BIOLOGICAL AND MEDICAL PHYSICS,
BIOMEDICAL ENGINEERING

BIOLOGICAL AND MEDICAL PHYSICS, BIOMEDICAL ENGINEERING

The fields of biological and medical physics and biomedical engineering are broad, multidisciplinary and dynamic. They lie at the crossroads of frontier research in physics, biology, chemistry, and medicine. The Biological and Medical Physics, Biomedical Engineering Series is intended to be comprehensive, covering a broad range of topics important to the study of the physical, chemical and biological sciences. Its goal is to provide scientists and engineers with textbooks, monographs, and reference works to address the growing need for information.

Books in the series emphasize established and emergent areas of science including molecular, membrane, and mathematical biophysics; photosynthetic energy harvesting and conversion; information processing; physical principles of genetics; sensory communications; automata networks, neural networks, and cellular automata. Equally important will be coverage of applied aspects of biological and medical physics and biomedical engineering such as molecular electronic components and devices, biosensors, medicine, imaging, physical principles of renewable energy production, advanced prostheses, and environmental control and engineering.

Editor-in-Chief:
Elias Greenbaum, Oak Ridge National Laboratory, Oak Ridge, Tennessee, USA

Editorial Board:

Masuo Aizawa, Department of Bioengineering, Tokyo Institute of Technology, Yokohama, Japan

Olaf S. Andersen, Department of Physiology, Biophysics & Molecular Medicine, Cornell University, New York, USA

Robert H. Austin, Department of Physics, Princeton University, Princeton, New Jersey, USA

James Barber, Department of Biochemistry, Imperial College of Science, Technology and Medicine, London, England

Howard C. Berg, Department of Molecular and Cellular Biology, Harvard University, Cambridge, Massachusetts, USA

Victor Bloomfield, Department of Biochemistry, University of Minnesota, St. Paul, Minnesota, USA

Robert Callender, Department of Biochemistry, Albert Einstein College of Medicine, Bronx, New York, USA

Britton Chance, Department of Biochemistry/Biophysics, University of Pennsylvania, Philadelphia, Pennsylvania, USA

Steven Chu, Lawrence Berkeley National Laboratory, Berkeley, California, USA

Louis J. DeFelice, Department of Pharmacology, Vanderbilt University, Nashville, Tennessee, USA

Johann Deisenhofer, Howard Hughes Medical Institute, The University of Texas, Dallas, Texas, USA

George Feher, Department of Physics, University of California, San Diego, La Jolla, California, USA

Hans Frauenfelder, Los Alamos National Laboratory, Los Alamos, New Mexico, USA

Ivar Giaever, Rensselaer Polytechnic Institute, Troy, New York, USA

Sol M. Gruner, Cornell University, Ithaca, New York, USA

Judith Herzfeld, Department of Chemistry, Brandeis University, Waltham, Massachusetts, USA

Mark S. Humayun, Doheny Eye Institute, Los Angeles, California, USA

Pierre Joliot, Institute de Biologie Physico-Chimique, Fondation Edmond de Rothschild, Paris, France

Lajos Keszthelyi, Institute of Biophysics, Hungarian Academy of Sciences, Szeged, Hungary

Robert S. Knox, Department of Physics and Astronomy, University of Rochester, Rochester, New York, USA

Aaron Lewis, Department of Applied Physics, Hebrew University, Jerusalem, Israel

Stuart M. Lindsay, Department of Physics and Astronomy, Arizona State University, Tempe, Arizona, USA

David Mauzerall, Rockefeller University, New York, New York, USA

Eugenie V. Mielczarek, Department of Physics and Astronomy, George Mason University, Fairfax, Virginia, USA

Markolf Niemz, Medical Faculty Mannheim, University of Heidelberg, Mannheim, Germany

V. Adrian Parsegian, Physical Science Laboratory, National Institutes of Health, Bethesda, Maryland, USA

Linda S. Powers, University of Arizona, Tucson, Arizona, USA

Earl W. Prohofsky, Department of Physics, Purdue University, West Lafayette, Indiana, USA

Andrew Rubin, Department of Biophysics, Moscow State University, Moscow, Russia

Michael Seibert, National Renewable Energy Laboratory, Golden, Colorado, USA

David Thomas, Department of Biochemistry, University of Minnesota Medical School, Minneapolis, Minnesota, USA

Lorenzo Pavesi
Philippe M. Fauchet
(Eds.)

Biophotonics

With 185 Figures

Springer

Professor Lorenzo Pavesi
University of Trento, Department of Physics, Laboratory of Nanoscience
Via Sommarive 14, 38050 Povo, Italy
E-mail: pavesi@science.unitn.it

Philippe M. Fauchet
University of Rochester, Department of Electrical and Computer Engineering
160 Trustee Road, Rochester, NY 14627-0231, USA
E-mail: fauchet@ece.rochester.edu

ISBN 978-3-540-76779-4   e-ISBN 978-3-540-76782-4

Biological and Medical Physics, Biomedical Engineering ISSN 1618-7210

Library of Congress Control Number: 2008920259

© 2008 Springer-Verlag Berlin Heidelberg

This work is subject to copyright. All rights are reserved, whether the whole or part of the material is concerned, specifically the rights of translation, reprinting, reuse of illustrations, recitation, broadcasting, reproduction on microfilm or in any other way, and storage in data banks. Duplication of this publication or parts thereof is permitted only under the provisions of the German Copyright Law of September 9, 1965, in its current version, and permission for use must always be obtained from Springer-Verlag. Violations are liable to prosecution under the German Copyright Law.

The use of general descriptive names, registered names, trademarks, etc. in this publication does not imply, even in the absence of a specific statement, that such names are exempt from the relevant protective laws and regulations and therefore free for general use.

Cover: eStudio Calamar Steinen

Printed on acid-free paper

9 8 7 6 5 4 3 2 1

springer.com

To Anna
To Melanie

Preface

This book is the brainchild of the fourth Optoelectronic and Photonic Winter School on "Biophotonics" which took place from 25th February 2007 to 3rd March 2007 in Sardagna, a small village on the mountains around Trento in Italy. This school, held every two years, have been promoted to trace the very fast developing technologies and the tremendous progress in photonics which have occurred and will occur in the near future. It is a common view that its current explosive development will lead to deep paradigm shifts in the near future. Identifying the plausible scenario for the future evolution of photonics presents an opportunity for constructive actions and scouting killer technologies.

This book gathers together distinguished authors to give an overview of the latest developments in biophotonics. Optical technologies have demonstrated a long tradition in the fields of life science and health care. The microscope has opened for us a view into the world of cells and bacteria and has evolved into a modern powerful tool for basic biological research on cell processes. Modern surgical microscopes have become key tools in neurosurgery, as well as in ophthalmology and surgery. Image-guided systems make use of computer tomography and MRI data in navigated surgery. Endoscopes have opened an easy access to the inner parts of the human body in minimal invasive surgery. Optical methods for gene sequencing and biochips open new routes for diagnosis and treatment of cancer. Fluorescence methods have replaced radioactive methods in screening processes in drug development. The role of optical technologies as the key enabling factor in health care and life sciences will grow tremendously in future. It is therefore a good time to write a book on biophotonics with the aim to introduce students and young researchers to this field and make a point of the importance of this science and technology. In particular, each chapter consists of a review, an introduction to the subject and a presentation of the current state of the art.

The first three chapters of the book are by Bassi, Giacometti and coworkers on photosynthesis, on the application of photosynthesis to biofuel production and on the role of carotenoids in photoprotection. Then, the chapter by

Diaspro and coworkers introduces the readers to non-linear microscopy. Another kind of microscopy, spectral self-interference microscopy, is presented in the chapter by Ünlü and coworkers. The following chapter, also by Ünlü and coworkers, deals with resonant cavity biosensors for sensing and imaging. A different way to engineer biosensors is the use of optical microcavities: these are reviewed in the chapter by Fauchet and coworkers.

The success and application of optical coherence tomography are described in the chapter by Boppart. Other coherent laser techniques for medical diagnostics are described in the chapter by Kemper and von Bally.

Two chapters follow on the use of fluorescence labelling for sensing and quantitative analysis. Caiolfa and coworkers detail both the principle and applications of luminescence probes in quantitative biology. Ligler reviews fluorescent-based optical biosensors.

The next chapter, by Seitz, discusses optical biochips and their applications to lab-on-a-chip. An example of the merging of microelectronics with biology is given in the chapter by Charbon on the use of CMOS camera for bioimaging applications.

Going back to basic optics, the chapter by Chiou and coworkers shows the use of optical forces in trapping and manipulating bio-objects, such as red blood cells or proteins. Optics can also be used in surgery. Pini and coworkers discuss the application of lasers in ophthalmology and vascular surgery in their chapter.

The final two chapters are examples of how photonics is used in common clinical practice. In the chapter by Pentland, the photobiology of the skin and the phototherapy are discussed, while in the chapter by Wilson the use of light in therapy is introduced and a few examples are given in detail.

We thank all the authors who presented very interesting state-of-the art lectures and chapters. Last but not least, we express our gratitude also to all those who have contributed to the success of the fourth Optoelectronic and Photonics Winter School on Biophotonics: the organizing committee (L. Pavesi, P. Fauchet, M. Anderle, L. Gonzo, R. Antolini, M. Ferrari, S. Iannotta, G. Righini), the staff of the University of Trento and all the students, whose participation was lively and stimulating. This year the winter school was also the first edition of the AIV school on nanotechnology. We acknowledge the generous contribution by the Società Italiana di Scienze e Tecnologie through its president Mariano Anderle. Further support came from the University of Trento, the Department of Physics, FBK-irst, LaserOptronic and Spectra-Physics, Crisel-Instruments and LOT Oriel group Europe. Particular thanks are also due to Alessandro Pitanti and his brother for their help in the editing the Latex files.

Trento, Rochester *Lorenzo Pavesi*
May 2008 *Philippe M. Fauchet*

Contents

1 Light Conversion in Photosynthetic Organisms
S. Frigerio, R. Bassi, and G.M. Giacometti 1
1.1 Introduction... 1
1.2 Chloroplast Structure 2
1.3 Pigments and Light Absorption 3
1.4 Photosynthetic Apparatus 4
    1.4.1 Photosystem II....................................... 6
    1.4.2 Photosystem I 7
    1.4.3 Cytochrome b_6f 8
    1.4.4 ATP Synthase 8
1.5 Cyclic Phosphorylation 9
1.6 Photoinhibition 10
References 13

2 Exploiting Photosynthesis for Biofuel Production
C. Govoni, T. Morosinotto, G. Giuliano, and R. Bassi 15
2.1 Biological Production of Vehicle Traction Fuels: Bioethanol
    and Biodiesel 17
    2.1.1 Bioethanol 17
    2.1.2 Biodiesel 18
    2.1.3 Biofuels Still Present Limitations Preventing
           Their Massive Utilization 18
2.2 Hydrogen Biological Production by Fermentative Processes 19
    2.2.1 Hydrogen Production by Bacterial Fermentation 20
2.3 Hydrogen Production by Photosynthetic Organisms 21
    2.3.1 Cyanobacteria 22
    2.3.2 Eukaryotic Algae 23
2.4 Challenges in Algal Hydrogen Production 23
    2.4.1 Oxygen Sensitivity of Hydrogen Production 23
    2.4.2 Optimization of Light Harvesting in Bioreactors 25
References 27

3 In Between Photosynthesis and Photoinhibition: The Fundamental Role of Carotenoids and Carotenoid-Binding Proteins in Photoprotection
G. Bonente, L. Dall'Osto, and R. Bassi............................ 29
3.1 When Light Becomes Dangerous for a Photosynthetic Organism .. 29
3.2 Acclimation 31
3.3 State 1–State 2 Transitions.................................... 32
3.4 Carotenoids Play a Fundamental Role in Many Photoprotection Mechanisms.. 33
3.5 Analysis of Xanthophyll Function In Vivo 36
3.6 Nonphotochemical Quenching 38
3.7 Feedback Deexcitation of Singlet-Excited Chlorophylls: qE 39
3.8 ΔpH - Independent Energy Thermal Dissipation (qI) 40
3.9 Chlorophyll Triplet Quenching................................ 41
3.10 Scavenging of Reactive Oxygen Species 41
3.11 Conclusions 43
References 44

4 Non-Linear Microscopy
D. Mazza, P. Bianchini, V. Caorsi, F. Cella, P.P. Mondal, E. Ronzitti, I. Testa, G. Vicidomini, and A. Diaspro 47
4.1 Introduction... 47
4.2 Chronological Notes on MPE 48
4.3 Principles of Confocal and Two-Photon Fluorescence Microscopy.. 49
    4.3.1 Fluorescence... 49
    4.3.2 Confocal Principles and Laser Scanning Microscopy....... 50
    4.3.3 Point Spread Function of a Confocal Microscope 52
4.4 Two-Photon Excitation 55
4.5 Two-Photon Optical Sectioning 59
4.6 Two-Photon Optical Setup 60
4.7 Second Harmonic Generation (SHG) Imaging................... 63
4.8 Conclusions 65
References 66

5 Applications of Optical Resonance to Biological Sensing and Imaging: I. Spectral Self-Interference Microscopy
M.S. Ünlü, A. Yalçin, M. Doğan, L. Moiseev, A. Swan, B.B. Goldberg, and C.R. Cantor... 71
5.1 High-Resolution Fluorescence Imaging 71
5.2 Self-Interference Imaging..................................... 71
5.3 Physical Model of SSFM 73
    5.3.1 Classical Dipole Emission Model 73
5.4 Acquisition and Data Processing 75
    5.4.1 Microscope Setup 75
    5.4.2 Fitting Algorithm 76

5.5	Experimental Results		77
	5.5.1	Monolayers of Fluorophores on Silicon Oxide Surfaces: Fluorescein, Quantum Dots, Lipid Films	77
	5.5.2	Conformation of Surface-Immobilized DNA	79
5.6	SSFM in 4Pi Configuration		82
5.7	Conclusions		84
References			85

6 Applications of Optical Resonance to Biological Sensing and Imaging: II. Resonant Cavity Biosensors
M.S. Ünlü, E. Özkumur, D.A. Bergstein, A. Yalçin, M.F. Ruane, and B.B. Goldberg 87

6.1	Multianalyte Sensing		87
6.2	Resonant Cavity Imaging Biosensor		88
	6.2.1	Detection Principle	88
	6.2.2	Experimental Setup, Data Acquisition, and Processing	90
	6.2.3	Experimental Results	91
	6.2.4	Spectral Reflectivity Imaging Biosensor	92
6.3	Optical Sensing of Biomolecules Using Microring Resonators		94
	6.3.1	Basics on Microring Resonators	94
	6.3.2	Setup and Data Acquisition	95
	6.3.3	Data Analysis and Discussion	96
6.4	Conclusions		98
References			98

7 Biodetection Using Silicon Photonic Crystal Microcavities
P.M. Fauchet, B.L. Miller, L.A. DeLouise, M.R. Lee, and H. Ouyang .. 101

7.1	Photonic Crystals: A Short Introduction		101
	7.1.1	Electromagnetic Theory	101
	7.1.2	One-Dimensional and Two-Dimensional PhC	103
	7.1.3	Microcavities: Breaking the Periodicity	105
	7.1.4	Computational Algorithms	106
7.2	One-Dimensional PhC Biosensors		107
	7.2.1	Preparation and Selected Properties of Porous Silicon	107
	7.2.2	Sensing Principle	109
	7.2.3	One-Dimensional Biosensor Design and Performance	111
	7.2.4	Fabrication of One-Dimensional PhC Biosensors	112
7.3	Selected Biosensing Results		114
	7.3.1	DNA Detection	114
	7.3.2	Bacteria Detection	114
	7.3.3	Protein Detection	115
	7.3.4	IgG Detection	117
7.4	Two-Dimensional PhC Biosensors		118
	7.4.1	Sample Preparation and Measurement	118
	7.4.2	Sensing Principle	119
	7.4.3	Selected Biosensing Results	120

8 Optical Coherence Tomography with Applications in Cancer Imaging

7.5 Conclusions 124
References 124

8 Optical Coherence Tomography with Applications in Cancer Imaging
S.A. Boppart 127
8.1 Introduction 127
8.2 Principles of Operation 128
8.3 Optical Sources for Optical Coherence Tomography 133
8.4 Fourier-Domain Optical Coherence Tomography 133
8.5 Beam Delivery Instruments for Optical Coherence Tomography ... 135
8.6 Spectroscopic Optical Coherence Tomography 136
8.7 Applications to Cancer Imaging 138
    8.7.1 Cellular Imaging for Tumor Cell Biology 138
    8.7.2 Translational Breast Cancer Imaging 140
8.8 Optical Coherence Tomography Contrast Agents 141
8.9 Molecular Imaging using Optical Coherence Tomography 145
8.10 Conclusions 147
References 149

9 Coherent Laser Measurement Techniques for Medical Diagnostics
B. Kemper and G. von Bally 151
9.1 Introduction 151
9.2 Electronic Speckle Pattern Interferometry (ESPI) 152
    9.2.1 Double Exposure Subtraction ESPI 152
    9.2.2 Spatial Phase Shifting (SPS) ESPI 153
9.3 Endoscopic Electronic Speckle Pattern Interferometry (ESPI) ... 156
    9.3.1 Proximal Endoscopic ESPI 156
    9.3.2 Distal Endoscopic ESPI 158
9.4 Microscopic (Speckle) Interferometry 161
9.5 Digital Holographic Microscopy 164
    9.5.1 Principle and Measurement Setup 164
    9.5.2 Nondiffractive Reconstruction 166
    9.5.3 Resolution and Numerical Focus 170
    9.5.4 Digital Holographic Phase Contrast Microscopy of Living Cells 171
9.6 Discussion and Conclusions 173
References 173

10 Biomarkers and Luminescent Probes in Quantitative Biology
M. Zamai, G. Malengo, and V.R. Caiolfa 177
10.1 Fluorophores and Genetic Dyes 177
    10.1.1 Small Organic Dyes and Quantum Dots 177
    10.1.2 Fluorescent Proteins 178

10.2	Microspectroscopy in Quantitative Biology: Where and How	183
	10.2.1 Fluorescence Correlation Spectroscopy	183
	10.2.2 Fluorescence Lifetime Imaging (FLIM)	188
	10.2.3 Glossary of Molecular Biology	194
References		195

11 Fluorescence-Based Optical Biosensors
F.S. Ligler 199
- 11.1 Introduction 199
- 11.2 Biological Recognition Molecules and Assay Formats 200
- 11.3 Displacement Immunosensors 203
- 11.4 Fiber Optic Biosensors 204
  - 11.4.1 Fiber Optics for Biosensor Applications 205
  - 11.4.2 Optrode Biosensors 207
  - 11.4.3 Evanescent Fiber Optic Biosensors 207
- 11.5 Bead-Based Biosensors 209
- 11.6 Planar Biosensors 210
- 11.7 Critical Issues and Future Opportunities 212
- References 214

12 Optical Biochips
P. Seitz 217
- 12.1 Taxonomy of Optical Biochips 217
  - 12.1.1 Basic Architecture of Optical Biochips 217
- 12.2 Analyte Classes for Optical Biochips 220
  - 12.2.1 DNA (DNA Fragments, mRNA, cDNA) 220
  - 12.2.2 Proteins (Antigens) 221
  - 12.2.3 Specific Organic Molecules 221
  - 12.2.4 Cell Gene Products (cDNA, Proteins) 221
  - 12.2.5 Tissue 222
- 12.3 Optical Effects for Biochemical Sensors 222
  - 12.3.1 Spectral Absorption 222
  - 12.3.2 Phase Shift 223
  - 12.3.3 Fluorescence 223
  - 12.3.4 Luminescence 223
  - 12.3.5 Raman Scattering 224
  - 12.3.6 Nonlinear Optical (NLO) Effects 224
- 12.4 Preferred Sensing Principles for Optical Biochips 224
  - 12.4.1 Evanescent Wave Sensing 225
  - 12.4.2 Fluorescence Sensing 228
- 12.5 Readout Methods for Evanescent Wave Sensors 229
  - 12.5.1 Angular Scanning 229
  - 12.5.2 Wavelength Tuning 230
  - 12.5.3 Grating Coupler Chirping 230
- 12.6 Substrates for Optical Biochips 230
- 12.7 Realization Example of an Optical Biosensor/Biochip: WIOS 231

12.8	Outlook: Lab-on-a-Chip Using Organic Semiconductors	232
	12.8.1 Basics of Organic Semiconductors	233
	12.8.2 Organic LEDs	233
	12.8.3 Organic Lasers	234
	12.8.4 Organic Photodetectors and Image Sensors	234
	12.8.5 Organic Photovoltaic Cells	234
	12.8.6 Organic Field Effect Transistors and Circuits	235
	12.8.7 Monolithic Photonic Microsystems Using Organic Semiconductors	235
12.9	Conclusions and Summary	236
References		236

13 CMOS Single-Photon Systems for Bioimaging Applications
E. Charbon 239

13.1	Introduction	239
13.2	Spectroscopy	240
13.3	Lifetime Imaging	242
13.4	Time-of-Flight in Bio- and Medical Imaging	244
13.5	System Considerations	245
13.6	Conclusions	247
References		248

14 Optical Trapping and Manipulation for Biomedical Applications
A. Chiou, M.-T. Wei, Y.-Q. Chen, T.-Y. Tseng, S.-L. Liu,
A. Karmenyan, and C.-H. Lin 249

14.1	Introduction	249
14.2	Theoretical Models for the Calculation of Optical Forces	252
	14.2.1 The Ray-Optics (RO) Model	252
	14.2.2 Electromagnetic (EM) Model	255
14.3	Experimental Measurements of Optical Forces	255
	14.3.1 Axial Optical Force as a Function of Position along the Optical Axis	255
	14.3.2 Transverse Trapping Force Measured by Viscous Drag	257
	14.3.3 Three-Dimensional Optical Force Field Probed by Particle Brownian Motion	257
	14.3.4 Optical Forced Oscillation	261
14.4	Potential Biomedical Applications	265
	14.4.1 Optical Forced Oscillation for the Measurement of Protein–Protein Interactions	266
	14.4.2 Protein–DNA Interaction	267
	14.4.3 Optical Trapping and Stretching of Red Blood Cells	269
14.5	Summary and Conclusion	271
References		271

15 Laser Tissue Welding in Minimally Invasive Surgery and Microsurgery
R. Pini, F. Rossi, P. Matteini, and F. Ratto 275
15.1 Introduction... 275
15.2 Laser Welding in Ophthalmology............................ 281
    15.2.1 Laser Welding of the Cornea 281
    15.2.2 Combing Femtosecond Laser Microsculpturing
            of the Cornea with Laser Welding 285
    15.2.3 Laser Closure of Capsular Tissue 287
15.3 Applications in Microvascular Surgery 289
15.4 Potentials in Other Surgical Fields 291
    15.4.1 Laser Welding of the Gastrointestinal Tract 291
    15.4.2 Laser Welding in Gynaecology 291
    15.4.3 Laser Welding in Neurosurgery 292
    15.4.4 Laser Welding in Orthopaedic Surgery 292
    15.4.5 Laser Welding of the Skin 292
    15.4.6 Laser Welding in Urology.............................. 293
15.5 Perspectives of Nanostructured Chromophores for Laser Welding.. 293
References 296

16 Photobiology of the Skin
A.P. Pentland 301
16.1 Basics of Skin Structure: Cell Types, Skin Structures,
      and Their Function 301
16.2 Effects of Light Exposure on Skin 304
16.3 Sun Protection and Sunscreens 309
16.4 Phototherapy: Use of Light for Treatment for Skin Disease 311
References 313

17 Advanced Photodynamic Therapy
B.C. Wilson... 315
17.1 Introduction.. 315
17.2 Basic Principles and Features of "Standard PDT" 316
17.3 Novel PDT Concepts 319
    17.3.1 Two-Photon PDT 319
    17.3.2 Metronomic PDT 322
    17.3.3 PDT Molecular Beacons............................... 323
    17.3.4 Nanoparticle-Based PDT 325
17.4 PDT Dosimetry Using Photonic Techniques 327
17.5 Biophotonic Techniques for Monitoring Response to PDT 330
17.6 Biophotonic Challenges and Opportunities in Clinical PDT........ 331
17.7 Conclusions 333
References 333

Index 335

List of Contributors

Gert von Bally
Center for Biomedical Optics
and Photonics,
University of Münster,
Robert-Koch-Straße 45,
48129 Münster,
Germany
Ce.BOP@uni-muenster.de

Roberto Bassi
Département de Biologie - Case 901,
Faculté des Sciences de Luminy,
Université Aix-Marseille II, LGBP,
163, Avenue de Luminy
13288 Marseille Cedex 09,
France
and
Dipartimento Scientifico
e Tecnologico,
Università di Verona
Strada Le Grazie 15,
37134 Verona,
Italy
bassi@sci.univr.it

David A. Bergstein
Department of Electrical
and Computer Engineering,
Boston University,
8 St. Mary's St.,
02215 Boston, MA, USA
bdave@bu.edu

Paolo Bianchini
LAMBS-MicroSCoBiO Research
Center, Department of Physics,
University of Genoa,
Via Dodecaneso 33,
16146 Genoa, Italy

Giulia Bonente
Laboratoire de Génétique et de
Biophysique des Plantes,
Faculté de Sciences de Luminy,
Université de Marseille,
Marseille Cedex 9,
France
bonente@sci.univr.it
and
Dipartimento Scientifico e
Tecnologico, Università di Verona,
Strada le Grazie 15,
37134 Verona,
Italy

Stephen A. Boppart
Beckman Institute for Advanced
Science and Technology,
Departments of Electrical
and Computer Engineering,
Bioengineering, and Medicine
University of Illinois at
Urbana-Champaign
Mills Breast Cancer Institute

Carle Foundation Hospital
Urbana, IL 61801
boppart@uiuc.edu

Valeria R. Caiolfa
Department of Molecular Biology
and Functional Genomics,
San Raffaele Scientific Institute
Milano, Italy
valeria.caiolfa@hsr.it
and
IIT Network Research,
Unit of Molecular Neuroscience,
San Raffaele Scientific Institute,
Milano, Italy

Charles R. Cantor
Department of Biomedical
Engineering, 44 Cummington St.,
02215 Boston, MA, USA
crcantor@bu.edu
and
Center for Advanced Biotechnology,
36 Cummington St.,
02215 Boston, MA, USA

Valentina Caorsi
LAMBS-MicroSCoBiO Research
Center, Department of Physics,
University of Genoa,
Via Dodecaneso 33,
16146 Genoa, Italy

Francesca Cella
LAMBS-MicroSCoBiO Research
Center, Department of Physics,
University of Genoa,
Via Dodecaneso 33,
16146 Genoa, Italy

Edoardo Charbon
Quantum Architecture Group
(AQUA), EPFL,
1015 Lausanne, Switzerland
edoardo.charbon@epfl.ch

Yin-Quan Chen
Institute of Biophotonics
Engineering, National Yang-Ming
University Taipei, Taiwan, ROC

Arthur Chiou
Institute of Biophotonics
Engineering, National Yang-Ming
University Taipei, Taiwan, ROC
aechiou@ym.edu.tw

Luca Dall'Osto
Dipartimento Scientifico e
Tecnologico, Università di Verona,
Strada le Grazie 15,
37134 Verona,
Italy
dallosto@sci.univr.it

Lisa A. DeLouis e
Center for Future Health,
University of Rochester,
Rochester, NY, USA
and
Department of Dermatology,
University of Rochester
Rochester, NY,
USA

Alberto Diaspro
LAMBS-MicroSCoBiO Research
Center, Department of Physics,
University of Genoa,
Via Dodecaneso 33,
16146 Genoa,
Italy

Mehmet Doğan
Department of Physics,
Boston University,
590 Commonwealth Ave.,
02215 Boston, MA, USA
mdogan@bu.edu

Philippe M. Fauchet
Center for Future Health,
University of Rochester,
Rochester, NY, USA
and

Department of Electrical
and Computer Engineering,
University of Rochester,
Rochester, NY, USA
and
Department of Biomedical
Engineering,
University of Rochester,
Rochester, NY, USA
and
The Institute of Optics,
University of Rochester,
Rochester, NY, USA

Sara Frigerio
Université Aix-Marseille II, LGBP,
Faculté des Sciences de Luminy,
Département de Biologie - Case 901,
163, Avenue de Luminy
13288 Marseille Cedex 09,
France

Giorgio Giacometti
Dipartimento di Biologia,
Università di Padova,
Via Ugo Bassi 58 B,
35131 Padova, Italy

Giovanni M. Giuliano
Ente per le Nuove tecnologie,
l'Energia e l'Ambiente (ENEA),
Unità Biotecnologie,
Centro Ricerche Casaccia,
C.P. 2400, Roma 00100,
Italy

Bennett B. Goldberg
Department of Physics,
Boston University,
590 Commonwealth Ave.,
02215 Boston, MA,
USA
goldberg@bu.edu
and

Department of Electrical
and Computer Engineering,
Boston University,
8 St. Mary's St.,
02215 Boston, MA, USA
and
Department of Biomedical
Engineering, 44 Cummington St.,
02215 Boston, MA, USA

Chiara Govoni
Dipartimento Scientifico e
Tecnologico, Università di Verona,
Strada Le Grazie 15,
37134 Verona, Italy

Artashes Karmenyan
Institute of Biophotonics
Engineering, National Yang-Ming
University Taipei, Taiwan, ROC

Björn Kemper
Center for Biomedical Optics
and Photonics,
University of Münster,
Robert-Koch-Straße 45,
48129 Münster, Germany
bkemper@uni-muenster.de

Mindy R. Lee
Center for Future Health,
University of Rochester,
Rochester, NY, USA
and
The Institute of Optics,
University of Rochester,
Rochester, NY, USA

Frances S. Ligler
Center for Bio/Molecular Science
& Engineering,
Naval Research Laboratory,
Washington, DC 20375, USA
fligler@cbmse.nrl.navy.mil

Chi-Hung Lin
Institute of Biophotonics
Engineering, National Yang-Ming
University Taipei, Taiwan, ROC

Shang-Ling Liu
Institute of Biophotonics
Engineering, National Yang-Ming
University Taipei, Taiwan, ROC

Gabriele Malengo
Department of Molecular Biology
and Functional Genomics,
San Raffaele Scientific Institute,
Milano, Italy

Paolo Matteini
Istituto di Fisica Applicata,
Consiglio Nazionale delle Ricerche
Via Madonna del Piano 10,
50019 Sesto Fiorentino,
Italy

Davide Mazza
LAMBS-MicroSCoBiO Research
Center, Department of Physics,
University of Genoa,
Via Dodecaneso 33,
16146 Genoa, Italy
mazza@fisica.unige.it

Benjamin L. Miller
Center for Future Health,
University of Rochester,
Rochester, NY, USA
and
Department of Biomedical
Engineering, University of Rochester,
Rochester, NY, USA
and
Department of Dermatology,
University of Rochester
Rochester, NY, USA

Lev Moiseev
Center for Advanced Biotechnology
36 Cummington St., 02215 Boston,
MA, USA
leva1m@bu.edu

Partha P. Mondal
LAMBS-MicroSCoBiO Research
Center, Department of Physics,
University of Genoa,
Via Dodecaneso 33,
16146 Genoa, Italy

Tomas Morosinotto
Dipartimento di Biologia,
Università di Padova,
Via Ugo Bassi 58 B,
35131 Padova, Italy

Huimin Ouyang
Department of Electrical
and Computer Engineering,
University of Rochester,
Rochester, NY, USA

Emre Özkumur
Department of Electrical and
Computer Engineering,
Boston University,
8 St. Mary's St.,
02215 Boston, MA, USA
eozkumur@bu.edu

Alice P. Pentland
Department of Dermatology,
University of Rochester,
601 Elmwood,
Rochester, NY 14642, USA
Alice_Pentland@urmc.
rochester.edu

Roberto Pini
Istituto di Fisica Applicata,
Consiglio Nazionale delle Ricerche
Via Madonna del Piano 10,
50019 Sesto Fiorentino, Italy
R.Pini@ifac.cnr.it

Fulvio Ratto
Istituto di Fisica Applicata,
Consiglio Nazionale delle Ricerche,
Via Madonna del Piano 10,
50019 Sesto Fiorentino, Italy

Emiliano Ronzitti
LAMBS-MicroSCoBiO Research
Center, Department of Physics,
University of Genoa,
Via Dodecaneso 33,
16146 Genoa, Italy

Francesca Rossi
Istituto di Fisica Applicata,
Consiglio Nazionale delle Ricerche
Via Madonna del Piano 10,
50019 Sesto Fiorentino, Italy

Michael F. Ruane
Department of Electrical
and Computer Engineering,
Boston University,
8 St. Mary's St.,
02215, Boston, MA, USA
mfr@bu.edu

Peter Seitz
Swiss Center for Electronics
and Microtechnology,
CSEM SA,
Zurich Switzerland,
and
Institute for Microtechnology,
University of Neuchâtel,
Neuchâtel, Switzerland
peter.seitz@csem.ch

Anna Swan
Department of Electrical
and Computer Engineering,
Boston University,
8 St. Mary's St.,
02215 Boston, MA, USA
swan@bu.edu

Ilaria Testa
LAMBS-MicroSCoBiO Research
Center, Department of Physics,
University of Genoa,
Via Dodecaneso 33,
16146 Genoa, Italy

Te-Yu Tseng
Institute of Biophotonics
Engineering, National Yang-Ming
University, Taipei, Taiwan, ROC

M. Selim Ünlü
Department of Electrical
and Computer Engineering,
Boston University,
8 St. Mary's St.,
02215 Boston, MA,
USA
selim@bu.edu
and
Department of Physics,
Boston University,
590 Commonwealth Ave.,
02215 Boston, MA, USA
and
Department of Biomedical
Engineering,
44 Cummington St.,
02215 Boston, MA,
USA

Giuseppe Vicidomini
LAMBS-MicroSCoBiO Research
Center, Department of Physics,
University of Genoa,
Via Dodecaneso 33,
16146 Genoa,
Italy

Ming-Tzo Wei
Institute of Biophotonics
Engineering, National Yang-Ming
University, Taipei, Taiwan, ROC

Brian C. Wilson
Division of Biophysics and
Bioimaging, Ontario Cancer
Institute, Toronto,
ON, Canada
and
Department of Medical Biophysics,
University of Toronto, Toronto,
ON, Canada, M5G 2M9
wilson@uhnres.utoronto.ca

Ayça Yalçin
Department of Electrical
and Computer Engineering,
Boston University,
8 St. Mary's St.,
02215 Boston, MA, USA
ayca@bu.edu

Moreno Zamai
Department of Molecular Biology
and Functional Genomics,
San Raffaele Scientific Institute,
Milano, Italy
and
IIT Network Research,
Unit of Molecular Neuroscience,
San Raffaele Scientific Institute,
Milano, Italy
moreno.zamai@hsr.it

1
Light Conversion in Photosynthetic Organisms

S. Frigerio, R. Bassi, and G.M. Giacometti

1.1 Introduction

The sun is the ultimate source of free energy that drives all of the processes in living cells. The radiant energy of the sun is captured and converted to chemical energy by photosynthesis. The flux of carbon through the biosphere begins with photosynthesis. Photosynthetic organisms produce carbohydrates and molecular oxygen from carbon dioxide and water:

$$6H_2O + 6CO_2 + light \rightarrow (CH_2O)6 \text{ pedice} + 6O_2 \tag{1.1}$$

The carbohydrates produced by photosynthesis serve as the energy source for other non-photosynthetic (heterotrophic) organisms. In this process, carbohydrates are recycled to carbon dioxide and water by the combined action of cellular catabolic processes.

The *fixation* of carbon dioxide into sugars requires free energy in the form of ATP and reducing power in the form of NADPH. The light reactions of photosynthesis respond to this need: the visible component of solar radiation is captured and its energy is converted into ATP and NADPH through a complex series of redox reactions and membrane-mediated energy conversions.

The following reactions, referred to as *dark phase* of photosynthesis, as, in principle, they do not need solar radiation to be carried out if NADPH and ATP are provided, drive the reduction of CO_2 to the carbohydrate GAP (glyceraldehyde-3-phosphate). The series of reactions are altogether indicated as the Calvin-Benson cycle [1].

The enzyme directly responsible for the addition of CO_2 to ribulose 1,5 - bisphosphate is RubisCO [2]. The whole carboxylation cycle requires three CO_2 molecules to generate a molecule of GAP, which is then used for the synthesis of more complex sugars or other compounds. The energy required for this cycle is nine molecules of ATP and six of NADPH, as summarized in the following equation:

$$3CO_2 + 9ATP + 6NADPH \rightarrow GAP + 9ADP + 8P_i + 6NADP^+ \tag{1.2}$$

To carry out these reactions, in particular those of the light phase, plants need a specific apparatus, which is in part shared by other photosynthetic organisms like cyanobacteria and green algae. The photosynthetic apparatus is located within the cells in dedicated organelles called chloroplasts. The origins of chloroplasts is most probably due to an event of endosymbiosis of a cyanobacterium by an eukaryotic cell [3].

1.2 Chloroplast Structure

Chloroplasts are characterized by two different membrane systems: a double membrane envelope, which encloses a soluble fraction called stroma; within the stroma are located thylakoids, an extended and morphologically complex membrane system carrying the photosynthetic apparatus. This inner membranous system defines an internal space called thylakoid lumen. The membrane can be divided in two main regions: the *grana* region, where the membranes form stacked structures of flat vesicles and the *stroma lamellae*, which create connections between stacked vesicles, insuring continuity to the lumenal space (Fig. 1.1). Thylakoid membranes are composed principally of glycerol-lipids, in particular monogalactosyldiacyl-glycerol and digalactosyldiacyl-glycerol, which distinguish photosynthetic membranes from all the others [4].

The prokaryotic origin of chloroplast is supported by the presence of a circular genome, which encodes almost 90–120 sequences, while many others (∼3,400) or more proteins, predicted as chloroplastic, are encoded by nuclear genes [5].

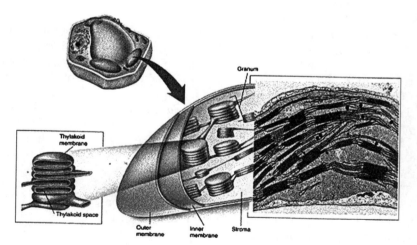

Fig. 1.1. Chloroplast structure. Localization in a typical mesophyll cell, schematic representation and electron micrograph, showing thylakoid membranes divided in grana and stroma lamellae

1.3 Pigments and Light Absorption

The capacity of plants to absorb light covers a wide range of wavelengths, between 350 and 700 nm, which include almost 60% of incident sunlight spectrum. All the pigments involved in light absorption, which are grouped in to two big classes, chlorophylls and carotenoids, are located within the thylakoids. Chlorophylls, named a or b, because of two different chemical forms, are organo-metallic compounds in which a substituted porphyrin ring structure coordinates a magnesium atom in its center. A phytol chain, necessary for their localization within the lipid membrane, constitutes one of the lateral chains of the porphyrin. These pigments strongly absorb red and blue light, while they scatter green light, thus giving reason for plant leafs color [6].

The absorption of a photon drives chlorophylls to an excited state, converting electromagnetic energy into electronic excitation; red light determines excitation to the first singlet excited state, while blue light causes excitation to the second singlet state (Fig. 1.2).

The photosynthetic apparatus makes the job of converting such molecular excitation into different forms of stable free energy. Electronic excitation of molecules decays to ground state in the nanosecond time scale, and the reactions that convert the excitation energy must be fast enough to successfully compete with decay.

The higher singlet-excited states, as well as the excited vibrational states, decay by internal conversion to the lowest vibrational state of the lowest excited singlet, in the time range of femtosecond, which is much faster than the

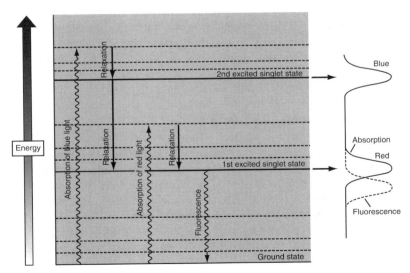

Fig. 1.2. Light absorption and energy decay scheme

rate of any subsequent process; therefore, the system equilibrate at the first singlet-excited state before anything else can happen.

Relaxation to ground state may occur through different pathways: (1) internal conversion, where the excitation energy is dissipated as heat; (2) fluorescence emission, where the energy is emitted as a photon; (3) energy transfer, where excitation energy is transferred to a nearby acceptor molecule (Forster energy transfer) [7]. In the photosynthetic apparatus, the most important is the latter: reiterated energy transfer between neighbor chlorophyll molecules allows excitation energy to reach a particular site (reaction center), where a redox reaction takes place, converting the excitation into chemical energy.

For efficient conversion of light energy into chemical energy, a highly efficient energy transfer through neighboring pigments is essential. However, the ideal conditions for high efficiency in energy transfer is not always satisfied in the photosynthetic apparatus; if the energy transfer rate is not competing successfully with the other relaxation mechanisms, the excitation energy is lost and a fluorescence emission or heat dissipation takes place.

Another important phenomenon must be taken into account. The lowest singlet-excited state of chlorophylls can also undergo intersystem crossing to the triplet state. This excited state cannot transfer its energy to the reaction center and therefore cannot contribute to photochemistry. Worse than that, chlorophyll triplet state can easily activate molecular oxygen to its singlet state, producing a highly reactive and dangerous molecule.

The second class of pigments is represented by carotenoids, linear polyenes [8] involved in facing the problem of singlet oxygen production, besides contributing to absorb light and transfer it to the nearby chlorophylls: they can quench chlorophyll triplet-excited states and also scavenge singlet oxygen. In both cases, this leads to the dissipation of the absorbed photon but it prevents oxidative damage to the membrane and photosynthetic apparatus components.

1.4 Photosynthetic Apparatus

Light absorption and energy conversion take place in specific and highly organized structures embedded in the thylakoid membrane. All together they constitute the photosynthetic apparatus. There are four different complexes: two Photosystems (PSI and PSII), the cytochrome b_6f, and the F-ATPase. These are composed of several protein subunits, cofactors, and pigments, while additional electron carriers can move within the lipid bilayer of the membrane or in the aqueous medium (Fig. 1.3). In the last years, high resolution structures for the four complexes, even though from different species, have become available, thus allowing a deeper comprehension of photosynthetic mechanisms [9–13].

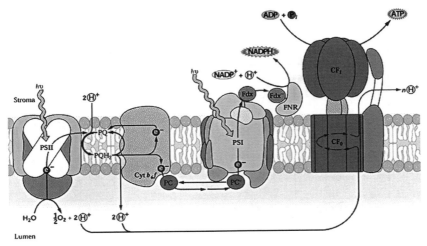

Fig. 1.3. Photosynthetic apparatus: Both the photosystems, cytochrome b_6f, and F-ATPase are shown, in addition to movable subunits as plastoquinone (PQ), plastocyanin (PC), and ferredoxin (Fdx)

As pointed out earlier, the task of the photosynthetic apparatus is that of providing reducing power and ATP to drive CO_2 fixation. The reducing power in the form of NADPH is generated through a redox reaction in which water is the ultimate electron donor:

$$2NADP^+ + 2H_2O + light \rightarrow 2NADPH + O_2 + 2H^+ \tag{1.3}$$

The energy of a single photon is not sufficient to drive an electron from water to $NADP^+$, and two photons must be used to make the job. For this reason, organisms using water as electron donor, i.e., oxygenic organisms such as higher plants, algae and cyanobacteria, make use of two photosystems that can operate in series.

Photosystem II is able to extract electrons from water and brings them to plastoquinone, a quinone molecule freely diffusible in the membrane lipid moiety. Photosystem I makes use of another photon to bring the electron at the level of $NADP^+$.

The connection between the two photosystems is operated by cytochrome b_6f complex, which transfers electrons from the plastoquinone pool to plastocyanine, a soluble Cu-protein freely diffusible in the thylakoids lumen. This electron transfer is in favor of potential gradient, and there is no need of light energy to drive it. Actually, the role of this complex is not limited to simply catalyze an electron transfer, as this would lead to a significant loss of energy. On the contrary, cytochrome b_6f, using the electron transfer energy, acts as a proton pump, which transfers protons against gradient from the stromal to the lumenal side of the membrane. The action of cytochrome b_6f is all the way analogue to that of complex III (or cytochrome b_1c) of the mitochondrial

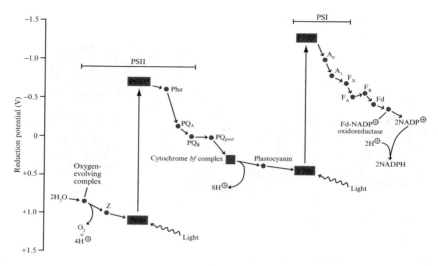

Fig. 1.4. Electron transport chain

respiratory chain, and plays an essential role in building up the proton gradient necessary to drive ADP phosphorylation by ATP synthase (Fig. 1.4). Photosystems are composed of proteins and pigments, chlorophylls and carotenoids, responsible for light harvesting; PSI and PSII are different in several aspects, in particular for the supramolecular organization, but it is possible to identify the same two functional moieties in both: (a) a core complex, responsible for charge separation and the first steps of the electron transport and (b) an antenna complex responsible of increasing the light harvesting and transferring of absorbed energy to the reaction center.

Both the core complexes are composed of protein encoded by *psa* and *psb* genes, respectively, for PSI and PSII, which bind pigments and surround the so called "special chlorophyll pair," where the charge separation actually occurs.

The core complex is surrounded by antenna proteins, encoded by the multigenic family of *light-harvesting complexes (lhc)*, which primarily increase the light-harvesting capacity, but are also involved in regulation and photoprotection [14]. PSI antenna complex is composed of Lhca1-4 as major subunits and Lhca5-6 as minor [15]; on the contrary, the major component of PSII antenna complex is a heterotrimer of Lhcb1-3 subunits, while remaining Lhcb proteins, Lhcb4-6, are altogether classified as minor antennas and they are generally found as monomers [16].

1.4.1 Photosystem II

Photosystem II (PSII), which is mainly located in grana thylakoids, is a light driven water-plastoquinone reductase. It performs the thermodynamically most demanding and most dangerous reaction: splitting water into its elementary components, molecular oxygen and protons.

The reaction center (RCII) of PSII is composed of the heterodimer constituted by the two proteins D1 (PsbA) and D2 (PsbD), which coordinate all the essential cofactors for its electron transport chain [9].

When the central chlorophylls special pair (P_{680}) is excited by a photon absorbed by the antenna apparatus, a charge separation occurs as an electron is transferred from chlorophyll to pheophytin, in one of the fastest electron transfer reaction (2–20 ps) ever observed in nature. The cation P_{680}^+ that is formed on this chlorophyll pair is one of the most oxidizing centers that can be formed in a living cell. The subsequent electron transfer is to Q_A,, a plastoquinone molecule tightly bound to the D2 protein near the stromal surface. This electron transfer is slower than the previous one but still fast enough (~400 ps) to compete with recombination. This step involves a significant loss of energy that contributes to stabilize the charge separation.

Because of its very high oxidation potential, P_{680}^+ is able to extract an electron from a tyrosine residue of the D1 protein (Tyr161, also named TyrZ) in about 20 ns; the electron hole at the donor side of PSII ends its run at the Mn cluster, near the lumenal surface of the membrane. In the mean time, in the acceptor side at the other membrane surface, the electron standing on Q_A finds its way to Q_B, a plastoquinone loosely bound on the D1 protein, also thanks to the assistance of a non-heme iron, sitting in between the two. This is the last reaction in the electron transfer chain within PSII reaction center and its rate is the limiting step (200–300 µs).

A second turnover, driven by a further excitation from the antenna, brings a second electron to Q_B with the very same reaction sequence. When Q_B receives the second electron, two protons are up-taken from the stroma and the plastoquinol leaves its binding site on D1. This is then reoccupied by an oxidized plastoquinone from the membrane pool. A two-photons-two-electrons gated mechanism brings about the reduction of a plastoquinone molecule at the acceptor side of the complex.

In a four-photons-four-electrons gated mechanism, two more photons and two more electrons injected in the transport chain are required to reduce a second PQ and oxidize two water molecules with the evolution of one oxygen molecule. The so-called "oxygen evolving complex" (OEC) sitting in the lumenal side of PSII, often indicated also as the "water splitting enzyme," can be considered as the heart of the PSII activity, and most of the today research efforts are directed to elucidate the detailed mechanism for its activity [17].

1.4.2 Photosystem I

In photosystem I (PSI), which is located almost exclusively in stroma lamellae, the excitation of the chlorophyll special pair P_{700} in the reaction center of PSI (RCI) initiates a series of electron transfer reactions that culminate in reduction of $NADP^+$ to NADPH. RCI is activated by the absorption of red light that promotes it to an excited state (P_{700}*) at about -0.6 V with a jump of about 1 V with respect to the ground state P_{700} ($E°' = +0.4$ V).

This generates a reductant strong enough to reduce ferredoxin, a soluble iron-sulfur protein able to donate electrons to $NADP^+$ by the action of the stromal soluble enzyme ferredoxin-$NADP^+$ reductase. P_{700}^+ formed is rereduced by an electron carried by the reduced plastocyanine (Fig. 1.4). Therefore, PSI which is situated mainly in the nonstacked, stroma lamellae regions of the thylakoid membrane, functions as a plastocyanine–ferredoxin reductase. The core of the RCI is composed of two major subunits, PsaA and PsaB, which contain all the pigments of the complex (including most of the antenna chlorophylls and carotenoids) and all the electron carriers. Upon absorption of a photon, charge separation occurs and an electron is transferred to a primary acceptor A_o (chlorophyll a monomer) from which it proceeds to the intermediate acceptor A_1 (phylloquinone) and subsequently to the second intermediate acceptor A_2 also indicated as F_x. This is a (4Fe–4S) cluster, as well as the subsequent electron acceptors F_A and F_B, which transfer the electron to the ferredoxin bound in its docking site on the stromal surface of the PSI complex. P700, A_o, A_1, and F_X are associated with the PSI intrinsic core subunits PsaA and PsaB, while the terminal electron acceptors F_A and F_B are bound to an extrinsic low molecular mass subunit PsaC [11]. It should be noticed that the electron transfer kinetics within RCI is very fast and not completely resolved yet.

1.4.3 Cytochrome b_6f

The electron transfer bridge between the two photosystems is realized by the cytochrome b_6f complex. This is assembled as a functional dimer with a twofold symmetry axis, each monomer being composed of four subunits: the cytochrome b_6 carrying two b-type heme groups classified as high and low potential, the cytochrome f carrying a single c-type heme group, the Rieske iron-sulphur protein carrying a $(Fe_2\text{-}S_2)$ cluster and the subunit IV [13].

The reduced plastoquinoles, freely diffusing inside the lipid bilayer, bind to the complex and electrons are transferred through cytochrome b_6f to the acceptor, plastocyanin. The transfer of two electrons from plastoquinol to the single electron carrier plastocyanin (summarized in the reaction below)

$$QH_2 + 2\text{plastocyanin}(Cu^{2+}) \rightleftharpoons Q + 2\text{plastocyanin}(Cu^+) + 2H^+ \qquad (1.4)$$

is realized through a complex series of redox reactions, which involves the semiquinone form Q^-, named "Q cycle," with the effect of transferring 4 H^+ instead of only two from the stroma to the lumen. This increases the efficiency of converting reduction potential energy into transmembrane proton gradient. Reduced plastocyanin moves in the lumen and reaches the PSI where it will donate its electron to P_{700}^+ (see earlier).

1.4.4 ATP Synthase

The last complex in the photosynthetic apparatus is the ATP synthase (F-ATPase). This kind of enzyme is ubiquitous in energy-transducing membranes, and the structure is highly conserved [12]. It presents a transmembrane domain and a stromal portion. F-ATPase in the chloroplast synthesizes ATP, using the proton-motive force generate by H^+ accumulation in thylakoid lumen. The diffusion of protons through a channel (Fo) determines the rotation of F-ATPase inner subunits inducing cyclic conformational changes in the F1 subunits, which provide the energy for ADP phosphorylation and release of ATP in the stroma; four protons are needed for the synthesis of one molecule of ATP [18].

The equilibrium along the electron transport chain, between reduced and oxidized species, is crucial for plant life. Excess energy absorbed by chlorophylls drives the formation of extremely dangerous reactive oxygen species (ROS), which can oxidize all the biological component of a cell. For this reason, absorbed energy can be redistributed among photosystems. When excess light is absorbed by PSII, thanks to a phosphorylation process, a population of LHCII trimers moves through the membrane and attach to PSI, thus increasing PSI light absorption capacity [19]. This phenomenon is called "state 1–state 2 transition."

1.5 Cyclic Phosphorylation

High electron transport rates drive the formation of reduced ferredoxin (Fd) in excess with respect to the ferredoxin-$NADP^+$ reductase activity. To avoid over-reduction of the electron transport chain and block of proton pumping and ATP synthesis, a mechanism, alternative to liner electron transport and involving PSI, can be activated, the so-called cyclic phosphorylation [20].

In this case, reduced ferredoxin can diffuse in the stroma back to the cytochrome b_6f and transfer electrons to the heme exposed in this region, through the action of a Fd-NADP-reductase [21]. Translocation of the electron through cytochrome to plastocyanin and then again to PSI leads to the generation of protons in the lumen, which can be used for ATP synthesis, even without NADPH production.

This alternative pathway exists in parallel with linear electron transport; the latter is limited by the oxidation rate of PSI acceptors, thus leading, under constant light intensity, to reach a kind of steady-state. Before the attainment of this steady-state, if PSII is inhibited, few plastocyanin molecules move to PSI, so all the electrons that reach Ferredoxin will be used for $NADP^+$ reduction, thus inhibiting cyclic electron flow [22]. With the increase of excitation pressure, the oxidation of Fd becomes the limiting step, so the cyclic phosphorylation is activated. In addition, the dark phase of photosynthesis is also involved in this alternative pathway, as it has been recently observed that in dark-adapted plants cyclic electron transport is activated first, as a consequence of a scarce NADPH requirement due to Calvin cycle inhibition [23].

The two alternative electron flows can exist simultaneously because they are structurally separated [24]. Because PSI and PSII are differentially located within thylakoids, the position of the cytochrome b_6f complex assumes a different role: the population located near to PSII participates to linear electron flow, while the population close to PSI is involved in cyclic electron transport. This separation is driven to the extreme level in the case of C4 plants, like maize, where PSII, thus the linear electron flow, is located in mesophyll, while PSI, thus cyclic electron flow, is located in bundle sheath [25]. The size of the populations located close to PSII vs. PSI can be regulated by phosphorylation events [26].

1.6 Photoinhibition

In describing the function of PSII, the attention has been focused on the reaction center, where the excitation energy coming from the antenna apparatus is converted into chemical energy through a series of redox reactions. The composition of the reaction center and its electron transfer cofactors has been briefly discussed, but PSII core is much more complex and contains a number of other protein subunits as always happens with enzymatic complexes that must be finely regulated.

In fact, a PSII core monomer contains at least 16 integral subunits, 3 lumenal subunits, 36 chlorophylls a (two loosely attached), 7 *all-trans* carotenoids, 2 heme groups (b and c type) belonging to the cytochrome b559, 1 non-heme iron, 2 quinones, 2 pheophitins, and it is organized in functional dimers characterized by a pseudo twofold symmetry [9].

Such a great complexity is partially justified by the difficult task the PSII core must accomplish of splitting water bringing its electrons at the level of plastoquinone in the membrane, but it also strongly suggests a need for a fine regulation of its function. There are, indeed, several good reasons for PSII activity to be regulated: the electron transport turnover depends on the frequency of P_{680} excitation, which, in turn, depends on the light fluence reaching the light harvesting apparatus. This must be large enough to ensure good photosynthetic activity even when the environment light intensity is very dim. But the photon fluence can vary very quickly by several orders of magnitude. Thus, a plant, which is tuned for optimal electron transport in dim light, will find itself overexcited when exposed to full sunlight. Over excitation of PSII brings oxidative damage to the protein-pigment complexes, thus impairing the photosynthetic activity. This is a well-known phenomenon called "photoinhibition" and represents the most important limiting factor to crop productivity [27].

Therefore, a regulation is needed at the level of the excitation transfer from the antenna to the reaction center. Indeed several sophisticated control mechanisms have been set up by evolution for this process. Nevertheless, at a light fluence typical of midday full sun, a significant amount of photoinhibition takes place, which severely limits energy conversion.

Although some kind of photoinhibition has been shown to take place also at the level of PSI [28], the main phenomenon regards PSII, where an inactivation mechanism inherent to the same electron transfer activity takes place, with low quantum yield, even at low light irradiance [29]. At increasing light intensities, inactivation of PSII occurs with higher frequency and in extreme cases may bring to irreversible oxidative damage of the complex.

In the last decades, considerable efforts were directed to the elucidation of the molecular mechanisms of this important phenomenon. It is long known that the D1 protein (PsbA) of the PSII reaction center is rapidly turned over and that its turnover rate increases with the light intensity [30]. Why is that so? It has been hypothesized that D1 is the protein subunit, which is preferentially damaged when the reaction center is over excited, and it needs to be frequently substituted to maintain the reaction center active. In fact, the water splitting photochemistry of PSII produces various radicals and active oxygen species, which cause irreversible damage to PSII. However, damaged PSII reaction centers do not usually accumulate in the thylakoid membrane because of a rapid and efficient repair mechanism. It has been calculated that this repair mechanism is at least as important as that of DNA, so that the green kingdom in the biosphere could not survive in its absence.

It is commonly believed that the design of PSII allows protection for most of its protein and pigment components, with the oxidative damage being mainly targeted to a single subunit, the reaction center D1 protein. Repair of PSII via turnover of the D1 protein is a complex process (Fig. 1.5)

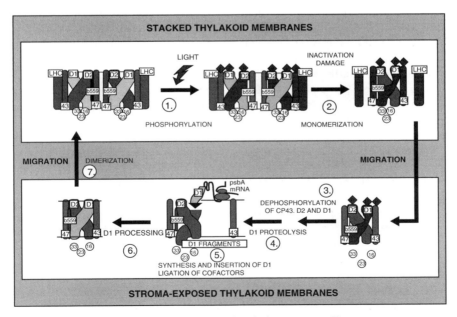

Fig. 1.5. Repair cycle of photosystem II

that involves reversible phosphorylation of PSII proteins and changes in the oligomeric structure of the complex [31]. This is combined with the shuttling of the complex between grana and stroma-exposed thylakoid domains, partial PSII disassembly, and highly specific proteolysis of the damaged D1 protein. Replacement of degraded D1 protein with a new copy requires a complex coordination between its degradation, resynthesis, insertion, and assembly into the PSII core. Although enhanced during higher irradiances, turnover of D1 occurs at all light intensities, and can be easily monitored in pulse-chase experiments, an efficient tool to study the phenomenon. It is now clear that most of the regulation of gene expression required for PSII repair cycle is exerted at the translational and posttranslational levels by the redox conditions in the thylakoid membrane and in the stroma.

D1 protein together with D2 constitutes the scaffold for all the electron transfer cofactors within RCII. It is therefore not surprising that its sacrificial high turnover produces profound effects on the photosynthetic activity as a whole.

There are intrinsic difficulties in understanding the rational of the proposed model: why should the D1 subunit be preferentially damaged by excess irradiation? How the specific proteases (belonging to the FtsH family) can actually recognize the damaged protein, considering that oxidative damage is probably a random process with no specific targets? The common view proposes a damage-induced conformational change on the D1 protein that would trigger its proteolytic degradation and some experimental evidences are brought about to support this view [32].

Let us take into consideration the main mechanism for oxidative damage inside the PSII core. When the frequency of excitation of P_{680} is higher than the rate of utilization of the reduction equivalents by PSI or, ultimately by CO_2 fixing activity (Calvin-Benson cycle), an over-reduction of the plastoquinone pool is produced and the electrons coming from RCII cannot find the physiological exit from the complex. In these conditions, i.e., when the electron acceptors are fully reduced and cannot receive further electrons, charge recombination at P_{680} produces its triplet form $^{T}P_{680}$ that is able to activate oxygen molecules to their singlet state $^{1}O_2$.

Normally carotenoids take care to avoid oxygen activation in the antenna by scavenging the chlorophyll triplet states by thermal dissipation. In the reaction center, there are actually carotene molecules, but none of them are close enough to P_{680} for the chlorophyll triplet to be transferred to the carotene and dissipated. If they were, they would immediately become oxidized by P_{680}^{+}. Singlet oxygen is the main responsible of direct or indirect oxidative damage to both pigments and proteins of PSII but it is not completely clear why the action of activated form of oxygen should be limited in their target to the D1 subunit.

A possible alternative model comes from experiments on PSII from cyanobacteria [33]. In this paper, it is shown that, under conditions of strong over-reduction of the PSII electron acceptors, massive chlorophyll bleaching from singlet oxygen is observed in a mutant that is not able to quickly degrade the D1 protein. On the contrary, in a mutant strain in which degradation of D1 takes place quickly, chlorophylls bleaching is very little, if any. On the basis of these and other experimental evidences, it is proposed that triggering of D1 degradation is not necessarily associated to an oxidative damage, which makes it a substrate for the proteolytic activity but, provided that the D1 subunit is maintained in the right conformation, the protein is simply turned over and replaced by the so-called "repair cycle" at a rate which is regulated by the redox state of the chloroplast. When the excitation supply exceeds the rate at which carbon dioxide can be fixed by the Calvin-Benson cycle, the reduction level increases and the DegP-FtsH proteases are activated to degrade D1. This mechanism ensure a protection to PSII before the oxidative damage by activated species of oxygen takes place.

It is like the case of a prudent car driver who decides to substitute its car tires after a given number of kilometers rather than waiting for them to blow up.

In this view, continuous degradation and resynthesis of D1 is the price the chloroplast must pay to preserve a very complex and delicate machinery, which is able to perform the redox chemistry that is fundamental for all forms of life on the biosphere.

References

1. A.A. Benson, M. Calvin, Ann. Rev. Plant Physiol. Plant Mol. Biol. **1**, 25–42 (1950)
2. H.M. Miziorko, G.H. Lorimer, Ann. Rev. Biochem. **52**, 507–535 (1983)
3. J. Goksoyr, Nature **214**, 1161
4. D.J. Murphy, Biochim. Biophys. Acta. **864**, 33–94 (1986)
5. F. Abdallah, F. Salamini, D. Leister, Trends Plant Sci. **5**, 141–142 (2000)
6. J.P. Thornber, R.S. Alberte, in *Encyclopedia of Plant Physiol; New Series. 5*, ed. by A. Trebst, M. Avron (Springer, Berlin Heidelberg New York, 1977) pp. 574–582
7. T. Forster, Ann. Phys. Leipzig. **2**, 55–75 (1948)
8. G. Britton, FASEB J. **9**, 1551–1558 (1995)
9. K.N. Ferreira, T.M. Iverson, K. Maghlaoui, J. Barber, S. Iwata, Science **303**, 1831–1838 (2004)
10. Z. Liu, H. Yan, K. Wang, T. Kuang, J. Zhang, L. Gui, X. An, W. Chang, Nature **428**, 287–292 (2004)
11. A. Ben Shem, F. Frolow, N. Nelson, Nature **426**, 630–635 (2003)
12. J.P. Abrahams, A.G.W. Leslie, G.R. Lutter, J.E. Walker, Nature **370**, 621–628 (1994)
13. D. Stroebel, Y. Choquet, J.L. Popot, D. Picot, Nature **426**, 413–418 (2003)
14. S. Jansson, Trends Plant Sci. **4**, 236–240 (1999)
15. P.R. Chitnis, Annu. Rev. Plant Physiol. Plant Mol. Biol. **52**, 593–626 (2001)
16. S. Caffarri, R. Croce, L. Cattivelli, R. Bassi, Biochemistry **43**, 9467–9476 (2004)
17. P. Joliot, Photosynth. Res. **76**, 65–72 (2003)

18. P.D. Boyer, Ann. Rev. Biochem. **66**, 717–749 (1997)
19. J.F. Allen, J. Bennett, K.E. Steinback, C.J. Arntzen, Nature **291**, 25–29 (1981)
20. D.I. Arnon, R.K. Chain, Proc Natl Acad Sci USA **72**, 4961–4965 (1975)
21. Z. Zhang, L. Huang, V.M. Shulmeister, Y.I. Chi, K.K. Kim, L.W. Hung, A.R. Crofts, E.A. Berry, S.H. Kim, Nature **392**, 677–684 (1998)
22. P. Joliot, A. Joliot, Proc. Natl. Acad. Sci. USA **99**, 10209–10214 (2002)
23. P. Joliot, A. Joliot, Proc. Natl. Acad. Sci. USA **102**, 4913–4918 (2005)
24. P. Joliot, A. Joliot, Biochim. Biophys. Acta. **1757**, 362–368 (2006)
25. R. Bassi, O. Machold, D. Simpson (1985) Carlsberg Res Commun **50**, 145–162 (1985)
26. O. Vallon, L. Bulte, P. Dainese, J. Olive, R. Bassi, FA Wollman, Proc Natl Acad Sci USA **88**, 8262–8266 (1991)
27. S.B. Powles, K.S. Chapman, F.R. Whatley, Plant Physiol **69**, 371–374 (1982)
28. Y. Hihara, K. Sonoike, in *Advances in Photosynthesis and Respiration*, vol. 11, (Kluver Academic Publishers, Dordrecht, 2001) pp. 507–525
29. B. Andersson, J. Barber, (1996) in Kluwer Academic Publishers, Dordrecht.
30. J.B. Marder, A.K. Mattoo, P. Goloubinoff, M. Edelman, in *Biosynthesis of the Photosynthetic Apparatus. UCLA Symposia* vol. 14, ed. by J.P. Thornber, L.A. Staehelin, R.B. Hallick (Liss, A.R., New York, 1984) pp. 309–311
31. R. Barbato, G. Friso, F. Rigoni, F.D. Vecchia, G.M. Giacometti, J. Cell Biol. **119**, 325–335 (1992)
32. E-M Aro, T. Hundal, I. Carlberg, B. Andersson, Biochim. Biophys. Acta. **1019**, 269–275 (1990)
33. E. Bergantino, A. Brunetta, E. Touloupakis, A. Segalla, I. Szabo, G.M. Giacometti, J. Biol. Chem. **278**, 41820–41829 (2003)

2

Exploiting Photosynthesis for Biofuel Production

C. Govoni, T. Morosinotto, G. Giuliano, and R. Bassi

During the recent "Energy Outlook and Modeling Conference" of March 2006 in Washington DC, it was estimated the world current energy consumption in 100 "Quads," where one Quad corresponds to about 25 million of oil equivalent tons (MTep, http://www.eia.doe.gov/oiaf/aeo/conf/). Nowadays, over 85% of world energy demands are met by the combustion of fossil fuels: coal, oil, and natural gas. Current oil reserves are estimated to be about 1,277 billions of barrels and, assuming a stable consumption, they would be sufficient for next 42 years. However, in the middle-term oil utilization is expected to rise of 1.6% every year thus making reserves exhaustion even faster (http://www.eia.doe.gov/oiaf/aeo/conf/pdf/petak.pdf). It is thus clear that to sustain our lifestyle to find alternative and renewable sources of energy is a striking and urgent need.

Besides the problem of fossil reserves depletion, the massive combustion of fossil fuels in the past decades also had a high environmental impact. In fact, this leads to releases of large amounts of carbon dioxide and other pollutants in the atmosphere. However, these emissions are reassimilated by natural processes, which cannot keep the pace of the present CO_2 production rates. In fact, every year the forests are able to fix about 1 billion of tons of carbon in organic matter and further 2 billion of tons are fixed in the ocean every year by the sea photosynthesis, but the CO_2 emissions caused by the human activity are about 6 billion of tons. So the balance is positive and every year 3 billion of tons increase the global level of CO_2 in the atmosphere (Fig. 2.1).

The alteration of this equilibrium caused an accumulation of carbon dioxide in the atmosphere, which is estimated to have risen of about 15% since 1750. The main consequence of the accumulation of CO_2 in the atmosphere has been proposed to be the "greenhouse effect" some gases like carbon dioxide, in fact, are able to retain the infrared radiation into the atmosphere and cause a global temperature increase of the planet. The correlation between CO_2 and temperature increase is clear, although a cause/effect relationship has not been clearly established, yet.

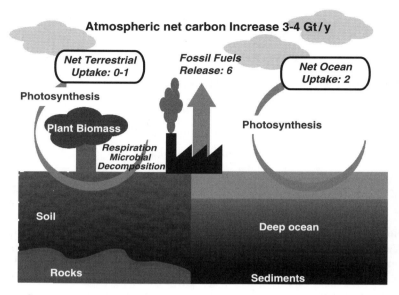

Fig. 2.1. Simplified global carbon cycle. Biogeothermal carbon cycle is schematized, distinguishing between terrestrial and aquatic processes. Photosynthesis, which fixes atmospheric CO_2 is indicated in *black*, while processes responsible of carbon dioxide release (respiration, microbial decomposition, and fossil fuels combustion) are indicated in *italic*. The net uptake activity of terrestrial and ocean activities are *circled*. Plants biomass, soil, rock, deep ocean, and marine sediments are indicated in *white* and are sites of long-term carbon storage.

This temperature increase has two major consequences. The first is desertification of subtropical areas, which caused a massive reduction of forests. As trees are photosynthetic organisms, which fix the atmospheric carbon dioxide into biomass, their massive reduction is causing further enhancement of carbon dioxide accumulation. Although a transitory effect, even more important is that desertification implies oxidation to CO_2 of the organic mass contained in the soil (1–2%). The second effect consists into the thawing out of the polar areas. In these regions, a large amount of CO_2 has been fixed as frozen organic matter in the underground during several million years. Because of the global warming, this organic matter is being thawed out and mineralized by bacteria and fungi causing a further CO_2 evolution.

The public conscience that the search of energy sources alternative to fossil fuels is fundamental for future prospects is generally increasing, and research efforts are devoted to this field in several countries. The attention is focused on renewable resources, such as sun and wind, which can be considered as no exhaustible. Furthermore, their massive exploitation will drastically reduce chemical, radioactive, and thermal pollution, and they therefore stand out as a viable source of clean and limitless energy.

Among the possible renewable energy sources, an interesting emerging option is the exploitation of higher plants or algae and their capacity to harvest solar light and convert it into chemical energy. This option is particularly interesting over other methods of solar radiation exploitation because these organisms are able to convert CO_2 into biomass, through the photosynthesis, and thus their utilization will also contribute to reduce carbon dioxide levels.

A further advantage is that, unlike fossil fuels, photosynthetic organisms are rather uniformly distributed over earth surface. Thus, their exploitation would not require large plants, and energy will be produced locally in small bio-power plants. This different spatial distribution of power plants has alone a positive effect on energy efficiencies: in fact, as a relevant percentage of energy is lost only during its transport from production to utilization, the reduction in distance would reduce significantly energy production needs.

Possible applications of plants or algae for energy production are described, with particular attention on their possible exploitation for biofuels and bio-hydrogen production.

2.1 Biological Production of Vehicle Traction Fuels: Bioethanol and Biodiesel

Any fuel derived recently from living organisms or from their metabolic by-product is defined as a biofuel. Unlike other natural resources like oil, coal, and nuclear fuels, biofuels are a renewable energy source. They present the advantage of a cleaner combustion in comparison to conventional fuels, and thus have a lower impact on atmospheric CO_2 levels.

Liquid biofuels are mainly developed for vehicle traction. At present the most promising options are bioethanol and biodiesel.

2.1.1 Bioethanol

Bioethanol can be utilized as a fuel in combustion engines in different ways: (1) as hydrous ethanol (95% by volume), containing small percentage of water, it can be used as a gasoline substitute in cars with modified engines. (2) As anhydrous (or dehydrated) ethanol, free from water or at least 99% pure, it can be blended with conventional fuels in ratios between 5 and 85% (E85). As a 5% component, it can be used in all recently produced engines without modification. Higher blends require modified engines installed in the so-called flexible fuel vehicles. (3) Finally, bioethanol is also used to manufacture ethyl-tertiary-butyl-ether (ETBE), a fuel additive for conventional gasoline.

Bioethanol is produced from starch plants (like grain, corn, and tubers-like cassava), sugar plants (sugar beet or sugar cane), and – although not yet on a large scale – from cellulose. Carbohydrates present in these plants are used as a substrate for the microbial fermentation, and ethanol is subsequently enriched by distillation/rectification and dehydration. The CO_2 liberated during its combustion was first fixed from the atmosphere by the crops during their growth, and thus the use of bioethanol in place/addition of fossil fuels would then reduce the net release of greenhouse gases in the atmosphere. The potential impact of this strategy is large if we consider that road transport accounts for 22% of all greenhouse gas emissions (http://:www.foodfen.org.uk). Bioethanol would have a positive environmental impact also because it is biodegradable and not toxic for living organisms as fossil fuels. A further advantage is that bioethanol can be easily integrated into the existing fuel distribution system. In fact, as mentioned, in quantities up to 5%, bioethanol can be blended with conventional fuel without the need of engine modifications and adaptation of forecourts and of transportation system.

Finally, its increased use would also have positive a socio-political impact, by reducing the dependence from the oil producing countries and boosting the rural economy.

2.1.2 Biodiesel

Biodiesel is an alternative fuel, which can be produced from vegetal resources. It is made by the *trans*-esterification of vegetal lipids generating lipid methyl esters (the chemical name for biodiesel) and glycerine, which is a valuable by-product used in soaps and other products.

In USA, biodiesel is mainly produced from soybean oil, while in Europe from rapeseed oil. In addition to these crops, biodiesel can be produced using some microalgae, which are able to accumulate large amount of triglycerides in their cells. Biodiesel contains no petroleum oil, but it can be blended at any level with petroleum diesel, and in this form it can be used in recent compression-ignition (diesel) engines with little or no modifications. Thus its introduction, even if massive, would not need modification in the fuel distribution systems. It also shares other advantages with bioethanol. As it is produced from vegetal sources, its utilization would have no net impact on the concentration of carbon dioxide in the atmosphere. Furthermore, biodiesel is also biodegradable, nontoxic, and essentially free of sulphur and aromatics.

2.1.3 Biofuels Still Present Limitations Preventing Their Massive Utilization

Although bioethanol and biodiesel present big advantages for the environment and they could rather easily substitute at least partially fossil fuels, there are still two major limitations to the massive use of biofuels: energetic yield and costs.

First, the energetic yield is calculated as the ratio between all the energetic inputs (used for the cultivation of the crop, transport, and transformation) and all the outputs (fuel, but also all the by-products, as glycerine in the case of biodiesel). The energetic yield of biodiesel from soya, for example, is about 3.2, meaning that for every unit of energy employed in the production of biodiesel, 3.2 units are harvested at the end of the process. Performances with bioethanol from maize are even less encouraging being its energetic yield about 1 or even less [1, 2].

Second, considering the present oil price, biofuels are still not economically competitive. In fact, biodiesel is estimated to be economically convenient over the limit of 80$/barrel (and even higher for bioethanol). So, at least on a short term, biofuels introduction in the commercial networks needs to be financially supported by governments willing to stimulate their utilization. However, in a medium timespan perspective, when oil reserves will be exhausting and the oil price will consequently rise, biodiesel will be a major option for vehicle traction and investments will be paid back.

2.2 Hydrogen Biological Production by Fermentative Processes

Hydrogen has some very interesting properties, which makes it the potential energy source of the future. It is easily converted into electricity, and it has very high energy content per weight unit. Moreover, its utilization has the lowest possible environmental impact. It produces electricity by reacting with oxygen and yielding water as the only waste. However, these advantages are effective only if hydrogen is produced by an efficient and clean process. Unfortunately, this is currently not the case, and hydrogen is mainly derived from fossil fuels and its production generates considerable amounts of CO_2 in addition to other air pollutants such as sulphur dioxide and nitrogen oxides (http://www.fao.org/docrep/w7241e/w7241e05.htm).

Thus, hydrogen's positive impact on the energy system depends on the development of new methods for its efficient production. One major option is the water electrolysis, which, however, requires an efficient, clean, and cheap method for electricity generation, which is at present not available.

For this reason, other options are receiving an increasing attention and, among them, the biological H_2 production. In fact, some living organisms are able to produce H_2 in anaerobic conditions thanks to metabolic reactions coupled with fermentation, nitrogen fixation, or photosynthesis. This is a very promising approach, as it has the potential to provide cheap hydrogen without using fossil fuels and emitting pollutants.

Living organisms can synthesize hydrogen from protons and electrons, using special enzymes named hydrogenases. Hydrogenases fall into two major classes: Ni-Fe and Fe-Fe hydrogenase identified from the metal content in their active site [3]. They differ in the composition and in the structure of the metal center active site as well as in their polypeptide sequence, but also show some conserved properties, namely the presence of CN and/or CO ligands in the active site. In general, Fe-Fe hydrogenases were shown to be more efficient catalysers, and therefore they are more promising for practical exploitations [4]. Both Fe-Fe and Ni-Fe hydrogenases have a strong sensitivity to O_2 presence, probably because of its binding to the metallocatalytic sites. For this reason, all hydrogen producing organisms identified till very recently are active in anaerobic conditions only. Exceptions to this pattern are still under evaluation.

Several organisms can produce hydrogen and differ for the source of reducing power they use for this biosynthesis. Some bacterial strains can exploit organic compounds by fermentation, while other organisms, such as green algae, are able to use directly the solar radiation harvested by the photosynthetic apparatus. However, we must be aware that, despite these strong potential advantages, several bottlenecks need to be resolved before practical application can be economically sustainable and further research is still needed.

2.2.1 Hydrogen Production by Bacterial Fermentation

Several bacteria are able to produce hydrogen from the fermentation of carbohydrates in anaerobic conditions. As substrates for this reaction, several sources could be used. However, the use of human and industrial wastes is particularly interesting as it attains two objectives at the same time: processing the waste and produce environmental-friendly energy.

Anaerobic bacteria use organic substances as source of electrons and energy, converting them into hydrogen. The reactions involved in hydrogen production (2.1 and 2.2) are rapid and suitable for treating large quantities of wastewater in large scale fermenters.

$$\text{Glucose} + 2H_2O = 2\text{Acetate} + 2CO_2 + 4H_2 \quad \Delta G = -184.2 \, \text{kJ} \quad (2.1)$$

$$\text{Glucose} = \text{Butyrate} + 2CO_2 + 2H_2 \quad \Delta G = -257.1 \, \text{kJ} \quad (2.2)$$

The produced molecules, acetate and butyrate, still have a residual chemical energy, which remains unexploited. This is because the complete oxidation of carbohydrates to H_2 and CO_2 is not a spontaneous process as it has a ΔG of only $-46 \, \text{kJ mol}^{-1}$ hydrogen. There are no known fermentation pathways that can achieve a conversion efficiency larger than 4 mol H_2/mol hexose [5].

To achieve a further decomposition of the remaining organic substances, an external energy input is necessary. This can be derived, as recently described, from the application of an external small electric potential, which stimulates the fermentation of acetate to H_2 in modified microbial fuel cells [6].

Another valuable option is to couple the first fermentation step by anaerobic bacteria with the second one performed by photosynthetic sulphur bacteria, which use the residual organic acids to produce hydrogen (Fig. 2.2). This last step, which completes the fermentation of organic wastes to hydrogen, can be performed in these organisms, even if energetically unfavorable, as they employ light energy to drive the reaction.

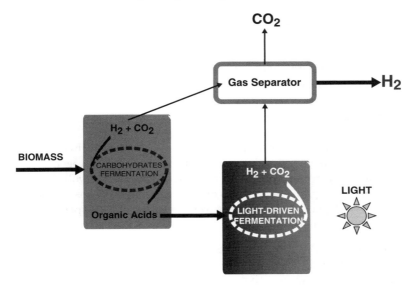

Fig. 2.2. Hydrogen production by dark and light fermentation of organic wastes. Carbohydrates from several sources such as human or industrial wastes are fermented from anaerobic bacteria into organic acids, carbon dioxide, and hydrogen. The former are decomposed by photosynthetic bacteria, using light energy to drive the reaction. Gases yields from both fermentation processes are separated by a gas separator, yielding pure hydrogen.

Clearly, more research and development is required but, when developed, H_2 fermentations from organic wastes would be relatively low cost comparable to methane fermentation (in the range of $4–8/MBTU[1]) as they could also use similar mixed-tank reactors, without need for the building of new installations. Thus, dark H_2 fermentation of wastes is close to be competitive with fossil fuel-derived H_2, providing a first approach to large scale biohydrogen production.

2.3 Hydrogen Production by Photosynthetic Organisms

In addition to the mentioned sulphur bacteria, other photosynthetic microorganisms are able to produce hydrogen. However, they do not use light energy only to drive a fermentative reaction, but also to directly decompose water to hydrogen and oxygen. This possibility is clearly very attractive from the

---

[1] *British thermal unit*. One Btu is equal to the amount of heat required to raise the temperature of one pound of liquid water by 1°F at its maximum density, which occurs at a temperature of 39.1°F. One Btu is equal to ~251.9 calories or 1,055 J.

point of view of environmental impact because this would be the perfect way to produce energy, using only two renewable resources, solar radiation and water.

The photosynthetic microorganisms able to produce hydrogen are cyanobacteria and green algae.

2.3.1 Cyanobacteria

Cyanobacteria can use two different enzymes to generate hydrogen gas [7]. The first is *nitrogenase*, which catalyzes the fixation of nitrogen into ammonia, producing hydrogen as a by-product. Hydrogen photo-evolution catalyzed by nitrogenases requires anaerobic conditions, as hydrogenases. As oxygen is a by-product of photosynthesis, cyanobacteria have developed two different strategies to achieve anaerobiosis:

1. Some evolved the capacity of building structures with reduced oxygen-permeability, called heterocysts, thus separating physically oxygen evolution from nitrogenase activity [8].
2. Nonheterocystous cyanobacteria instead separate temporally oxygen evolution from nitrogenase activity activating the latter only during dark periods.

Whatever is the case, hydrogen production by nitrogenase enzyme is interesting but presents an important limitation. As the reaction is coupled to the nitrogen fixation, it is energetically demanding, making the whole process poorly efficient.

The other hydrogen-metabolizing enzymes in cyanobacteria belong to the class of hydrogenases. One class, *uptake hydrogenase* (encoded by *hupSL gene*), however, catalyzes the oxidation of hydrogen molecules to produce protons and electrons [9]. Uptake hydrogenase enzymes are found in the thylakoid membrane of heterocystis from filamentous cyanobacteria, and they transfer the electrons from hydrogen for to the respiratory chain. Nitrogenases of this type are not responsible for net hydrogen production in vivo, rather, allow for partial recuperation of energy invested in N_2 fixation and employed for hydrogen production. In the perspective of the production of hydrogen molecules, their activity is not desirable and should be inhibited.

The second type of hydrogenases are *reversible, or bi-directional hydrogenases* (encoded by *hoxFUYH*) that can either consume or produce hydrogen. The biological role of these bidirectional or reversible hydrogenases is poorly understood and thought to be involved in the ions level control of the organism. Reversible hydrogenases are associated with the cytoplasmic membrane and likely function as electron acceptor from both NADH and H_2 [10]. If opportunely regulated, these enzymes can be exploited for hydrogen production.

2.3.2 Eukaryotic Algae

Gaffron and coworkers over 60 years ago discovered that green algae, when illuminated after an anaerobic incubation in the dark, have the ability to produce H_2 [11, 12]. This hydrogen production is directly coupled with water oxidation through Photosystem II (PSII) and the photosynthetic electron transport chain. Thus, H_2 is synthesized directly from water, using sunlight as a source of energy, with a totally clean process, making very promising the perspective of exploiting green algae for a large scale H_2 production. Moreover, differently from the cyanobacteria, H2 production is achieved by Fe-Fe hydrogenases, thus in a more efficient way.

2.4 Challenges in Algal Hydrogen Production

The exploitation of algae for hydrogen production is probably the method with the best perspectives in the long term among those that are analyzed here. However, it has two major limitations that prevent its large scale exploitation: (1) the great sensibility of the process to the oxygen presence; (2) the inefficient light energy distribution in the bioreactor.

2.4.1 Oxygen Sensitivity of Hydrogen Production

As mentioned, hydrogen is always produced in anaerobic conditions because all hydrogenases are sensitive to oxygen. This limitation becomes deleterious when oxygenic photosynthetic organisms such as algae are discussed: in fact, they use water as electron donor to produce protons and oxygen. The latter is thus an unavoidable by-product of photosynthesis and even the incubation in anaerobic conditions is sufficient to avoid inhibition of hydrogen production.

Different strategies are possible to obtain a significant level of hydrogen production by photosynthetic organisms: one is to separate temporally or spatially oxygen evolving process from the hydrogen production. Alternatively, modified hydrogenase enzymes, less sensitive to O_2, can be searched.

Some cyanobacteria, as mentioned, can maintain a state of controlled anaerobiosis in heterocystis; however, it is difficult to imagine modifying green algae to make them able to build such complex structures. In the case of algae, thus, the approach of a temporal separation of oxygen and hydrogen evolutions is more promising. In 2000, Melis et al. [13] demonstrated in *Chlamydomonas reinhardtii* that, when sulphur in the growth medium is limiting, rate of oxygenic photosynthesis declines without affecting significantly the mitochondrial respiration. In fact, under illumination Photosystem II (PSII) polypeptides, in particular the D1 subunit, have a very fast turnover. Under sulphur deprivation, when the protein biosynthesis is inhibited, PSII is thus among the first protein complexes affected, causing the decrease of photosynthetic activity and oxygen evolution. On the contrary, the mitochondrial complexes, which have a lower turnover rate, can maintain longer their function. In sealed cultures, the imbalance between photosynthesis and respiration results in a net consumption of oxygen by the cells and the establishment of anaerobiosis, which stimulates the H_2 production.

Sulphur deprivation, however, is a stress condition for the cells and after a few hours H_2 production yield decreases. Therefore, it is necessary to restore a normal level of sulphur in the culture to allow the cells to recover, re-establish photosynthesis and reconstitute energy reserves.

However, the process is reversible and after a recovery delay the sulphur deprivation treatment can be repeated for further induction of hydrogen production. In fact, it has been demonstrated that phases of S deprivation and H_2 production can be alternated to phases where sulphur is supplied and H_2 is not produced. The maximum energy yield obtained by this method is approximately 10%, so close to the best photovoltaic cells. The cost is lower, however, as solar cells need to be built while algal cells reproduce. However, the biological systems can reach maximum yields only in a transitory way, while in the long-term average yields are still too low (0.1–1%).

In addition to the still insufficient yields, the implementation of this method of hydrogen production on a large scale system is complicated by the need of changing the growing medium to add or remove sulphur. For this reason, it is important to search more efficient methods to induce anaerobiosis, as PSII is the only responsible of oxygen evolution, and the regulation of its expression is a relatively simple method to induce or suppress anaerobiosis. For this reason in the perspective of a large scale application, it would be more practical to have *Chlamydomonas* strains where one or more genes essential for the PSII assembly are under the control of an inducible and easily controlled promoter. If the presence or absence of PSII can be readily controlled in the culture, it would be possible to control the succession of anaerobiosis and aerobiosis.

The oxygen sensitivity of the hydrogen production is mainly due to the sensitivity of the enzyme hydrogenase itself. In fact, these enzymes are inactivated by the presence of very low oxygen concentrations. The exact reason for this sensitivity is not fully understood at molecular detail and is the subject of intense research. However, as different hydrogenases have variable structural organization but share a similar active site, the nature of this latter component is probably causing oxygen sensitivity.

The most efficient method to obtain large scales hydrogen production would be to find an oxygen resistant hydrogenase. To this aim two different approaches are possible: first, to search for hydrogenase mutants with an higher selectivity for hydrogen. In particular, the idea of reducing the oxygen diffusion to the active site by modulating the molecular size of the gas channels within the molecule looks promising. The second possible approach is to find organisms that can synthesize hydrogen even in presence of oxygen and thus naturally have oxygen resistant hydrogenases. In this respect, it is interesting the finding that such organisms may actually exist in nature, as demonstrated by the example of *Thermotoga neapolitana*, which was shown to produce hydrogen even in the presence of 4–6% oxygen [14]. The gene encoding the hydrogenase of this organism has been isolated and sequence analysis suggested that its superior resistance could be due the fact that this enzyme is generally found as a trimer, while all the other hydrogenases known are monomers. Thanks to this larger molecular size, so that its active site can be buried and less accessible to oxygen molecules. However, this explanation alone is insufficient as a homologous organism, *T. maritima* has an hydrogenase with similar size but it is not equally resistant to oxygen. In a close future perspective, the study of this hydrogenase would suggest possible ways to further improve oxygen resistance, while the expression of this resistant hydrogenase in *Chlamydomonas* cells would allow understanding the possible impact of a oxygen-resistant hydrogenase on hydrogen yields.

2.4.2 Optimization of Light Harvesting in Bioreactors

The first step in hydrogen production is light absorption by photosynthetic complexes. When algae grow in photobioreactors, this process has a poor efficiency because of the bad distribution of light. In fact, to have good production yields, biomass concentration need to be high, leading to increased optical density of the culture where light can only penetrate for a few millimeters, as shown in Fig. 2.3. As a consequence, the external cell layers are exposed to a very intense light leading to activation of photo-protective mechanisms to dissipate energy in excess, to avoid formation of oxygen reactive species and consequent cellular damages. This cells population, thus, is not efficient in the light utilization because absorbed energy is dissipated rather than used for metabolic reactions. On the contrary, cells located at more internal level in the culture are exposed to very low light, insufficient for sustaining intense metabolic activity.

This unequal light distribution has another negative consequence: the connective fluxes in the medium cause algal cells to stir into the reactor where they can move from dark to strong light within few seconds. The fast variation in light intensity leads to an input of reducing power into the photosynthetic chain at very different rates, without allowing the time for acclimation of photosynthetic apparatus [15]. These conditions are highly stressful because, without the delay needed for the activation of photo-protective mechanisms, the energy absorbed in excess generates oxidative stress and leads to PSII photoinhibition.

To overcome these problems and increase the productivity of photobioreactors, two type of interventions are needed:

1. To reduce the antenna size to lower the optical density of the culture and allow for a better light distribution within the bioreactor without reducing the cellular density
2. To strengthen the algal mechanisms of oxidative stress resistance and of thermal energy dissipation

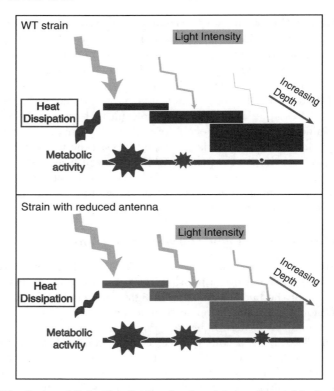

Fig. 2.3. The limits of light distribution in photobioreactors yields. In photobioreactors, a high cell concentration should be maintained to obtain a reasonable yield. This leads to unequal light distribution. In fact, more exposed cellular layers (*left*) absorb a very high light, which can be used only partially for metabolism. A large fraction is also dissipated thermally to avoid establishing oxidative stress in cells. Internal layers (*right*) are illuminated by a very poor light, insufficient to support intense metabolic activity. If a strain with reduced antenna is used in the photobioreactor, with an equal cellular concentration, the optical density of the culture is lower and light is more efficiently distributed. External cells absorb a lower proportion of available light, which thus reaches more internal layers supporting metabolic activity.

Among the different mechanisms, it is important to increase the capacity to respond to fast alterations in light intensity.

The only strategy attempted so far to increase the light distribution in photobioreactors has been the use of mutants lacking all the antenna proteins, a class of proteins that are responsible of a large fraction of light harvesting in photosynthetic eukaryotes. With this approach, light distribution has been indeed improved. However, cells suffered for strong photo-sensitivity and did not survive when exposed to full sunlight. This puzzling phenotype (lower antenna size is expected to decrease the number of photons funnelled to RCII-reaction centre II- and thus reduce sensitivity) is explained by the observation that antenna proteins (named Lhc) are not only involved in light harvesting but also have a major role in photoprotection [16]. As a consequence, the deletion of all antenna proteins reduces the capacity for photoprotection to a level making insignificant the advantage obtained in terms of light distribution.

However, among Lhc polypeptides, individual members proteins are specialized in light harvesting, while others in photoprotection [17]. Thus, to obtain mutants optimized to bioreactor conditions, the first step consists in identifying members of *Chlamydomonas* Lhc protein family essential for photoprotection vs. those that can be deleted to lower the optical density.

To this aim, we apply two different approaches: in vivo approach consists in the creation of a library of *Chlamydomonas* insertional mutants. In these mutants genes are randomly knocked out by the insertion of a resistance cassette. These mutants are screened for a smaller antenna size, by analyzing their fluorescence yield properties and Chl a/b ratio. Strains with a reduced antenna content, in fact, would have lower cell fluorescence yield and higher Chl a/b ratio, as Chl b is specifically bound to Lhc proteins.

An in vitro approach is complementary: individual members of Lhc family are studied one by one. Each Lhc protein identified in the genome [18] is characterized by expressing in bacteria the corresponding polypeptides and refolded in vitro to obtain the native pigment protein complex. By this method, each protein will be characterized individually for its capacity of light harvesting and photoprotection.

References

1. A.E. Farrell, R.J. Plevin, B.T. Turner, A.D. Jones, M. O'Hare, D.M. Kammen, Science **311**, 506–508 (2006)
2. C.N. Hunter, J.D. Pennoyer, J.N. Sturgis, D. Farrelly, R.A. Niederman, Biochemistry **27**, 3459–3467 (1988)
3. P.M. Vignais, B. Billoud, J. Meyer, FEMS Microbiol. Rev. **25**, 455–501 (2001)
4. M.W. Adams, Biochim. Biophys. Acta **1020**, 115–145 (1990)
5. R.K. Thauer, K. Jungmann, K. Decker, Bacteriol. Rev. **41**, 100–180 (1977)
6. H. Liu, S. Grot, B.E. Logan, Environ. Sci Technol. **39**, 4317–4320 (2005)
7. D. Dutta, D. De, S. Chaudhuri, S.K. Bhattacharya, Microb. Cell Fact. **4**, 36 (2005)
8. B. Bergman, J.R. Gallon, A.N. Rai, L.J. Stal, FEMS Microbiol. Reviews **19**, 139–185 (1997)
9. P. Tamagnini, E. Leitao, F. Oxelfelt, Biochem. Soc. Trans. **33**, 67–69 (2005)
10. G. Boison, H. Bothe, A. Hansel, P. Lindblad, FEMS Microbiol. Letters **174**, 159–165 (1999)
11. H. Gaffron, Nature 204–205 (1939)
12. H. Gaffron, J. Rubin, J. Gen. Physiol. **26**, 219–240 (1942)
13. A. Melis, L. Zhang, M. Forestier, M.L. Ghirardi, M. Seibert, Plant Physiol. **122**, 127–136 (2000)

14. S.A. Van Ooteghem, A. Jones, L.D. Van Der, B. Dong, D. Mahajan, Biotechnol. Lett. **26**, 1223–1232 (2004)
15. A.V. Vener, P.J. van Kan, P.R. Rich, I.I. Ohad, B. Andersson, Proc. Natl. Acad. Sci. USA **94**, 1585–1590 (1997)
16. D. Elrad, K.K. Niyogi, A.R. Grossman, Plant Cell **14**, 1801–1816 (2002)
17. P. Horton, A. Ruban, J. Exp. Bot. **56**, 365–373 (2005)
18. D. Elrad, A.R. Grossman, Curr. Genet. **45**, 61–75 (2004)

3

In Between Photosynthesis and Photoinhibition: The Fundamental Role of Carotenoids and Carotenoid-Binding Proteins in Photoprotection

G. Bonente, L. Dall'Osto, and R. Bassi

3.1 When Light Becomes Dangerous for a Photosynthetic Organism

During operation of oxygenic photosynthesis, highly reactive molecules such as excited chlorophylls are placed in an environment, the chloroplast. This is probably the very spot where oxygen concentration is the highest on earth. There are two major sources of reactive oxygen species (ROS) in the photosynthetic apparatus. The first source is electron transport, where electrons are extracted from water and transported to $NADP^+$. The powerhouses of photosynthetic electron transport are the PSI and PSII reaction centers, which transfer one electron at time. Now, relatively stable oxygen forms are, respectively, the most oxidized one (O_2) and the most reduced one (H_2O). All the other intermediate redox states are highly reactive and are, in fact, called ROS. It can easily be understood that during a multistep electron transport chain, the occasions in which O_2 can be exposed to reduction by a single electron thus yielding superoxide are easily produced. Moreover, in PSII, the site where electrons are extracted one by one from water, intermediate redox states may be produced.

The second major source of ROS is the process of light energy absorption by chlorophylls and transfer of the excitation energy between the many chlorophylls forming the antenna system. Plants are particularly prone to photo-oxidative damage, and for the same reasons they are effective at photosynthesis, namely because the primary pigment, chlorophyll (Chl) is a very efficient sensitizer. In fact, chlorophyll has a long living singlet state (5 ns) thus allowing for intersystem crossing and formation of triplet chlorophyll-excited states, which can react with O_2, a triplet in its ground state, to yield $^1O_2^*$. This reactive species as well as others deriving from its reaction with water and organic molecules cause oxidative damage to occur in proteins, lipid, and pigments, leading to photoinhibition of photosynthesis and, ultimately, to photobleaching of pigments.

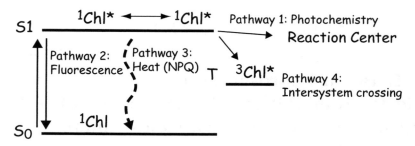

Fig. 3.1. The singlet chlorophyll (S1) deexcitation pathways: Pathway 1, energy utilization in photosynthesis (photochemical quenching); pathway 2, the energy reemitted as fluorescence; pathway 3, thermal dissipation pathway (NPQ, nonphotochemical quenching); pathway 4, the singlet chlorophyll conversion into triplet (intersystem crossing)

When does this happens and how? When a photon is absorbed by a chlorophyll molecule, the molecule is promoted to its first singlet excited state. This excited state does not live long because it is usually dissipated through different pathways (Fig. 3.1); the most important (for photosynthesis) is photochemistry, i.e., charge separation and electron transport. If this pathway is active, most of the energy is used through this way and Chl lifetime is below 1 ns. The second pathway is fluorescence: the photon is reemitted and Chl goes back to its ground state. This pathway is less important and fluorescence is usually very low (below 1%); however, it is very useful for scientists as it tells us in any moment the level of excited states in the photosynthetic system. The third path is heat dissipation, which can be extremely variable. Normally, it changes in a complementary way with respect to photochemistry. When the photochemistry is high, there is very little heat dissipation and the yield of the photosynthetic process is high. When some factor (water, temperature, CO_2) limits photochemistry, then mechanisms are activated for unharmful dissipation of the excitation energy into heat. The reason for this careful control is actually to keep as low as possible the number of excitons entering the last pathway of deexcitation: intersystem crossing (IC). Intersystem crossing is the conversion, through electronic spin inversion, of chlorophyll single-excited state into triplet-excited state.

As light increases, chlorophyll-excited states concentration stays low as far as photochemistry keeps the pace. When photochemistry is limited, heat dissipation is activated, after saturation of heat dissipation pathway, no further prevention is available and the probability of intersystem crossing occur increases, followed by photodamage. Photodamage can be limited by another set of mechanisms collectively indicated as "ROS scavenging." Singlet oxygen seems to be the main species produced in photoinhibited membranes, other reactive species are involved, as superoxide anion (O_2^-), which is produced at both PSII oxygen evolving complex and PSI (Mehler reaction), hydrogen peroxide (H_2O_2), and hydroxyl radical ($\cdot OH$), species which are easily produced when the electron transport carriers are over-reduced. A major target is membrane lipids, which undergo peroxidation. This proceeds by a free radical chain reaction mechanism. The level of lipid peroxidation, in fact, is one of the most valuable indicators for photooxidative damage.

This article reviews several photoprotective processes that occur within chloroplasts of eukaryotic photosynthetic organisms, starting with medium and long-term response strategies to photooxidative stress, as state transition and acclimation. Following, a particular emphasis will be laid on the molecular mechanisms, which prevents ROS formation or participate in ROS detoxification inside the chloroplast. In these mechanisms, carotenoids and carotenoids-binding proteins, which belong to the Lhc family, have a key role.

3.2 Acclimation

When photosynthetic organisms are exposed to different light conditions with respect to those of growth, acclimation processes are activated, which consist into remodeling of the PSII composition. In case of acclimation to high light, e.g., we observe a strong decrease of LHCII trimeric complexes yielding a lower cross section for each PSII reaction center complex, while PSI complexes act as a pivot reference and do not change their composition [1]. An extreme example of such acclimation has been described in the green alga *Dunaliella salina*, where the chlorophyll antenna size of PSII has been reported to be as small as 60 chlorophyll molecules under high light conditions, while as large as 460 chlorophyll molecules under low light growth [2]. The light-harvesting antenna reduction seems to be a common strategy, across evolutionary distant organisms, adopted to optimize photosynthesis and avoid ROS production in stressing conditions. Besides the ratio between antenna and reaction center complexes, the stoichiometry of electron transport components with respect to light-harvesting components also changes to sustain an higher electron transport rate. Also, enzymes involved in metabolic sinks, such as the Calvin cycle, increase their relative abundance and activity [1]. The mechanism for stoichiometric adaptation of PSII antenna size is posttranslational: accumulation of zeaxanthin induced by the excess light [3] binds to Lhc proteins into a specific allosteric site called L2 [4], thus inducing a conformational change [5, 6], which leads to monomerization and degradation of LHCII but not of minor Lhcb components [7]. Increase in the stoichiometry of others electron transport components, such as cytochrome $b_6 f$ complex, is likely to be controlled by transcriptional activation [8].

Although acclimation to constant light conditions has been studied to some extent, the resulting architecture of the photosynthetic apparatus can only reach a compromise condition reflecting the average conditions over a period of several days, as shown by the need of at least three days excess light period to elicit adaptative proteolysis of LHCII [9].

In fact, over a single day, light intensity and temperature change dramatically from dawn to midday and sunset, thus requiring regulation mechanisms operating over a short time-span (minutes, hours) and thus cannot be activated by the synthesis/degradation of photosynthetic components, whose everyday operation would be energetically unsustainable. These short-term processes will be described later.

3.3 State 1–State 2 Transitions

State transitions are a protective mechanism, which acts through a reversible protein modification in a few minutes timescale [10, 11] to balance the energy pressure between the two photosystems, physically displacing LHCII antenna complexes from PSII to PSI [12].

Higher plants and green algae oxygenic photosynthesis works through the synchronized action of two photosystems: PSII reaction center, whose maximum absorption peak is at 680 nm, and PSI, where it is at 700 nm.

The coordinated and energy collection by the two photosystems is a necessary condition for the proceeding of photosynthesis and the generation of a strongly reducing molecule, which is able to transfer its electron to $NADP^+$.

The kinetically limiting step in the electron transport chain between PSII and PSI involves the oxidation of liposoluble hydrogen carrier plastoquinone (PQH_2), which once reduced at the Q_b site PSII, diffuses in the thylakoid membrane to cytochrome b_6f complex. In turn, cytochrome b_6f is maintained to an oxidized state by the activity of PSI, which transfers electrons to $NADP^+$. Thus, over-excitation of PSII leads to over reduction of PQ, leaving PSII without a suitable electron acceptor. PSII undergoes charge recombination and closure of energy traps thus increasing Chl lifetime and singlet oxygen production.

To avoid this problem, the PQH_2 acts as a signal activating chloroplastic protein kinase [13] to allow phosphorylation of LHCII antenna complex associated to PSII (State I). Phosphorylated LHCII undergoes conformational change, disconnection from PSII, and migration to the stroma lamellae where it associates to PSI (State II). This increases light harvesting, and thus electron transport capacity of PSI thus compensate for the uphill electron carriers over-reduction. Consequence is the oxidation of PQH2 to PQ and downregulation of the LHCII kinase. Dephosphorylation of LHCII and its reassociation to PSII reaction centers is ensured by chloroplastic phosphatases. This mechanism is photoprotective as it balances a potentially dangerous over-excitation on PSII, source of ROS, and it regulates light-harvesting activity depending on light intensity and light spectral quality changes consequent to rapid transitions from shade to full sunlight. In fact, the absorption spectrum of PSI is red-shifted with respect to PSII.

In the green alga *C. reinhardtii* state transitions are the main photoprotective strategy. This organism is able to shift from PSII to PSI up to 80% of its LHCII complexes, while in higher plants this parameter is not more than 15% [4].

3.4 Carotenoids Play a Fundamental Role in Many Photoprotection Mechanisms

The role of carotenoids in photoprotection of photosynthetic systems is extremely important as shown by the early experiments of [15] with carotenoidless bacterial strains and by the effect of herbicide norflurazon on plants [16]. Carotenoids are 40 carbon atoms polyisoprenoid compounds, made by the condensation of eight isoprene units. In this class of compounds, we can distinguish between carotenes, linear or cyclic hydrocarbons, and xanthophylls, their oxygenated derivatives. Carotenoids are located in the photosynthetic membranes, within the chloroplast, both free in the lipid bilayer (1–2%) and bound to the subunits of PSI and PSII in specific binding sites whose occupancy is necessary for the folding of these pigment–protein complexes. In higher plants, β-carotene binds to reaction center subunits of both PSI (PsaA and PsaB) and PSII (PsbA,PsbB,PsbC,PsbD). Xanthophylls (lutein, violaxanthin, neoxanthin, and zeaxanthin) are, instead, bound to Lhca and Lhcb proteins forming the outer antenna complexes of PSI and PSII, respectively. In *C. reinhardtii*, the additional xanthophyll species loroxanthin is present, also bound to Lhc complexes. Location of carotenoid species in both PSI and PSII supramolecular complexes is shown in Fig. 3.2.

Together with β-carotene, xanthophylls act both as photoreceptors, absorbing light energy, which is used in photosynthetic electron transport, and as photoprotectants of the photosynthetic apparatus from excess light energy as well as from the reactive oxygen species that are generated during oxygenic photosynthesis.

Carotenoids exert their photoprotective role through several mechanisms: (a) Chlorophyll triplet quenching: this action protects thylakoid lipids from peroxidation by quenching triplet chlorophyll species, which prevents production of reactive oxygen species (ROS) [17–19]; (b) ROS scavenging: incomplete quenching of ^{3}Chl* may yield into ROS production, which can be scavenged by reaction with carotenoids; (c) Activation of heat dissipation of excess energy (NPQ): specific xanthophylls (lutein, zeaxanthin) are needed for operation of nonphotochemical quenching of excess light energy (NPQ), thus decreasing probability for IC and ^{3}Chl* formation. All these mechanisms are cooperatively activated for protection of photosynthetic apparatus when environmental conditions promote photooxidative stress.

It is worth nothing that all these mechanisms are rapidly activated: carotenoids detoxify short-living reactive species such as singlet oxygen ($^1O_2$),

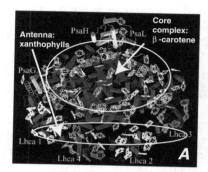
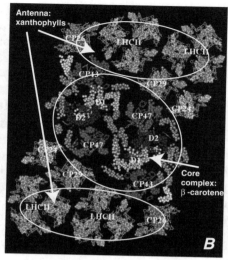

Fig. 3.2. Location of carotenoid species in both PSI (**a**) and PSII (**b**) supramolecular complexes. (**a**) PSI image by [58], based on PSI crystal structure by [59]; (**b**) PSII image by [60]

whose lifetime is ∼200 ns, while the activation of heat dissipation of excess energy (NPQ) requires less than a minute. In fact, carotenoids are involved in many of the mechanisms that allows survival of the photosynthetic cell when light intensity and photosynthetic electron transport rate undergo sudden changes, thus providing a more efficient response in conditions of repetitive excess light exposure, such as under sun flecks deriving from the overcasting of the leaf canopy; particularly, those mechanisms that are localized in the hydrophobic bilayer of the photosynthetic membrane, hosting reaction centers, and antenna proteins. Additional mechanisms located on the soluble phase of the chloroplast stroma are also of great importance for photoprotection and are based on the activity of superoxide dismutase, ascorbate peroxydase [20]. Their role has been the object of excellent reviews and will not be discussed here.

The study of xanthophyll function in plants and algae has been carried on in the most recent years through an integrative approach, using genetics, physiology, biochemistry, and molecular biology [21]. First of all, it can be asked why plants do actually need xanthophylls at all. In fact all the main photoprotective functions of xanthophylls (namely, triplet quenching and ROS scavenging) are well-known properties of β-carotene [22]. Thus, the extreme conservation of carotenoid composition in plant species has little or no explanation on the basis of published chemical properties of these molecules. Nevertheless, the fact that all plant species and to a large extent green algae contain, be-

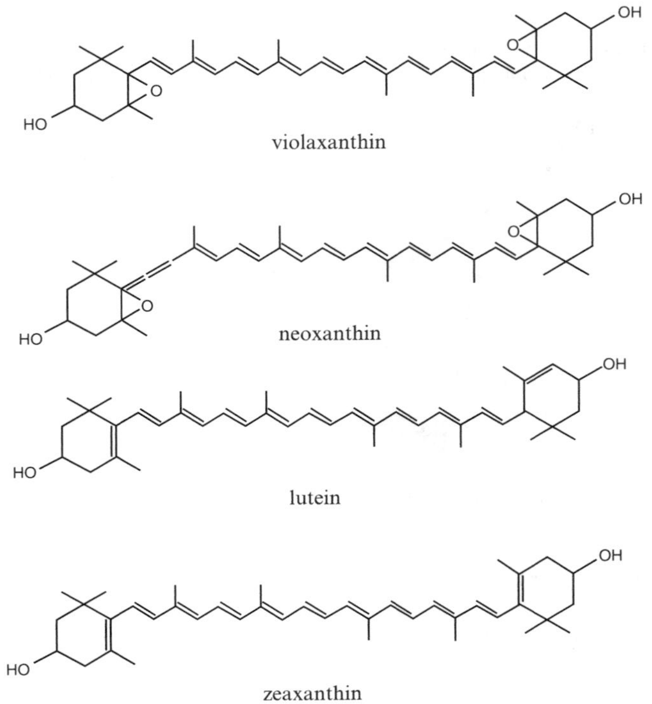

Fig. 3.3. The three xanthophylls, namely lutein, violaxanthin, and neoxanthin, contained by all plant species and to a large extent green algae in low light, and the xanthophyll zeaxanthin, synthesized in high light

side β-carotene, the same three xanthophylls, namely lutein, violaxanthin, and neoxanthin (Fig. 3.3) in low light; and that violaxanthin is deepoxidized to zeaxanthin when light is in excess through the green lineage [23] is a clear biological demonstration that each of these xanthophylls have a specific function. As mentioned earlier, xanthophylls in the thylakoid membrane are bound to Lhc proteins. Moreover, each Lhc protein has several (2–4) xanthophyll-binding sites. The affinity of each site for xanthophyll species is summarized in Fig. 3.4 for the major LHCII protein, which constitute by far the most abundant xanthophyll–protein binding complex in the chloroplast. Detailed biochemical and spectroscopic analysis has shown that the binding site L1 is occupied by lutein in all Lhc proteins, site L2 can bind lutein or violaxanthin, site N1 is specific for neoxanthin, and site V1 bind violaxanthin [24–26]. Site L2 can exchange violaxanthin with zeaxanthin, produced in excess light, and undergoes a conformational change, which increases heat dissipation and decrease the lifetime of the ^{1}Chl*-excited states [5, 6, 27]. As mentioned earlier, xanthophylls have a light harvesting function whose importance can be evaluated by spectroscopic methods. Figure 3.5 shows the deconvolution of

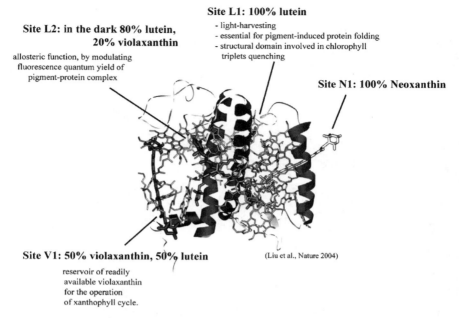

Fig. 3.4. Occupancy of the xanthophyll-binding sites in the major LHCII antenna complex. The affinity of each site for xanthophyll species is summarized in the figure. From: [45]

1-T spectra and of fluorescence excitation spectra of a typical antenna protein, LHCII. This procedure [28] allows identification of xanthophyll spectral contributions, while the ratio between the amplitude of each component in 1-T vs. fluorescence excitation spectra yields the efficiency of energy transfer to Chl a. On these basis, it can be obtained that xanthophylls contribute to light harvesting by less than 10%, the balance being the effect of Chl a and Chl b absorption. This simple consideration suggests that the major function of xanthophylls is not light harvesting but photoprotection.

3.5 Analysis of Xanthophyll Function In Vivo

Genetic has been used as the main strategy for the understanding of the specific function of each carotenoid species in photosynthesis. Single and multiple mutants targeting gene products catalyzing enzymatic steps in the carotenoid biosynthesis pathway have been produced by several laboratories yielding collection of mutants with altered xanthophyll composition. Most used mutants have been *npq1* (without zeaxanthin), *npq2* (without violaxanthin and neoxanthin), *lut2* (without lutein), and *aba4* (without neoxanthin), which have been useful in elucidating the basic function of the different xanthophyll species. Ideally, the physiologist would like to work on:

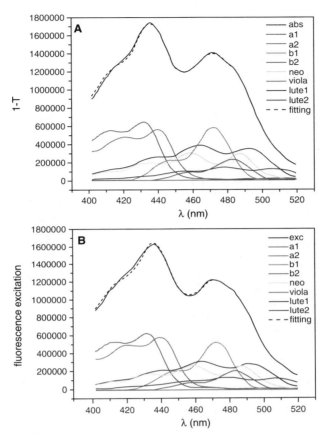

Fig. 3.5. Deconvolution of 1-T spectra (**a**) and of fluorescence excitation spectra (**b**) of a typical antenna protein, LHCII. The ratio of the amplitude of each spectral form in (**b**) vs. (**a**) yields the efficiency of excitation energy transfer from each pigment to Chl a, the final sensitizer of photosynthesis. The figure obtained is that light absorption by xanthophylls (vs. Chls, 350–750 nm) = 16.8%; contribution to Chl a fluorescence by xanthophylls (vs. Chls) in trimeric LHCII = 9.9%

1. Genotypes lacking one specific carotenoid
2. Genotypes retaining only one carotenoid species
3. Genotypes lacking one or more carotenoid-binding proteins

Although not complete, a considerable collection is now available, whose analysis will gave us a detailed functional survey. Mutant collections were obtained both in higher plants (*A. thaliana*) and in the green algae model organism *C. reinhardtii*. Table 3.1 includes several genotypes available, with indications relative to laboratories that contributed to their isolation. These mutant plants were fundamental for the initial work. Major findings are summarized in the following paragraphs.

Table 3.1. State of the art in genetics of carotenoids biosynthesis and function. Mutants affected in carotenoids or in carotenoids-binding proteins composition are listed

Arabidopsis thaliana mutants		
MUTANT	MUTATION EFFECT	Ref.
npq1	without zeaxanthin	Niyogi. K.K. 1998
npq2	constitutive zeaxanthin. without violaxanthin	Niyogi. K.K. 1998
lut2	without lutein	Pogson. B. 1998
npq2 lut2	zeaxanthin only	Havaux. M. 2004
npq1 lut2	without lutein and zeaxanthin	Niyogi. K.K. 2001
aba4-1	without neoxanthin	Dall'Osto. L. 2007
npq4	without PsbS	Li. X.P. 2000
npq1 npq4	without PsbS and zeaxanthin	Havaux. M. 1999
lut2 npq4	without PsbS and lutein	Dall'Osto. L. unpublished
aba4-1 npq1	without neoxanthin and zeaxanthin	Dall'Osto. L. 2007
aba4-1 lut2	without neoxanthin and lutein	Dall'Osto. L. 2007
lhcb3-	Knock out Lhcb3 (LHCII)	SALK line. unpublished
lhcb5 as	antisense Lhcb5 (CP26)	Andersson.J. 2001
lhcb5-	Knock out Lhcb5 (CP26)	SALK line. unpublished
lhcb6-	Knock out Lhcb6 (CP24)	Kovacs.L. 2006
lhcb4 as	antisense Lhcb4 (CP29)	Andersson.J. 2001
lhcb1+2 as	antisense Lhcb1+2 (LHCII)	Andersson.J. 2003
ch1	without chlorophyll b. without antennas	Espineda. C. 1999
ch1 npq1	without chlorophyll b and zeaxanthin	Dall'Osto. L. unpublished
ch1 npq2	without chlorophyll b. constitutive zeaxanthin. without violaxanthin	Dall'Osto. L. unpublished
ch1 npq4	without chlorophyll b and PsbS	Dall'Osto. L. unpublished
ch1 lut2	without chlorophyll b and lutein	Dall'Osto. L. unpublished
ch1 npq1 lut2	without chlorophyll b. lutein and zeaxanthin	Dall'Osto. L. unpublished
ch1 lhcb5-	without chlorophyll b and knock out Lhcb5 (CP26)	Dall'Osto. L. unpublished
ch1 npq1 lhcb5-	without chlorophyll b and zea. knock out Lhcb5 (CP26)	Dall'Osto. L. unpublished
Chlamydomonas reinhardtii mutants		
MUTANT	MUTATION EFFECT	Ref.
npq1	without zeaxanthin	Niyogi.K.K. 1997
lor1	without loroxanthin and lutein	Chunayev.A.S. 1991
npq1 lor1	without loroxanthin. lutein and zeaxanthin	Niyogi.K.K. 1997
npq2	constitutive zeaxanthin. without violaxanthin	Govindje 2002
npq2 npq1 lor1	without loroxanthin. lutein. constitutive zeaxanthin. without violaxanthin	Baroli.I. 2003
npq5	knock out Lhcbm1 (LHCII)	Elrad. D. 2005

3.6 Nonphotochemical Quenching

In strong light, the energy pressure on the photosynthetic apparatus dramatically increases thus causing increase of the fluorescence lifetime of Chl and oxidative stress (see earlier). In higher plants and some green algae, a mechanism is activated in these conditions for downregulation of chlorophyll-excited states concentration by opening a dissipation channel into heat of the excess energy absorbed over the rate that can be utilized by photochemistry, which is saturated in these conditions. Thus, quenching by reaction centers (photochemical quenching), when saturared, is integrated by an additional quenching effect originated outside reaction centers (NPQ, Nonphotochemical quenching) (Fig. 3.1). In doing so, the probability of chlorophyll intersystem crossing is decreased and the formation of reactive oxygen species down rated. NPQ can be measured through the decrease of the leaf fluorescence upon high light exposure. It was soon realized that NPQ involves several functional components: state transitions (see 3) account for a small reduction in leaf fluorescence (called qT); the formation of quenching species within PSII core complex also contribute, together with similar effect within antenna system (qI), while the largest contribution is provided by a quenching mechanism located in the antenna proteins, which is activated by the transmembrane pH gradient and is, therefore, called "Energy quenching" or qE.

3.7 Feedback Deexcitation of Singlet-Excited Chlorophylls: qE

qE is the most relevant NPQ component. It can quench up to 75% of $^1Chl^*$-excited states with an half time of 1–2 min. qE is triggered by trans-thylakoid ΔpH formation: absorption of excess photons causes the buildup of a high ΔpH, and the resulting decrease in lumen pH is essential for qE. Thus, qE acts as a feedback-regulated mechanism in photosyntheis, as it is modulated by the extent of *trans*-thylakoid ΔpH, generated by photosynthetic electron transport [29].

The need for stroma/lumen ΔpH and qE is clearly demonstrated by the inhibitory effect on qE of ionophore molecules as nigericin, which is able to abolish pH gradients. Moreover, DCCD (dicyclohexylcarbodiimide), a molecule able to covalently bind protonatable residues of proteins, is a powerful inhibitor of qE [30]. Recently, the target site for DCCD has been localized in the PsbS subunit [31], a PSII subunit whose deletion abolishes qE [32].

Xanthophylls have a key role in qE. The *lut2* mutant, lacking lutein, shows both a reduced qE amplitude and a slower induction kinetic and an even stronger effect is observed in the *npq1* mutant [3]. In this genotype, the enzyme violaxanthin deepoxidase (VDE) is knocked out. VDE is usually activated by lumen acidification, thus leading to deepoxidation of violaxanthin to zeaxanthin. This event of light-dependent, reversible deepoxidation of the violaxanthin pool is referred as "xanthophyll cycle" [33]. Lack of high light-induced Zea synthesis is thus in *npq1* mutant and is thus the reason for the strong reduction in qE. This general figure is supported by the phenotype of the double mutant *npq1lut2*, where qE is completely abolished leading to photooxidative stress [34], and of the *npq2* mutant, where zeaxanthin is constitutively accumulated, which exhibits qE similar to WT in amplitude but kinetically faster in onset and slower in recovery [3]. These evidences support a role for zeaxanthin as positive allosteric modulator of qE, expressed upon its reversible binding to Lhc proteins [35].

The mutation *npq4* of *A. thaliana*, the gene encoding the PSII subunit PsbS, is epistatic over carotenoid biosynthesis mutations [32], suggesting the step it catalyzes is upstream with respect to carotenoid function in qE mechanism. Conversely, qE amplitude depends on the stoichiometric amount of PsbS polypeptide and on the specific protonation of two acidic lumenal residues, E122 and E226, the sites of DCCD inhibition. Inactivation of one of these sites (*A. thaliana* point mutants E122Q and E226Q) halves qE, while the double mutant E122Q-E226Q gives equals the deletion mutant *npq4* [31,36]. All together these results suggest that qE mechanism is triggered in excess light by the protonation of PsbS, acting as a sensor of lumen pH, while the quenching event itself is activated downstream and involves the carotenoids lutein and zeaxanthin. Information on the quenching mechanism itself have been obtained by femtosecond transient absorption in vivo [37] suggest that the quenching reaction consists into the transient formation of a Chlorophyll–Zeaxanthin radical cation, which then recombines with heat dissipation of the excitation energy. This radical cation is proposed to trap excitation energy from PSII chlorophylls (Chl bulk) because of a high rate of formation (0,1–1 ps). *Npq4* and *npq1* mutants did not show any radical cation formation. A tentative model can be proposed based on the recent finding that (a) PsbS it is not a pigment-binding protein, thus leading to the conclusion that quenching is not located in this subunit [38]; (b) by the detection of the radical cations in purified minor chlorophyll-proteins (CP24, CP26, CP29) [61]; and (c) by the finding that zeaxanthin binding induces conformational change and quenching in the very same antenna proteins [27,39,40]. Thus, PsbS would act as a pH sensor, activated by protonation of E122 and E226 residues, triggering a conformational change in neighbor PSII antenna subunits (CP24, CP26, CP29), whose ability to undergo formation of the radical cation quenching species is modulated by their capacity for exchanging violaxanthin with zeaxanthin [41,42].

3.8 ΔpH - Independent Energy Thermal Dissipation (qI)

When zeaxanthin is synthesized in high light and replaces violaxanthin in site L2 of Lhc antenna proteins, a second type of quenching is produced, which is not dependent on the formation of radical cations but rather on energy transfer from Chl a to the short living Zea S1-excited state [43]. This kind of quenching, although weaker, is not restricted to the minor proteins CP24, CP26 [27], CP29 [39]; but is also observed in the major LHCII complex [6,43]. A further difference with respect to qE consists in its lack of nigericine sensitivity, once that zeaxanthin synthesis has occurred and is thus constitutively active in *npq2* mutant, even in the dark [27]. In WT plants, it is induced upon strong illumination that induces accumulation of zeaxanthin. Plants that have previously accumulated zeaxanthin becomes less sensitive to strong light because of a molecular shift in their antenna proteins, which are set to their short lifetime state through a zeaxanthin-induced protein conformational change that can be easily detected through isoelectric point shift and spectroscopy [27,44].

This mechanism, together with qE and qT, contributes to the quenching phenomenon on the whole called NPQ. Although qE is very fast ($t_{1/2}$ min=1 min) and qT somehow slower (8 min), qI appears to have long relaxation time (20 min–1 h) related to the release of Zea from Lhc proteins. The interplay of these mechanisms ensure photoprotection under changes of light intensity with different time constants.

3.9 Chlorophyll Triplet Quenching

The above-described mechanisms concur to avoid overexcitation of PSII. Nevertheless, by increasing light and/or decreasing temperature or metabolic sink activity, overexcitation eventually occurs. As described in section 1, IC (intersystem crossing) converts chlorophyll from its singlet-excited state into triplet, that, ultimately, can react with oxygen yielding ROS. However, carotenoids can largely prevent reaction with oxygen by quenching $^3Chl^*$. The transition energy level of carotenoid triplet state with nine or more conjugated double bonds is lower than that of chlorophyll. Carotenoid triplet states, thus, can directly receive energy from triplet chlorophyll (triplet chlorophyll quenching), thus quenching $^3Chl^*$. This reaction is located within Lhc proteins and prevents the formation of singlet oxygen. Indeed, xanthophylls bound to Lhc proteins are located in proximity to Chl for efficient quenching of $^3Chl^*$.

As energy of carotenoid triplet state is too low to be transferred to other acceptor molecules, it is directly dissipated as heat:

$$^3Chl^* +\, ^1Car \rightarrow\, ^3Car^* +\, ^1Chl$$
$$^3Car^* \rightarrow\, ^1Car + heat \quad (3.1)$$

In LHCII, the major PSII light-harvesting complex, chlorophyll triplet quenching is mainly catalyzed by lutein bound in site L1, through transfer of excitation energy from nearby chlorophylls [5]. Lutein appears to be the xanthophyll species most efficient in $^3Chl^*$ quenching as determined by direct measurements of the kinetics of carotenoid triplet formation upon excitation of Chl. Transient absorption spectroscopy of lutein vs. violaxanthin containing Lhcb1 has shown lower Car triplet yield and slower kinetics of Chl a to Car triplet transfer in the latter [45]. Consistently, increased photo-damage has been observed in vivo [45] because of formation of $^1O_2^*$ in chloroplasts and purified complexes. Thus, changes in the xanthophyll occupancy of sites L1 and L2 of Lhcb proteins, particularly LHCII, affect photoprotection.

In other Lhc proteins, such as Lhca4, also xanthophylls bound to site L2 are active in triplet quenching [18,19]. Lhca4 is a subunit of PSI characterized by red shifted spectra forms absorbing at >700 nm. Although the function of these spectroscopic features is the object of debate, analysis of recombinant Lhca4 WT vs. mutant missing red-forms showed that xanthophylls efficiency in triplet chlorophyll quenching was improved in the presence of red-forms. It is proposed that "red" chlorophylls, located near to xanthophyll in site L2 [46], act as a funnel for Chl triplet states to xanthophyll molecules bound to site L2, thus allowing 100% efficiency in triplet quenching [18].

3.10 Scavenging of Reactive Oxygen Species

Earlier, we have discussed the key role of carotenoids in mechanisms preventing ROS formation. However, these mechanisms can be saturated and ROS formed in vitro and in vivo [47,48]. Carotenoids are well suited as antioxidants as they are active in scavenging reactive oxygen species generated in the chloroplast. Carotenoids protect from oxidative damage by two general mechanisms: (1) quenching of singlet oxygen with dissipation of energy as heat; (2) scavenging of radical species thus preventing or terminate radical chain reactions.

Carotenoids are hydrophobic antioxidants located in the thylakoid membranes. Xanthophylls bound to the Lhc proteins can catalyze $^1O_2^*$ scavenging, while β-carotene perform this function in the PS II core complex on $^1O_2^*$ produced from interaction of $^3P680^*$ and O_2.

After reaction with ROS, the carotenoid is excited to a triplet state ($^3Car^*$), and then relaxes into its ground state (^1Car) by loosing the extra energy as heat.

$$^1O_2^* +^1 Car \rightarrow^3 O_2 +^3 Car^*$$
$$^3Car^* \rightarrow^1 Car + heat \qquad (3.2)$$

Carotenoids can act also as chain-breaking antioxidants in the peroxidation of membrane phospholipids [49] and, therefore, protect unsaturated-rich lipid membranes from rapid degradation.

It has been reported that *A. thaliana npq1* mutant, unable to synthesize zeaxanthin, has higher lipid peroxidation levels than WT; this phenotype is not related to the lower thermal energy dissipation ability (qE) in *npq1* because of the absence of zeaxanthin [47].

Differently on mechanisms like qE or qI, where carotenoids active in the process are bound to Lhc proteins, an important role in ROS scavenging and detoxification is played by carotenoids, which are free to diffuse in the lipid bilayer. In particular, zeaxanthin has been proposed to favor membrane thermostability and protecting from lipid peroxidation [50].

Zeaxanthin has a major role in scavenging: the *A. thaliana npq2lut2* double mutant has zeaxanthin as the only available xanthophyll, and is more resistant than WT to photooxidative stress and lipid peroxidation [7]. Nevertheless, zeaxanthin-enriched plants have a decreased growth [27] implying that lutein and neoxanthin play a role in the photoprotection against ROS, during normal growth in the absence of Zea.

Lutein has a synergistic effect in photoprotection together with zeaxanthin. *A. thaliana lut2* mutant produces more singlet oxygen than WT under photooxidative conditions; nevertheless, stress symptoms are partially rescued by enhanced zeaxanthin accumulation. Indeed, plants lacking both zeaxanthin and lutein (*npq1 lut2* double mutant) show much more photosensitivity and higher lipid peroxidation with respect to each single mutant [45]. The same effect can be observed in green algae: *C. reinhardtii* single mutants *npq1* and *lor1* are altered in photoprotective mechanisms both uphill (singlet and triplet chlorophyll quenching) and downhill (scavenging) ROS production, although they are still able to survive in high light. On the contrary, *npq1 lor1* double mutant shows a lethal phenotype in high light, because of its lower ability to detoxify singlet oxygen and superoxide anion, as it has been demonstrated by the exogenous addition of pro-oxidants chemicals [51]. A similar synergistic effect of zeaxanthin has been shown for neoxanthin. The *A. thaliana aba4-1* mutant specifically lacks neoxanthin, and is only slightly more sensitive than WT; however, the double mutants *aba4-1 npq1* is prone to photoxidative stress with respect to *npq1*, suggesting that Zea compensate for lack of Neo, whose specific function appears to be the scavenging of superoxide anion (O_2) [52] mainly produced in the Mehler reaction.

A role in chloroplast protection from ROS has been recently reported for the small amphiphilic lipid tocopherol. Early *in vitro* experiments have demonstrated that this compound can efficiently terminate chain reactions of polyunsaturated fatty acid free radicals and quench singlet oxygen [53–55]. It was found that zeaxanthin-supplemented human cells, in the presence of either α-tocopherol or ascorbic acid, were significantly more resistant to photoinduced oxidative stress. The authors postulated that the underlying mechanism responsible for the synergistic action is based on prevention of zeaxanthin depletion, because of its free radical degradation. α-tocopherol would be the final radical scavenger, thus preventing carotenoid consumption. This model may be applied also in plants; *A. thaliana vte 1* mutant is impaired in tocopherol biosynthesis. When *vte1* mutation is coupled with xanthophyll cycle mutation in double mutant *npq1 vte1*, a higher PSII photoinhibition with respect to that of single mutants is reported [56]. These data together with the observation that *npq1* plants accumulate higher tocopherol level than WT [47, 57], while *vte1* plants accumulate zeaxanthin led to the hypothesis that zeaxanthin and tocopherol have overlapping functions in protecting from photodamage by ROS.

3.11 Conclusions

Oxygen, essential for animals, is produced by PSII in photosynthetic organisms. These organisms use chlorophylls as sensitizers for light-absorption water photolysis leading to O_2 evolution. The chloroplast is thus a biological compartment where the highest O_2 concentrations coexists with reducing species produced by light-driven electron transport thus leading to high probability of ROS formation and photodamage. Mechanisms have evolved for preventing formation of photooxidant and detoxify them once they are formed. Photosynthesis heavily depends on their efficiency. First, target of photoprotection mechanisms is maintenance of the balance between light absorption and utilization in the ever-changing natural environment; second, the target is quenching of Chl triplets when balance is lost. Finally, when ROS are eventually formed, the last target consists in their scavenging. Carotenoids are the essential molecules for photoprotection in that they are involved in each of these three levels of photoprotection, namely in thermal energy dissipation (qE and qI), chlorophyll triplet quenching, and direct scavenging of reactive oxygen species.

References

1. M. Ballottari, L. Dall'Osto, T. Morosinotto, R Bassi, J. Biol. Chem. **282**, 8947–8958 (2007)
2. B.M. Smith, P.J. Morrissey, J.E. Guenther, JA Nemson, M.A. Harrison, J.F. Allen, A. Melis, Plant Physiol. **93**, 1433–1440 (1999)
3. K.K. Niyogi, A.R. Grossman, O. Bjrkman, Plant Cell **10**, 1121–1134 (1998)
4. T. Morosinotto, R. Baronio, R. Bassi, J. Biol. Chem. **277**, 36913–36920 (2002)
5. E. Formaggio, G. Cinque, R. Bassi, J. Mol. Biol. **314**, 1157–1166 (2001)
6. I. Moya, M. Silvestri, O. Vallon, G. Cinque, R Bassi, Biochemistry **40**, 12552–12561 (2001)
7. M. Havaux, L. Dall'Osto, S. Cuine, G. Giuliano, R Bassi, J. Biol. Chem. **279**, 13878–13888 (2004)
8. Y. Choquet, F. Zito, K. Wostrikoff, F.A. Wollman, Plant Cell **15**, 1443–1454 (2003)
9. D.-H. Yang, J. Webster, Z. Adam, M. Lindahl, B Andersson, Plant Physiol **118**, 827–834 (1998)
10. C. Bonaventura, J. Myers, Biochim. Biophys. Acta. **189**, 366–383 (1969)
11. J.F. Allen, Curr. Biol. **15**, R929–R932 (2005)
12. J.F. Allen, Biochim. Biophys. Acta. **1098**, 275–335 (1992)
13. N. Depege, S. Bellafiore, J.D. Rochaix, Science **299**, 1572–1575 (2003)
14. R. Delosme, J. Olive, F.A. Wollman, Biochim. Biophys. Acta. **1273**, 150–158 (1996)
15. W.R. Sistrom, M. Griffiths, R.Y. Stanier, J. Cellular and Comparative Physiology **48**, 473–515 (1956)
16. J.C. Gray, Trends Genet. **19**, 526–529 (2003)
17. E.J. Peterman, C.C. Gradinaru, F. Calkoen, JC Borst, R. van Grondelle, H. Van Amerongen, Biochemistry **36**, 12208–12215 (1997)
18. D. Carbonera, G. Agostini, T. Morosinotto, R Bassi, Biochemistry **44**, 8337–8346 (2005)
19. R. Croce, M. Mozzo, T. Morosinotto, A. Romeo, R Hienerwadel, R. Bassi, Biochemistry **46**, 3846–3855 (2007)
20. K. Asada, Annu. Rev. Plant Physiol. Plant Mol. Biol. **50**, 601–639 (1999)
21. K.K. Niyogi, Annu. Rev. Plant Physiol. Plant Mol. Biol. **50**, 333–359 (1999)
22. A. Telfer, S. Dhami, S.M. Bishop, D. Phillips, J Barber, Biochemistry **33**, 14469–14474 (1994)
23. B. ALE Demmig-Adams, K. Winter, A. Kruger, F-C Czygan, in *Photosynthesis. Plant Biology* Vol.8, ed. by W.R. Briggs, (New York, Alan R. Liss, 1989) pp. 375–391
24. R. Croce, R. Remelli, C. Varotto, J. Breton, R Bassi, FEBS Lett **456**, 1–6 (1999)
25. R. Croce, S. Weiss, R. Bassi, J. Biol. Chem. **274**, 29613–29623 (1999)
26. S. Caffarri, R. Croce, J. Breton, R. Bassi, J. Biol. Chem. **276**, 35924–35933 (2001)
27. L. Dall'Osto, S. Caffarri, R. Bassi, Plant Cell **17**, 1217–1232 (2005)
28. R. Croce, G. Cinque, A.R. Holzwarth, R. Bassi, Photosynth. Res. **64**, 221–231 (2000)
29. J.M. Briantais, C. R. Acad. Sci. Hebd. Seances Acad. Sci. D **263**, 1899–1902 (1966)

30. A.V. Ruban, R.G. Walters, P. Horton, FEBS Lett. **309**, 175–179 (1992)
31. X.P. Li, A.M. Gilmore, S. Caffarri, R. Bassi, T Golan, D. Kramer, K.K. Niyogi, J. Biol. Chem. **279**, 22866–22874 (2004)
32. X.P. Li, O. Bjorkman, C. Shih, A.R. Grossman, M Rosenquist, S. Jansson, K.K. Niyogi, Nature **403**, 391–395 (2000)
33. R.C. Bugos, H.Y. Yamamoto, Proc. Natl. Acad. Sci. USA **93**, 6320–6325 (1996)
34. K.K. Niyogi, C. Shih, C.W. Soon, B.J. Pogson, D DellaPenna, O. Bjorkman, Photosynth. Res. **67**, 139–145 (2001)
35. P. Horton, A.V. Ruban, M. Wentworth, Rev. Philos. Trans. R Soc. Lond. B Biol. Sci. **355**, 1361–1370 (2000)
36. X-P Li, A. Phippard, J. Pasari, K.K. Niyogi, Funct. Plant Biol. **29**, 1131–1139 (2002)
37. N.E. Holt, D. Zigmantas, L. Valkunas, X.P. Li, KK Niyogi, G.R. Fleming, Science **307**, 433–436 (2005)
38. P. Dominici, S. Caffarri, F. Armenante, S Ceoldo, M. Crimi, R. Bassi, J. Biol. Chem. **277**, 22750–22758 (2002)
39. M. Crimi, D. Dorra, C.S. Bosinger, E. Giuffra, A.R. Holzwarth, R. Bassi, Eur. J. Biochem. **268**, 260–267 (2001)
40. R. Croce, J. Breton, R. Bassi, Biochemistry **35**, 11142–11148 (1996)
41. R. Bassi, B. Pineau, P. Dainese, J. Marquardt, Eur. J. Biochem. **212**, 297–303 (1993)
42. T. Morosinotto, S. Caffarri, L. Dall'Osto, R. Bassi, Physiol. Plant. **119**, 347–354 (2003)
43. T. Polivka, D. Zigmantas, V. Sundstrom, E. Formaggio, G. Cinque, R. Bassi, Biochemistry **41**, 439–450 (2002)
44. N.E. Holt, G.R. Fleming, K.K. Niyogi, Biochemistry **43**, 8281–8289 (2004)
45. L. Dall'Osto, C. Lico, J. Alric, G. Giuliano, M Havaux, R. Bassi, BMC Plant Biol. **6**, 32 (2006)
46. T. Morosinotto, J. Breton, R. Bassi, R. Croce, J. Biol. Chem. **278**, 49223–49229 (2003)
47. M. Havaux, K.K. Niyogi, Proc. Natl. Acad. Sci. USA **96**, 8762–8767 (1999)
48. A. Krieger-Liszkay, J. Exp. Bot. **56**, 337–346 (2005)
49. B.P. Lim, A. Nagao, J. Terao, K. Tanaka, T. Suzuki, K. Takama, Biochim. Biophys. Acta. **1126**, 178–184 (1992)
50. M. Havaux, Trends Plant Sci. **3**, 147–151 (1998)
51. I. Baroli, B.L. Gutman, H.K. Ledford, J.W. Shin, B.L. Chin, M. Havaux, K.K. Niyogi, J. Biol. Chem. **279**, 6337–6344 (2004)
52. L. Dall'Osto, S. Cazzaniga, H. North, A. Marion-Poll, R. Bassi, Plant Cell **19**, 1048–1064 (2007)
53. A. Kamal-Eldin, L.A. Appelqvist, Lipids **31**, 671–701 (1996)
54. P. Di Mascio, T.P. Devasagayam, S. Kaiser, H. Sies, Biochem. Soc. Trans. **18**, 1054–1056 (1990)
55. M. Wrona, M. Rozanowska, T. Sarna, Free Radic. Biol. Med. **36**, 1094–1101 (2004)
56. M. Havaux, F. Eymery, S. Porfirova, P. Rey, P. Dormann, Plant Cell. **17**, 3451–3469 (2005)
57. M. Havaux, J.P. Bonfils, C. Lutz, K.K. Niyogi, Plant Physiol. **124**, 273–284 (2000)

58. M.K. Sener, C. Jolley, A. Ben Shem, P. Fromme, N. Nelson, R. Croce, K. Schulten, Biophys. J. **89**, 1630–1642 (2005)
59. A. Ben Shem, F. Frolow, N. Nelson, Nature **426**, 630–635 (2003)
60. A. Eckardt, Plant Cell. **13**, 1245–1248 (2001)
61. T.J. Avenson, T.K. Ahn, D. Zigmantas, K.K. Niyogi, Z. Li, M. Ballottari, R. Bassi, G.R. Fleming, J. Biol. Chem. **274**, 29613–29623 (2008)

4

Non-Linear Microscopy

D. Mazza, P. Bianchini, V. Caorsi, F. Cella, P.P. Mondal, E. Ronzitti,
I. Testa, G. Vicidomini, and A. Diaspro

4.1 Introduction

Non-linear interactions between light and matter have been extensively used for spectroscopic analysis of biological, natural and synthetic samples [1–5]. In the 1990s, the development in laser technology allowed to apply these principles to the light microscopy field [6–8]. In this context multi-photon excitation (MPE) fluorescence microscopy and second harmonic generation (SHG) imaging are representative of the continuing growth of interest in optical microscopy. Although other modern imaging techniques like scanning near-field microscopy [9], scanning probe microscopy [10] or electron microscopy [11] provide higher spatial resolution, light microscopy techniques have unique characteristics for the three-dimensional (3D) investigation of biological structures in hydrated states, including the direct observation of living samples [12–14]. Multi-photon microscopy relies on the property of fluorescent molecules to simultaneously absorb two or more photons [15]. In this context, the advances in fluorescence labelling and the development of new fluorescent/luminescent probes like the so-called quantum dots [16] and the visible fluorescent proteins (VFPs), that can be expressed permanently bound to proteins of interest by genetically modified cells, allow the study of the complex and delicate relationships existing between structure and function in the four-dimensions (x-y-z-t) biological systems domain [17, 18]. MPE shares with the confocal microscopy the intrinsical 3D exploration capability and provides some additional interesting features. First, MPE greatly reduces photo-interactions and permits to image living samples for long time periods. Second, it allows high sensitivity measurements due to the low background signal. Third, since most of the fluorescent molecules show a wide two-photon absorption spectrum, MPE allows simultaneous excitation of multiple fluorescent molecules with only one excitation wavelength, reducing the effects of chromatic abberations of the optical path. Fourth, two-photon microscopy can penetrate into thick and turbid media up to a depth of some $100\,\mu\text{m}$. Fifth, MPE can induce chemical rearrangement and photochemical reactions within a sub-femtoliter volume in solutions, culture cells and living tissues.

Another non-linear process that can be applied to the optical microscopy field is the generation of second harmonic signal. This phenomenon is related to the capability of non-centrosymmetrical matter to scatter light at the double of the illumination wavelength. Since highly organized biological matters such as myosin or collagen fiber are excellent SHG sources, SHG allows for 3D non-invasive imaging of biological matter in general.

It is important to remember here that other kind of non-linear microscopy has been developed in the recent years. One example is Coherent anti-Stokes Raman (CARS) microscopy; since this signal is directly derived from molecular vibrations that are characteristic of the chemical composition and molecular structure of the sample, this third-order non-linear technique allows for the 3D analysis of both the chemical and the morphological information about the sample, without needing of external labelling.

Furthermore, the development of non-linear microscopy favoured progresses of several investigative techniques as fluorescence correlation spectroscopy [19–22], image correlation spectroscopy [23, 24], fluorescence lifetime imaging [25–28], single molecule detection schemes [29–32], photodynamic therapies [33] and two-photon photoactivation and photoswitching of VFPs [34–36].

Finally, an exciting scenario that has opened in recent years is the non-linear optical nanoscopy [37], related to the possibility of breaking the optical resolution limit by combining the light coming from different sources, so that interference patterns, as in 4Pi microscopy [38], or sequential excitation and depletion of fluorescent molecules, as in STED microscopy [38], allow to investigate 3D samples at a nanometric level.

4.2 Chronological Notes on MPE

The TPE story starts in 1931, with the theory originally developed by Maria Göppert Mayer in her Ph.D. Thesis [15]. The keystone of TPE theory lies in the prediction that one atom (or molecule) can simultaneously absorb two photons in the same quantum event, within a temporal window of 10^{-16}–10^{-15} s. This temporal scale reflects the rarity of a two photon event: in the daylight an efficient fluorescent molecule undergoes a two-photon absorption once every 10 million years, while one photon absorption occurs once a second [39]. Therefore, to increase the probability of the two-photon event, a high flux of photons is necessary, or in other words a laser source. As for confocal microscopy [40], in fact, the development of laser technology has been a key factor in the experimental evidence of two-photon events and in the spreading of the technique. The first observation of a two-photon signal in CaF_2 dates back to 1961 [1], just one year later than the introduction of the first Ruby laser. Some years later, the third-order absorption (three-photon effect) was observed in naphthalene crystals [2]. In the same years second harmonic scattered signal was measured by illuminating quartz crystals with a ruby laser beam [41]. For many years the application of these non-linear processes was limited to the spectroscopic studies of inorganic samples. The first observation of TPE of organic dyes is dated to 1970 [42], while in 1976 Berns reported a two-photon effect as a result of focusing an intense laser beam onto chromosomes of living cells [43].

The application of these principles to the microscopy field required 20 more years. Even if the original idea of generating 3D microscopy images by means of non-linear effects was first suggested and attempted in the 1970s by Sheppard, Kompfner, Gannaway and Choudhury of the Oxford group [44, 45], the practical realization of a 'two-photon' microscope is related to the pioneering work of W. Denk in W.W. Webb Laboratories (Cornell University, Ithaca, NY), which was responsible for spreading the technique that revolutionized fluorescence microscopy imaging [6].

4.3 Principles of Confocal and Two-Photon Fluorescence Microscopy

4.3.1 Fluorescence

The term fluorescence is related to the capability of certain molecules to emit light (in a time scale of 10^{-9} s) when they are illuminated with a proper wavelength. More precisely, the energy required to prime fluorescence is the energy that is necessary to produce a molecular transition to an electronic excited state [46, 47]. In other words, if λ is the wavelength of the light delivered on the sample, the energy provided by photons $E = hc/\lambda$ (where $h = 6.6 \times 10^{-34}$ J s is the Plank's constant and $c = 3 \times 10^8$ m s^{-1} is the speed of light) should be equal to the molecular energy gap ΔE_g between the ground state and one vibrational or rotational level of the electronic excited state:

$$\Delta E_g = E = hc/\lambda. \tag{4.1}$$

Once the molecule has adsorbed the photon, it has several pathways for relaxing back to the ground state, including non-radiative phenomena, phosphorescence (associated to the forbidden transition to the triplet state) and fluorescence (Fig. 4.1). In fluorescence the internal conversion from the lower vibrational level of the excited state is not associated with light emission while the relaxation to the ground state is achieved by emitting one photon. For this reason, the fluorescence emission is generally shifted towards a longer wavelength than the one used for exciting the molecule. This phenomenon is known as Stokes' shift and it ranges from 50 to 200 nm depending on the fluorescent molecule in consideration. Conventional imaging techniques use ultraviolet or visible light for the excitation. In multi-photon excitation the jump between the ground state and the excited one is due to the simultaneous absorption of two or more photons [5]. As the sum of the energy of the absorbed photons

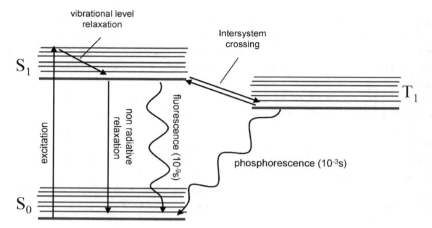

Fig. 4.1. Perrin–Jablonski diagram representing the possible pathways for one molecule to relax back to the ground state once excited

must match the molecular energy gap, multi-photon processes allow using longer excitation wavelength. Multi-photon microscopy generally use infrared (IR) light for the excitation, resulting in the increase of penetration depth in turbid media and in the decrease of the photo-toxicity in biological samples.

Because of the fact that the molecule is memoryless of the way in which excitation is accomplished, the two-photon induced fluorescence retains the same characteristics of conventional emission process. Therefore, in multi-photon excitation, the fluorescent emission is generally at shorter wavelengths than the one absorbed by the molecule.

4.3.2 Confocal Principles and Laser Scanning Microscopy

In conventional wide-field microscopy a large portion of the sample is entirely illuminated with a light source and viewed directly by eye or through any image collection system (charged coupled device (CCD), for instance). With this method the sample undergoes continuous excitation and all the points of the sample, both in-focus and out-of-focus, will contribute to the image. As a consequence, the out-of-focus contribution will appear blurred in the image, resulting in the decrease of the axial resolution and in the hazing of the collected image. In this context, as reported by Minsky in 1961, the maximum performance of a microscopy imaging system should be met if it would be possible to investigate point by point the observed sample in order to collect at each position the light scattered or emitted by that point alone, rejecting all the contributions from the other parts of the sample, especially from those belonging to different focal planes [48]. Even if it is not possible to eliminate every undesired ray because of multiple scattering, it is straightforward to remove all the rays that do not focus on the point of interest, by using a point-like source. This can be achieved by using the condenser lens to project a pinhole aperture (a small aperture in an opaque screen) on the focal plane or by using a laser source focused on the focal plane. In this way the amount of light delivered on the sample is reduced by order of magnitudes without affecting the focal brightness. However, even in this way, all the points along the focal axis will be able to scatter light (or to produce fluorescence signal) that will contribute to the image. The solution to this problem resides in placing a second pinhole aperture in the image plane that lies beyond the back aperture of the objective lens, so that all the light coming from out of focus scattering sources (or fluorescence sources) will be rejected. As shown in Fig. 4.2, the final result is a symmetrical set-up, made up of two lenses at the two sides of the specimen with two point-like apertures beyond them: in this way the role played by the lens on the excitation side is identical to the one on the detection side and their combined effect results in a relevant improvement of the axial response of the imaging system [49]. The strong symmetry of the system results also in the possibility (adopted by modern raster scanning confocal microscopes) of using the same lens both for the excitation and for the acquisition, in an *epi-fluorescence* scheme.

Because of the point-like nature of the focused beam, in order to acquire microscopic images of an extended portion of the sample, it is necessary to perform a scan. This is usually obtained with one of these methods.

The first one is based on raster scanning (point by point in a line and then line by line) of the image field. This operation requires a finite time in order

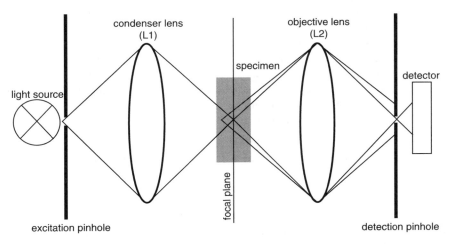

Fig. 4.2. Simplified view of a confocal microscope. A point-like source (a lamp with an excitation pinhole mask or a laser) is projected on the specimen by a condenser lens L_1. All the fluorescent molecules illuminated by the cone of light will emit. This emission is brought to the detector by the objective lens L_2. Right before the detector the collection pinhole allow to exclude all the light emitted by out of focus planes

to collect a sufficient signal from each point of the sample. In the first confocal microscopes the scanning was performed by mechanically moving the sample under the investigating beam. However, a faster and more effective way of performing the raster scanning is related to the use of galvanometric mirrors, which are capable to direct the imaging ray on the different points of the sample. With this method the emitted light is usually detected through a photomultiplier tube (PMT) and is displayed by memorizing the detected intensity at each point of the sample. The resulting imaging rate is typically of about one image per second, but modern confocal heads with resonant scanning mirrors can scan images up to 10 times faster.

A second possible approach to obtain confocal images consists in using a disk containing multiple sets of pinholes, namely a Nipkow spinning disk, placed in the image plane of the objective lens [50, 51]. A large parallel beam is pointed at the disk and the light passing through the pinholes is focused by the objective. The light emitted by the sample crosses again the pinholes and it is brought to a high-efficiency and high time-resolution CCD. When spinning the disk at a high rate it is possible to produce excitation several times per second, reaching an imaging rate comparable to video rate.

Despite the scanning architecture, both systems allow for optical sectioning: the sample is placed in a conjugate focal plane, and by moving the objective along the optical axis it is possible to scan different fields of view at different depths in the sample and to collect a series of in-focus optical sections that allow for 3D reconstruction.

The quality of the 3D reconstruction will be strictly dependent on the capability of the system to reject out-of-focus light and, therefore, to the pinhole aperture size: the use of smaller pinholes improves the discrimination of in-focus light and an improvement in axial resolution but will also result in a lower signal output. However, axial resolution and optical sectioning do not depend only on the pinhole size but also on the optical properties of the lenses (numerical aperture), on the excitation and emission wavelength and on the properties of the sample (its refractive index) as well as on the overall optical alignment of the instrument.

4.3.3 Point Spread Function of a Confocal Microscope

A simple way for schematizing an optical microscope is a linear space invariant (LSI) system. In this context it is necessary to determine the response of the system to an impulsive signal (in this specific case the image obtained when observing a point-like object). We will refer at this response as to the point spread function of the microscope (PSF). If the LSI approximation holds, it is possible to idealize any observed object as a collection of point-like sources, and the image can be properly described by the convolution product of the PSF with the object. Let us consider the schematic view of a confocal microscope that we depicted earlier: a monochromatic point-like source is focused onto a sample through a lens L_1 and the emitted radiation (also supposed to be monochromatic) is collected through the second lens L_2 by a point detector. If h_{ex} and h_{em} are, respectively, the impulse response of the lens L_1 and L_2, the radiation distribution delivered on the sample will be $U_{ex} = (h_{ex} \otimes \delta_s)(x) = h_{ex}$, where the point-like excitation source has been modelled with a Dirac delta.

The fluorescence emitted by each point x of the sample, $U_{em}(x)$, will be proportional to the product of the field intensity delivered on the sample and on the distribution of the fluorescent dye $D(x)$, $U_{em}(x) = U_{ex}(x)D(x)$. This radiation will be then brought by the second lens to the point-like detector, leading to the signal recorded by the detector, $U_{det}(x) = [(h_{em} \otimes U_{em})\delta_d](x)$, where the detector has been modelled by a delta function. The overall collected signal will be therefore

$$\begin{aligned}I_{tot} &= \int U_{det}(x)\,dx \\ &= \int \delta_d(x)\,dx \int h_{em}(x-y)U_{em}(y)\,dy \\ &= \int \delta_d(x)\,dx \int h_{em}(x-y)h_{ex}(y)D(y)\,dy \\ &= \int h_{ex}(y)D(y)\,dy \int \delta_d(x)h_{em}(x-y)\,dx \\ &= \int h_{ex}(y)h_{em}(-y)D(y)\,dy. \end{aligned} \quad (4.2)$$

If we consider the particular case of a point-like object, (4.2) provides the impulsive response of the system, i.e. the total PSF of the confocal microscope. By modelling the point-like object as a Dirac impulse δ_0 (4.2) becomes

$$I_{tot} = \int h_{ex}(y)h_{em}(-y)\delta_0(y)\,dy = h_{ex}(0)h_{em}(0). \quad (4.3)$$

If we consider the confocal epi-fluorescence scheme $L_1 = L_2$, and if we assume that $\lambda_{ex} = \lambda_{em}$,[1] we end up with $h_{ex} = h_{em} = h$. We can generally extend the previous formulas for an x-y-z scanning coupled to the imaging process. We therefore obtain for a general point $P(x,y,z)$, $I_{tot} = h^2(x,y,z)$, which is the general expression for the PSF. The mathematical expression for $h(x,y,z)$ can be formulated through the electromagnetic waves scalar theory [52] and through Fraunhofer diffraction, leading to

$$h(u,v) \propto \left| \int_0^1 J_0(v\rho)\,e^{-\frac{iu\rho^2}{2}} \rho\,d\rho \right|^2, \quad (4.4)$$

---

[1] The equivalence of the excitation and the emission wavelength is an approximation for fluorescence where generally $\lambda_{ex} \leq \lambda_{em}$ due to Stokes' shift. A more precise expression for the fluorescence case is to consider a weighted mean of the excitation and emission wavelength $\bar{\lambda} = \sqrt{2}\lambda_{em}\lambda_{ex}/\sqrt{\lambda_{em}^2 + \lambda_{ex}^2}$.

where J_0 is a Bessel function of the 0th order and u and v are dimensionless variables that depend on the lens parameters and wavelengths and are respectively proportional to the axial and the radial coordinates in the object space: $u \propto z$, $v \propto (x^2 + y^2)^{1/2}$.

By making use of the asymptotic expansions of the Bessel function we have therefore expressions for the points along the optical axis and in the focal plane:

$$h(0, v) \propto \left(\frac{2J_1(v)}{v} \right)^2 \qquad h(u, 0) \propto \left(\frac{\sin(u/4)}{u/4} \right)^2. \qquad (4.5)$$

The expressions above can be considered lateral and the axial PSF components of a conventional microscope, as only one lens is used. We can compare these relation with the confocal PSF:

$$I_{\text{tot}}(0, v) = h(0, v)^2 \propto \left(\frac{2J_1(v)}{v} \right)^4 \quad I_{\text{tot}}(u, 0) = h(u, 0)^2 \propto \left(\frac{\sin(u/4)}{u/4} \right)^4. \qquad (4.6)$$

In this context the calculation of the full width at half maximum (FWHM) of these expressions represents the system resolution. By limiting our attention to the axial resolution it can be shown that confocal imaging results in an improvement of a factor 1.4 [53,54]. In particular, the expression of the axial resolution for a pinhole aperture diameter smaller than 1 AU[2] results

$$r_z = \frac{0.64\lambda}{n - \sqrt{n^2 - \text{NA}^2}} \qquad (4.7)$$

where n is the refractive index of the medium and NA is the numerical aperture of the objective.

Despite the theoretical formalism for evaluating the ideal PSF of a confocal system shows the improvement in lateral and axial resolution, some consideration must be given about the real response of a system when imaging non-ideal samples. First it must be considered that the pinhole aperture is strictly related to the improvement in resolution: by opening the pinhole, the detector cannot be considered anymore as punctual and the resolution gets worse.

Therefore, in the observation of dim samples some compromise between the resolution and the intensity of the collected signal must be found. Table 4.1 reports the dependence of axial and lateral resolution on the pinhole size. The experimental results have been obtained by imaging sub-resolved fluorescent beads (Polyscience, diameter ($64 \pm 9\,\mu\text{m}$)) with a $\times 100$, 1.3 NA objective under 488 nm excitation.

Furthermore, the real PSF will depend on the physical and optical characteristics of the observed sample. In particular, the refractive index mismatch

---

[2] AU is the Airy-Unit, the diameter of the Airy disk.

Table 4.1. FWHM of confocal PSFs for different pinhole sizes

	Oil ($n = 1.5$)			
	Lateral (nm)		Axial (nm)	
	Pinhole 20 μm	Pinhole 50 μm	Pinhole 20 μm	Pinhole 50 μm
Experimental	186 ± 6	215 ± 5	489 ± 6	596 ± 4
Theoretical	180	210	480	560

Table 4.2. FWHM of confocal PSFs for different focusing depths and objective media

Depth (μm)	Air		Glycerol		Oil	
	Lateral (nm)	Axial (nm)	Lateral (nm)	Axial (nm)	Lateral (nm)	Axial (nm)
0	187 ± 8	484 ± 24	183 ± 14	495 ± 29	186 ± 6	489 ± 6
30	244 ± 10	623 ± 9	221 ± 5	545 ± 12	197 ± 10	497 ± 21
60	269 ± 11	798 ± 10	252 ± 7	628 ± 9	186 ± 12	196 ± 19
90	277 ± 5	1063 ± 24	268 ± 8	797 ± 26	191 ± 9	484 ± 12

between the objective immersion medium and the sample solution can play a crucial role as some spherical aberration effects can arise, leading to a loss of axial resolution and deforming the actual shape of the PSF [55]. In particular, the effects of spherical abberation get more important when focusing deep in the sample. Table 4.2 reports the broadening of the lateral and axial resolution when imaging planes at different depths of focus.

4.4 Two-Photon Excitation

Let us consider the special case of two-photon excitation. All the considerations made here can be easily extended to the more general case of multi-photon excitation. In TPE, two photons are absorbed by a fluorescent molecule in the same quantum event. The interaction may occur only if the global energy delivered by the photons is identical to the energy gap of the molecule between the ground state and some vibrational level of the excited electronic state, or in other words if the sum of the energy of the two photons (that do not have to be necessarily identical) is equal to the energy required for prime excitation with a conventional one-photon absorption (ref. Fig. 4.3). This means in terms of the wavelength of the two photons that prime the non-linear excitation, λ_1 and λ_2:

$$\lambda_{1P} = \left(\frac{1}{\lambda_1} + \frac{1}{\lambda_2}\right)^{-1}, \tag{4.8}$$

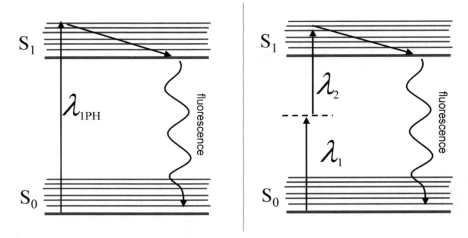

Fig. 4.3. Perrin–Jablonsky simplified diagram representing the differences between confocal and two-photon excitation microscopy. Fluorescence emission does not depend on the way in which excitation is performed

where λ_{1P} is the wavelength required to induce one-photon excitation. For practical purposes the wavelength of the two photons is chosen identical, so that

$$\lambda_{2P} = \lambda_1 = \lambda_2 = 2\lambda_{1P} \quad \text{and} \quad \Delta E_g = \frac{2hc}{\lambda_{2P}}. \tag{4.9}$$

Considering the two-photon process as non-resonant, we can assume the existence of a virtual intermediate state; the electron will reside in the virtual state for a time, τ_{virt}, that can be calculated using the time–energy uncertainty principle:

$$\Delta E_g \tau_{\text{virt}} \simeq \hbar/2 \quad \rightarrow \quad \tau_{\text{virt}} \simeq 10^{-15} - 10^{-16}\,\text{s}. \tag{4.10}$$

The two photons will have to be delayed no more than τ_{virt} to induce the non-linear interaction. It is clear that for a high flux of photons it is therefore required to have the two-photon interaction. In TPE the photons spatial and temporal concentration flux will have a crucial role. The shorter wavelength required for TPE allows using near IR to excite UV and visible electronic transitions. The rigorous development of multi-photon theory requires to apply perturbation quantum theory [56]. In particular, by solving the perturbation expansion of time-dependent Shrödinger equation and by using a Hamiltonian containing the dipole interaction of the molecule with the incoming radiation, it is possible to show that the first-order solution corresponds to one-photon interaction, while higher order solution are the n-photon ones [57]. Such a derivation allows to show that TPE depends on the square of the intensity delivered on the molecule; however, the same result can be obtained with classical or semi-quantic considerations [58].

Therefore, the fluorescence intensity emitted by one molecule can be considered proportional to the two-photon molecular cross section $\delta_2(\lambda)$ and to the square of the intensity delivered on the sample:

$$I_f(t) \propto \delta_2 I(t)^2 \propto \delta_2 P(t)^2 \left[\pi \frac{(\text{NA})^2}{hc\lambda}\right]^2, \tag{4.11}$$

where $P(t)$ is the laser power, and the intensity of the incoming radiation is calculated by using the paraxial approximation in an ideal optical system. Two-photon excitation is generally induced with ultra-fast pulsed lasers. The time averaged intensity emitted by one molecule over a time T will be therefore

$$\langle I_f(t) \rangle = \frac{1}{T} \int_0^T I_f(t)\, dt \propto \delta_2 \left[\pi \frac{(\text{NA})^2}{hc\lambda}\right]^2 \frac{1}{T} \int_0^T P(t)^2\, dt. \tag{4.12}$$

We can now consider $f_p = 1/T$ as the repetition rate of the pulsed laser and τ_p as the pulse width. If we consider that the emission of the laser is described by the profile

$$\begin{aligned} P(t) &= P_{\max} & \text{for} \quad & 0 < t < \tau_p \\ P(t) &= 0 & \text{for} \quad & \tau_p < t < T, \end{aligned} \tag{4.13}$$

then the mean square power delivered on the sample $P_{\text{ave}}^2 = 1/T \int_0^T P(t)^2\, dt$ becomes $P_{\text{ave}}^2 = P_{\max}^2 \tau_p f_p$ and (4.12) becomes

$$\langle I_f(t) \rangle \propto \delta_2 \frac{P_{\text{ave}}^2}{\tau_p f_p} \left[\pi \frac{(\text{NA})^2}{hc\lambda}\right]^2. \tag{4.14}$$

Writing the same relationship for a continuous wave (CW) laser, we obtain

$$\langle I_f(t) \rangle \propto \delta_2 P_{\text{ave}}^2 \left[\pi \frac{(\text{NA})^2}{hc\lambda}\right]^2. \tag{4.15}$$

Comparing (4.13) and (4.15) it is evident that the excitation of molecules with a pulsed laser is more efficient: in order to obtain the same fluorescence emission with a CW laser, it is necessary to use an average power $1/\sqrt{\tau_p f_p}$ higher relatively to the pulsed laser case.

From (4.13) it is possible to easily evaluate the probability n_a for a fluorophore to absorb two photons simultaneously during a single pulse:

$$n_a \propto \delta_2 \frac{P_{\text{ave}}^2}{\tau_p f_p^2} \left[\pi \frac{(\text{NA})^2}{hc\lambda}\right]^2. \tag{4.16}$$

The selection of the repetition rate and of the width of the beam pulse is related to the necessity of avoiding saturation of the fluorescence when n_a approaches unity. In other words, during the beam pulse the molecule must have enough time to relax back to the ground state as this is a prerequisite for the absorption of another pair of photons. If not, saturation effects arise, leading to a worsening of the axial and radial resolution of the system [59]. The application of (4.16) allows to choose for proper optical and laser parameters that maximize excitation out of the saturation regime. In particular, the typical values for the ultra-fast lasers used in two-photon microscopy have $\tau_p = 80-150$ fs and $f_p = 80-100$ MHz. The last considerations must be given about the two-photon cross section δ_2. Because of the fact that the quantum-mechanical selection rules for two-photon excitation differ from their one-photon counterpart, it is not easy to extend the data for conventional absorption to the non-linear case even if a simple "rule of thumb" can be considered. In fact, in general we can expect to have a two-photon excitation peak at a wavelength that is the double of the one-photon excitation maximum. However, the knowledge of the global trend of two-photon cross-section for a particular molecule requires the direct measurement. Figure 4.4 shows the TPE cross section for some of the most common fluorescent molecules. Because of the wide nature of the excitation spectra it can be noted that one single wavelength can be used for the simultaneous excitation of multiple dyes [60, 61]. It has been showed that endogenous fluorescent molecules like flavoproteins, NAD(P)H and tryptophane exhibit TPE fluorescence [61, 62] and also the fluorescent proteins like GFP (green fluorescent protein) are capable of undergoing TPE [63–65]. The two-photon cross-sections are generally expressed in GM (Goppert-Mayer, $1\,\mathrm{GM} = 10^{-58}\mathrm{m}^4\,\mathrm{s}$). GFP variants

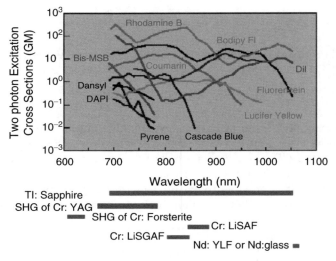

Fig. 4.4. Exemplary two-photon cross-sections for some typical fluorophores for biological imaging. The bars represent typical emissions of commonly used laser sources for TPE

have cross sections between 4 and 60 GM [66]. As comparison we can consider NADH that at its excitation maximum, 700 nm, shows a cross section of 0.02 GM [61]. Quantum dots can show cross sections up to 2,000 GM.

4.5 Two-Photon Optical Sectioning

In the following we will describe how two-photon (or multi-photon) excitation allows to spatially control excitation along the optical axis, limiting the excitation volume to a sub-femtoliter volume, thus allowing optical sectioning. We discussed above that the fluorescence signal emitted under two-photon excitation will depend on the second power of the excitation intensity. We can therefore consider that the intensity delivered point by point on the sample, I_{ex}, is proportional to the lens PSF, which in the paraxial approximation is described by (4.4). Therefore, we can expect that the fluorescent signal emitted by the sample $I_{\text{em}}(x, y, z)$ will be proportional to I_{ex}^2 and for the axial and radial components of I_{em} – introducing the dimensionless variables u, v – we get [67]

$$I_{\text{em}}(0, v) = h(0, v)^2 \propto \left(\frac{2J_1(v)}{v}\right)^4 \quad I_{\text{em}}(u, 0) = h(u, 0)^2 \propto \left(\frac{\sin(u/4)}{u/4}\right)^4.$$
(4.17)

These expressions are identical to the axial and radial parts of the total PSF of a confocal system (ref. (4.6)). However, it must be noted that while confocal microscopy achieves this result at the detection side, by means of the pinhole, in two-photon excitation the dependence of the fluorescence intensity on the inverse of the 4th power of the axial coordinate is achieved directly in the excitation step. This means that while in confocal microscopy a thick volume of the sample is excited and the selection of the in-focus contribution is obtained through the pinhole at the detection level, in two-photon excitation, only the molecules in a small volume (order of the femtoliter) around the focal point are properly excited and emit fluorescence [68,69]. This fact also implies that the contributes far-off the focal plane (depending on the NA of the lens) will not be affected by photobleaching [70,71] or phototoxicity [72–74] and do not contribute to the signal detected if a TPE architecture is used. In this way we have no need for a pinhole as all the fluorescent emission is supposed to be originated at the focal plane.

This means that TPE is intrinsically three-dimensional as it allows for optical sectioning by collecting images of the sample plane-by-plane and then reconstructing the three-dimensional map of the emitted fluorescence (ref. Fig. 4.5). It must be underlined that since all the fluorescence collected is necessarily originated at the focal point, the efficiency of the collection of the signal is much higher than in the confocal case. In TPE, in fact, over 80% of the signal arise from a 700–1,000 nm (depending on the NA of the objective) thick region around the focal plane [54, 75]. This results in a reduction in

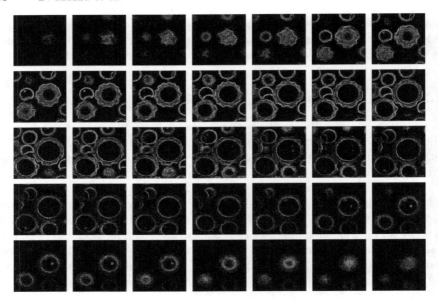

Fig. 4.5. Exemplary optical sectioning obtained with two-photon excitation at 720 nm. The sample is Colpoda cists, and the signal is produced by autofluorescence of the membranes

background and an increase in the image contrast. This compensates the decrease in resolution due to the longer wavelength compared to the one used in conventional excitation (4.7). However, the use of infrared wavelength instead of UV–visible allows deeper penetration as the long wavelength used in TPE (and in general in MPE) will be scattered less than the UV–visible light [55, 76]. Because of the high intensity delivered in the focal point (compared to conventional excitation), it must be noted that the laser power must be finely controlled in order to prevent photo-damage effects [73]. On the other side, the confinement of the excitation processes allows to photochemically modify the sample properties with a high 3D spatial control, opening new frontiers in the context on micro- and nano-surgery [77, 78]. Finally, the localization of the photobleaching effect allows to acquire 3D maps of the sample for longer times than in conventional and in confocal architecture, as when observing a specific slice of the sample, the out-of-focus contributes will not be affected by photobleaching [70].

4.6 Two-Photon Optical Setup

The basic setup for multi-photon imaging includes the following elements: a high-intensity ultra-fast infrared laser source (femto- or picosecond pulse width with a 80–100 MHz repetition rate); a laser beam scanning system (i.e. a pair of galvanometric mirrors); high numerical aperture objectives (NA 1.0–1.4) in order to deliver high peak intensity in the focal point; and a highly efficient detection system.

Laser source plays an essential role in the production of MPE signal. Because of relatively low cross-section of the non-linear processes, high photon flux is required, $>10^{24}$ photons cm^2 s^{-1} at the focal plane [79]: considering the spectral range of 600–1,100 nm this means a peak intensity in the MW cm^{-2}–GW cm^{-2}. By using an ultra-fast pulsed laser in combination with a high numerical aperture objective, this results in a mean laser power of 50 mW or less. This allows for a sufficient peak intensity to induce TPE, with a mean power levels that are biologically tolerable [72, 80].

The most common ultra-fast pulsed lasers are Ti:sapphire femtosecond laser sources [40], which can be tuned in wavelength between 700 and 1,050 nm allowing most of the common fluorescent molecules to be excited in TPE regime. Typical parameters for Ti:sapphire lasers include an average power of 700 mW–1 W, pulse width of 100–150 fs and repetition rate of 80–100 MHz. Other laser sources for MPE include Cr-LiSAF, pulse-compressed Nd-YLF in the femtosecond regime, and mode-locked Nd-YAG. Picosecond pulse width can also be reached by using a pulse stretcher. It must be noted that due to time–energy uncertainty principle, $\Delta E \Delta T \geq \hbar/2$, a shorter pulse width will return in a broader emission in terms of wavelengths: by considering $\Delta T = \tau_p$, $\Delta \lambda / \lambda = \Delta E / E$ and $E = hc/\lambda$, we obtain

$$\Delta \lambda = \frac{\lambda^2}{2\pi \tau_p c}, \qquad (4.18)$$

which for a central laser emission wavelength of 1,000 nm and pulse width $\tau_p = 100$ fs gives an uncertainty on the wavelength of about 5 nm. Furthermore, it must be considered that, when the laser light crosses the sample, a temporal broadening of the pulse occurs. Direct measurement of the pulse width at the focal plane is not an easy task [81–83]. For practical purposes, it can be assumed that at the focal volume 1.5–2 times broadening is obtained.

Some notes also must be given to the scanning system. Generally, a x-y raster scan is performed by means of galvanometric mirrors [84]. Therefore, the image acquisition rate is generally limited by the mechanical properties of the scanner. Fast scanning methods based on resonant scanners can be performed as well; however, it must be considered that fast scanning and subsequent high temporal resolution may require some compromises about spatial resolution and sensitivity, since by scanning faster the laser will illuminate for shorter times each point of the sample, resulting in a lower collected signal. Axial scanning can be performed by a variety of solutions, the most common being a single objective piezo nanopositioner or a galvanometric object table. Two-photon architecture usually allows to switch between confocal and TPE imaging retaining the x-y-z focal position [85]. Furthermore, the use of electro-optical (like electro-optic modulators, EOMs) or acusto-optical devices allow

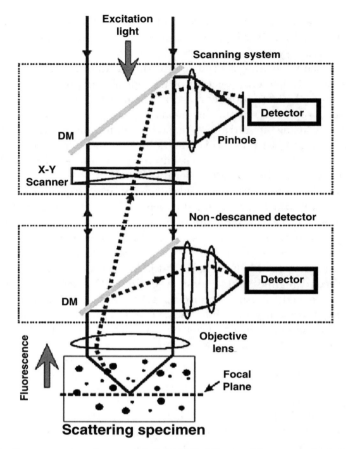

Fig. 4.6. Laser sources in use at LAMBS Microscobio research center of the University of Genoa for TPE

to rapidly modulate the laser intensity: in this way it is possible to define regions to be imaged with different powers during the very same scan.

As depicted in Fig. 4.6 it is possible to consider two popular approaches to TPE microscopy, the descanned and non-descanned modes. The first configuration uses the same optical pathway as used in confocal microscopy, and the light emitted by the sample is deflected back by the scanning mirrors and brought to the photomultipliers. The pinhole is usually kept open – or removed – as it is not necessary for TPE optical sectioning. In the non-descanned mode the microscope architecture is optimized for collecting the maximum TPE signal: pinholes are removed, the emitted light is collected through a dichroic mirror and it does not cross the scanning system. [61, 76, 86]

Despite the configuration used, the fluorescent light is brought by the detection system by a dichroic mirror. Several types of detectors can be used,

Tsunami-Millennia
(680-830/780-930),
Chameleon XR (720-980)$_{10}$

Fig. 4.7. Optical scheme of two-photon microscope in descanned and non-descanned configuration (Courtesy of M.Cannel and C. Soeller)

including PMTs, CCD cameras and avalanche photodiodes. PMTs are the most commonly used for their low cost, good sensitivity in the visible range (the quantum efficiency is about 20%–40% in the blue green range, dropping down to 1% in the red) and for the large photosensitive area that allows a good dynamic range; thus an efficient collection of the signal (Fig. 4.7). Avalanche photodiodes are extremely efficient in terms of sensitivity (quantum efficiency of about 70%–80%). Unfortunately, they are pretty expensive and their small photosensitive area requires the descanned architecture, limiting the global signal collection efficiency. CCD cameras are used for fast imaging, and are particulary useful in the context of tandem or multi-focal imaging. Finally, regarding the collection of the signal, it must be noted that due to the fact that the fluorescent signal is generated only in the focal point, we could think of an ideal TPE microscope with a detector that is able to collect all the light coming from the sample, as the fluorescence signal can be easily recognized by the scattered excitation light by means of the wavelength – i.e. with a low-pass filter.

4.7 Second Harmonic Generation (SHG) Imaging

In conventional microscopy, contrast methods consist of phenomena like absorption, reflection, scattering and fluorescence: in this conditions the specimen response is linearly dependent with the incoming light. This means that a linear relationship exists between the electric field strength $\tilde{E}$ of the light and the induced polarization of the object $\tilde{P}$. In particular, if we consider an incoming oscillating field with a non-resonant frequency (i.e. a frequency that is not absorbed by the molecules of the material), the optical response can be approximated to be a first-order response:

$$\tilde{P}(t) = \epsilon_0 \chi^{(1)} \tilde{E}(t), \tag{4.19}$$

where ϵ_0 is the dielectric constant of the vacuum and $\chi^{(1)}$ is the linear susceptibility of the specimen. Moving to the non-linear domain, high power intensity cause a large variety of unusual responses, with a non-linear dependency on the applied electric field. Second harmonic generation is one of these nonlinear optical effects in which the incident light is coherently scattered by the specimen at twice the optical frequency and at certain angles [87]. We can generally write the non-linear correspondent of (4.19) as

$$\tilde{P}(t) = \epsilon_0 \left(\chi^{(1)} \tilde{E}(t) + \chi^{(2)} \tilde{E}^2 + \chi^{(3)} \tilde{E}^3 ... \right), \quad (4.20)$$

where the first left term on the right is the linear scattering, the second is related to SHG, the third to third harmonics, etc. In the above equation, because the fields are vectors, the nonlinear susceptibilities are tensors. As each atom acts as an oscillating dipole that radiates in a dipole radiation pattern, the radiation phase among the enormous number of atoms must be matched to induce constructive interference and thus non-linear generation is allowed under phase-matching conditions (i.e. when the scattered light is in phase).

For the same reason, only molecules that exhibit a specific symmetry. In particular, only non-centrosimmetric materials can originate SHG. The SHG signal depends strictly on the relative orientation between the polarization of the incoming light and the direction of the symmetry constraints; a polarization analysis of the second harmonic signal can provide useful information about the orientation of molecules, impurities in crystal structures and characteristics of surfaces and optical interfaces. As for linear scattering, second harmonic generation is not associated with absorption and involves only virtual state transitions that are related to the imaginary part of the nonlinear susceptibilities, and so no energy is deposited in the specimen and no damage can be produced. This is one of the major advantages of applying SHG to the microscopy field and in particular to the investigation of biological samples [88, 89]. Furthermore, SHG imaging preserve the intrinsical capability of 3D investigation of matter, since the high photon flux required for generating this non-linear signal is achieved only within a femtoliter volume around the focal point of the lens. Since both SHG and TPE can be observed simultaneously from the same sample, the correlative analysis of these two signals provide additional insight about the specimen, allowing not only to identify the molecular source of the SHG, but also to probe radial and lateral symmetry within structures of interest.

Recognition of the SHG relies on the property that the emitted light has double wavelength of the incoming radiation. Therefore, by changing the color of the illumination laser we expect to observe an analogue shift in the emission wavelength. The emission of SHG signal is not isotropic, as it will be more efficient in the forward direction. However, in dense samples, multiple scattering allow for a detection in the backward direction too. Therefore, SHG can be usually collected both along the transmitted light and along the *epi*-pathways of the microscope.

A number of plant structures are capable of efficiently generating second harmonic signals. These structures include stacked membrane structures such as grana, starch granules, secondary cell wall, cellulose, cuticle and cuticular waxes, and silica deposits (bio-opals). Starch granules exhibit high conversion efficiency in SHG; in fact, a piece of potato tuber placed in an unfocused, ultra-fast laser beam can efficiently generate a bright SHG beam in the forward direction. For example, the SHG signal from a potato (Solanum tuberosum L.) starch granule is so strong that it is visible to the naked eye even under ambient room light. Cellulose is a linear molecule without branching. Neighboring cellulose chains may form hydrogen bonds leading to the formation of micro-fibrils (20–30 nm) with partially crystalline parts called micelles. All the highly organized structure may be responsible for strong SHG signal such as collagen or myosin fibers. SHG allows therefore for three-dimensional imaging and microscopical investigation of tissue organization with typical confocal resolution, without the need for any external labeller [90–92]. The possibility of selecting the wavelength in the IR range permits to perform practically non-invasive in vivo deep tissue imaging, being a promising tool for skin endoscopy.

4.8 Conclusions

The investigation of soft and living matter requires instruments capable to investigate the relationships between the 3D structure of the specimen and its functions and activity. Optical microscopy and especially the advances brought by confocal and non-linear microscopy in the last decades offers a powerful tool for this aim. Confocal microscopy allows to generate 4D (x-y-z-t) views of the specimen in a non-invasive way, preserving the vital condition of the sample, with a strong reduction of the out-of-focus haze, allows the multiple observation of different dyes by spectral fingerprinting and recent scanning heads provide a good tool for visualizing at video rate fast in vivo dynamics. The advances in confocal microscopy, together with the development of new fluorescent labellers and the spreading of visible fluorescent proteins have favoured the development and the spreading of two-photon and multi-photon imaging. Compared to confocal microscopy, we can mainly recognize two important properties by MPE.

1. The MPE excitation is confined both in the x-y directions and along the optical axis. Therefore, no fluorescence signal arises from elements out of the focal point. Consequently, optical sectioning can be performed without the use of a pinhole or deconvolution algorithms. Furthermore, photo-bleaching and photo-damage are also confined to the focal point, and the signal-to-noise ratio increases as well as the image contrast. The development of proprietary schemes, like the non-descanning acquisition, further improves the efficiency of the signal collection.

2. The use of near-infrared wavelengths allows to minimize biological damage as biological tissues in general poorly absorb infrared light. This combined with the in-focus photobleaching allows for long-term observation of living samples. Because of the wide two-photon cross section of most of the fluorescent molecules, it is possible to excite different dyes at once with only one laser line, reducing the average power delivered on the sample. Furthermore, the use of IR light permits deeper penetration into the sample (up to 0.5 mm) as both absorption and scattering are reduced with this wavelength. Finally, because of the fact that the separation between excitation and emission (excitation in the IR, emission in the UV–visible) is usually wider than in conventional excitation, it is considerably easier do discriminate between the actual emission and scattered/reflected excitation light.

The advances of TPE is strictly connected to the developments in another non-linear microscopy technique, second harmonic generation imaging. It offers a practically non-invasive tool for deep tissue imaging as it does not require fluorescent labelling of the samples and it retain the 3D investigation and optical sectioning properties of multi-photon excitation. Also in this case all the advantages using IR light for the illumination are conserved.

Since relevant generation of second harmonic signal is associated with the presence of tissue constituents as collagen, SHG may play an important role in skin disease diagnosis and in the medical research field in general, being a promising endoscopy tool.

The combination of MPE microscopy and SHG imaging provides both a deep insight into biological matter and offers possibilities in the nanomanipulation of living cells and tissues and in the 3D micro- and nano-surgery field. The range of application of these techniques is rapidly increasing in the biomedical, biotechnological and biophysics sciences and it is now facing the clinical applications.

References

1. W. Kaiser, C.G.B. Garrett, Phys. Rev. Lett. **7**, 229 (1961)
2. S. Singh, L.T. Bradley, Phys. Rev. Lett. **12**, 162 (1964)
3. D.M. Friedrich, J. Chem. Educ. **59**, 472 (1982)
4. D.M. Friedrich, W.M. McClain, Annu. Rev. Phys. Chem. **31**, 559 (1980)
5. P. R. Callis, Annu. Rev. Phys. Chem. **48**, 271 (1997)
6. W. Denk, J.H. Strickler, W.W. Webb, Science **248**(4951), 73 (1990)
7. A. Diaspro, IEEE Eng. Med. Biol. Mag. **15**, 29 (1996)
8. A.J. Koster, J. Klumperman, Nat. Rev. Mol. Cell. Biol. (Suppl), SS6 (2003)
9. U. Durig, J.K. Gimzewski, D.W. Pohl, Phys. Rev. Lett. **57**(19), 2403 (1986)
10. G. Binnig, C.F. Quate, Ch. Gerber, Phys. Rev. Lett. **56**(9), 9301 (1986)
11. E. Ruska, M. Knoll, Z. Tech. Phys. **12**, 389 (1931)
12. J.B. Pawley (ed.), *Handbook of Biological Confocal Microscopy*. (Plenum, New York, 2005)

13. A. Periasamy (ed.), *Methods in Cellular Imaging*. (Oxford University Press, New York, 2001)
14. A. Diaspro (ed.), *Confocal and Two-Photon Microscopy: Foundations, Applications and Advances*. (Wiley, New York, 2002)
15. M. Göppert-Mayer, Ann. Phys. **9**, 273 (1931)
16. J.K. Jaiswal, E.R. Goldman, H. Mattoussi, S.M. Simon, Nat. Methods **1**(1), 73 (2004)
17. D.J. Arndt-Jovin, M. Robert-Nicoud, S.J. Kaufman, T.M. Jovin, Science **230**(4723), 247 (1985)
18. J. Lippincott-Schwartz, E. Snapp, A. Kenworthy, Nat. Rev. Mol. Cell. Biol. **2**(6), 444 (2001)
19. K.M. Berland, P.T. So, E. Gratton, Biophys. J. **68**(2), 694 (1995)
20. P. Schwille, U. Haupts, S. Maiti, W.W. Webb, Biophys. J. **77**, 2251 (1999)
21. K.G. Heinze, M. Jahnz, P. Schwille, Biophys. J. **86**, 506 (2004)
22. Q. Ruan, M.A. Cheng, M. Levi, E. Gratton, W.W. Mantulin, Biophys. J. **87**, 1260 (2004)
23. P.W. Wiseman, J.A. Squier, M.H. Ellisman, K.R. Wilson, J. Microsc. **200**, 14 (2000)
24. P.W. Wiseman, F. Capani, J.A. Squier, M.E. Martone, J. Microsc. **205**, 177 (2002)
25. T. French, P.T. So, D.J. Weaver, T. Coelho-Sampaio, E. Gratton, E.W. Voss, J. Carrero, J. Microsc. **185**(Pt 3), 339 (1997)
26. K. König, S. Boehme, N. Leclerc, R. Ahuja, Cell. Mol. Biol. **44**, 763 (1998)
27. J. Sytsma, J.M. Vroom, C.J. De Grauw, H.C. Gerritsen, J. Microsc. **191**, 39 (1998)
28. M. Straub, S.W. Hell, App. Phys. Lett. **73**, 1769 (1998)
29. J. Mertz, C. Xu, W.W. Webb, Opt. Lett. **20**, 2532 (1995)
30. X.S. Xie, H.P. Lu, J. Biol. Chem. **274**, 15967 (1999)
31. G. Chirico, F. Cannone, S. Beretta, G. Baldini, A. Diaspro, Microsc. Res. Tech. **55**(5), 359 (2001)
32. G. Chirico, F. Cannone, G. Baldini, A. Diaspro, Biophys. J. **84**(1), 588 (2003)
33. J.D. Bhawalkar, N.D. Kumar, C.F. Zhao, P.N. Prasad, J. Clin. Laser Med. Surg. **15**(5), 201 (1997)
34. G.H. Patterson, J. Lippincott-Schwartz, Science **297**(5588), 1873 (2002)
35. G. Chirico, A. Diaspro, F. Cannone, M. Collini, S. Bologna, V. Pellegrini, F. Beltram, Chem. Phys. Chem. **6**(2), 328 (2005)
36. M. Schneider, S. Barozzi, I. Testa, M. Faretta, A. Diaspro. Biophys. J. **89**(2), 1346 (2005)
37. S.W. Hell, Nat. Biotechnol. **21**, 1347 (2003)
38. S.W. Hell, M. Schrader, H.T. van der Voort, J. Microsc. **187**(Pt 1), 1 (1997)
39. W. Denk, K. Svoboda, Neuron **18**(3), 351 (1997)
40. E. Gratton, M.J. van de Ven, in *Handbook of Confocal Microscopy*, ed. by J.B. Pawley (Springer, New York, 1995), p. s69
41. P.A. Franken, A.E. Hill, C.W. Peters, G. Weinreich, Phys. Rev. Lett. **7**, 118 (1961)
42. P.M. Rentzepis, C.J. Mitschele, A.C. Saxman, Appl. Phys. Lett. **17**, 122 (1970)
43. M.W. Berns, Biophys. J. **16**(8), 973 (1976)
44. J.N. Gannaway, C.J.R. Sheppard, Opt. Quant. Electron. **10**, 435 (1982)
45. C.J.R Sheppard, A. Choudhury, Opt. Acta **24**, 1051 (1977)

46. J.B. Birks, *Photophysics of Aromatic Molecules* (Wiley, London, 1970)
47. C.R. Cantor, P.R. Schimmel, *Biophysical Chemistry. Part II: Techniques for the Study of Biological Structure and Function* (Freeman, New York, 1980)
48. M. Minsky, Scanning **10**, 128 (1988)
49. C.R.J. Sheppard, in *Confocal and Two-Photon Microscopy: Foundations, Applications and Advances*, ed. by A. Diaspro (Wiley, New York, 2002), p. 1
50. M. Petran, M. Hadravsky, M.D. Egger, R. Galambos, J. Opt. Soc. Am. **58**, 661 (1968)
51. G.S. Kino, T.R. Corle, Phys. Today **42**, 55 (1989)
52. M. Born, E. Wolf, *Principles of Optics* (Pergamon, New York, 1993)
53. G.J. Brakenhoff, E.A. van Spronsen, H.T. van der Voort, N. Nanninga, Methods Cell. Biol. **30**, 379 (1989)
54. T. Wilson, C.R.J. Sheppard, *Theory and Pratice of Scanning Optical Microscopy* (Academic Press, London, 1984)
55. A. Diaspro, F. Federici, M. Robello, Appl. Opt. **41**(4), 685 (2002)
56. O. Nakamura, Microsc. Res. Tech. **47**, 165 (1999)
57. A. Esposito, F. Federici, C. Usai, F. Cannone, G. Chirico, M. Collini, A. Diaspro, Microsc. Res. Tech. **63**(1), 12 (2004)
58. A. Diaspro, C.J.R. Sheppard, in *Confocal and Two-Photon Microscopy: Foundations, Applications and Advances*, ed. by A. Diaspro (Wiley, New York, 2002), p. 39
59. G.C. Cianci, J. Wu, K.M. Berland, Microsc. Res. Tech. **64**(2), 135 (2004)
60. C. Xu, J. Guild, W.W. Webb, W. Denk, Opt. Lett. **20**, 2372 (1995)
61. P.T.C. So, C.Y. Dong, B.R. Masters, K.M. Berland, Annu. Rev. Biomed. Eng. **2**, 399 (2000)
62. D.W. Piston, B.R. Masters, W.W. Webb, J. Microsc. **178**, 20 (1995)
63. M. Chalfie, Y. Tu, G. Euskirchen, W.W. Ward, D.C. Prasher, Science **263**(5148), 802 (1994)
64. S.M. Potter, C.M. Wang, P.A. Garrity, S.E. Fraser, Gene **173**, 25 (1996)
65. M. Zimmer, Chem. Rev. **102**, 759 (2002)
66. G.A. Blab, P.H.M. Lommerse, L. Cognet, G.S. Harms, T. Schmidt, Chem. Phys. Lett. **350**, 71 (2001)
67. C.J.R. Sheppard, M. Gu, Optik **86**, 104 (1990)
68. O. Nakamura, Optik **93**, 39 (1993)
69. M. Gu, C.J.R. Sheppard, J. Microsc. **177**, 128 (1995)
70. G.H. Patterson, D.W. Piston, Ultramicroscopy **78**, 2159 (2000)
71. A. Diaspro, G. Chirico, C. Usai, P. Ramoino, J. Dobrucki, in *Handbook of Biological Confocal Microscopy*, 3rd edn., ed. by J. Pawley (Plenum, New York, 2006), p. 690
72. W. Denk, K.R. Delaney, A. Gelperin, D. Kleinfeld, B.W. Strowbridge, D.W. Tank, R. Yuste, J. Neurosci. Methods **54**(2), 151 (1994)
73. A. Hopt, E. Neher, Biophys. J. **80**, 2029 (2001)
74. A. Abbotto, G. Baldini, L. Beverina, G. Chirico, M. Collini, L. D'Alfonso, A. Diaspro, R. Magrassi, L. Nardo, G.A. Pagani, Biophys. Chem. **114**(1), 35 (2005)
75. T. Wilson, in *Confocal and Two-Photon Microscopy: Foundations, Applications and Advances*, ed. by A. Diaspro (Wiley, New York, 2002), p. 19
76. V.E. Centonze, J.G. White, Biophys. J. **75**(4), 2015 (1998)
77. K. König, H. Liang, M.W. Berns, B.J. Tromberg, Opt. Lett. **21**, 1090 (1996)

78. K. König, I. Riemann, P. Fischer, K.J. Halbhuber, Cell. Mol. Biol. **45**, 195 (1999)
79. K. König, J. Microsc. **200**, 83 (2000)
80. K. König, T. Krasieva, E. Bauer, U. Fiedler, M.W. Berns, B.J. Tromberg, K.O. Greulich, J. Biomed. Opt. **1**, 217 (1996)
81. P.E. Hanninen, S.W. Hell, Bioimaging **2**, 117 (1994)
82. J.B. Guild, C. Xu, W.W. Webb, App. Opt. **36**, 397 (1997)
83. R. Wolleschensky, T. Feurer, R. Sauerbrey, U. Simon, Appl. Phys. B **67**, 87 (1998)
84. R.H. Webb, Rep. Prog. Phys. **59**, 427 (1996)
85. A. Diaspro, G. Chirico, F. Federici, F. Cannone, S. Beretta, M. Robello, J. Biomed. Opt. **6**(3), 300 (2001)
86. B.R. Masters, P.T. So, E. Gratton, Biophys. J. **72**, 2405 (1997)
87. L. Moreaux, O. Sandre, S. Charpak, M. Blanchard-Desce, J. Mertz, Biophys. J. **80**(3), 1568 (2001)
88. P.J. Campagnola, M.D. Wei, A. Lewis, L.M. Loew, Biophys. J. **77**(6), 3341 (1999)
89. P.J. Campagnola, A.C. Millard, M. Terasaki, P.E. Hoppe, C.J. Malone, W.A. Mohler, Biophys. J. **82**(1), 493 (2002)
90. A. Zoumi, X. Lu, G.S Kassab, B.J Tromberg, Biophys. J. **87**(4), 2778 (2004)
91. W.R. Zipfel, R.M. Williams, R. Christie, A.Y. Nikitin, B.T. Hyman, W.W. Webb, Proc. Natl. Acad. Sci. USA **100**(12), 7075 (2003)
92. S. Rotha, I. Freund, J. Chem. Phys **70**, 1637 (1979)

5

Applications of Optical Resonance to Biological Sensing and Imaging: I. Spectral Self-Interference Microscopy

M.S. Ünlü, A. Yalçin, M. Doğan, L. Moiseev, A. Swan, B.B. Goldberg, and C.R. Cantor

5.1 High-Resolution Fluorescence Imaging

Fluorescence microscopy is an essential tool in modern biological research. It is a powerful method that allows noninvasive monitoring of specifically labeled targets within living cells, and simultaneous detection of multiple targets using different labels. The spatial resolution in fluorescence microscopy is limited because of the diffraction limit; the resolution in transverse direction is proportional to $\lambda/2NA = \lambda/2n\sin\theta$ (where n is the refractive index in the object space, and θ is the half-angle of the largest cone of rays that can enter or leave the optical system), whereas the longitudinal resolution is given by $2\lambda n/NA^2$. High-spatial resolution to detect fluorescent molecules below the diffraction limit can be achieved in several ways, such as by increasing the effective numerical aperture (as in 4Pi confocal microscopy) [1], introducing spatial variation in the excitation light creating finer spatial features in the image (as in standing wave microscopy) [2], using multiple-photon fluorescence absorption or emission mechanisms that lead to nonlinear effects in the light field (as in 2-photon microscopy) [3], and by selectively quenching the fluorescence from a focal spot to obtain a very small fluorescing volume (as in stimulated emission depletion microscopy) [4].

5.2 Self-Interference Imaging

Fluorophores, when immobilized on surfaces, are often quenched because of mechanisms of energy transfer, standing wave nodes in the excitation field, and destructive interference in the emission. Fromherz and coworkers noted that the intensity of the total fluorescence oscillates as a function of the fluorophore height above a reflecting substrate [5]. Their technique, fluorescence interference contrast microscopy (FLIC), is based on measuring the intensity of fluorophores located within $\sim\lambda$ of vertical distance from a reflecting surface. On the one hand, this proximity to the surface causes the entire emission

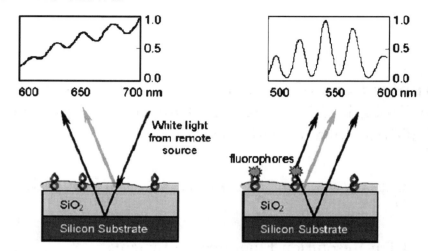

Fig. 5.1. The interferometric experimental configurations. (*Left*): WL reflection spectroscopy is based on spectral variations of reflection from thin transparent films. Interference of light reflected from the top surface and a buried reference surface results in periodic oscillations. (*Right*): The SSFM technique maps the spectral oscillations emitted by a fluorophore located on a layered reflecting surface into a precise position determination.

spectrum to oscillate as light undergoes constructive and destructive interferences. The fluorophore height can be determined from these oscillation curves; however, careful calibration of fluorescence intensity as a function of distance from the surface is required. On the other hand, for larger separations between the fluorophore and the reflecting surface (on the order of 10–20λ), light can go between constructive and destructive interferences multiple times even at the same height. This creates interference fringes in the emission spectrum, and spectral self-interference fluorescence microscopy (SSFM) [6–8] utilizes this principle to reveal height information.

Using a combination of a traditional reflection technique and an interferometric fluorescence spectroscopy technique, the average optical thickness of biological layers and the height of fluorescent markers can be measured with subnanometer accuracy. Figure 5.1 summarizes the two techniques schematically. The first technique, white light (WL) reflection spectroscopy, is based on spectral variations of reflection from thin transparent films. Interference of light reflected from the top surface and a buried reference surface results in periodic oscillations in the reflection spectrum. The principle is similar to the interference-based detection technique using color variations due to increased path length as a consequence of surface binding on optically coated silicon [9]. The spectral measurements result in very high accuracy (~ 0.2 nm) for measuring average optical density, comparable to ellipsometry, and provide a precise relative measure of the additional "optical" mass on the substrate. The second

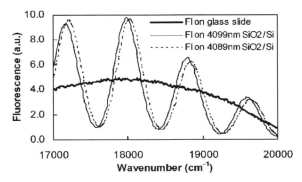

Fig. 5.2. Emission spectra of fluorescein immobilized on a glass slide and on top of a Si-SiO$_2$ chip with two different thicknesses of the oxide layer.

technique, SSFM, analyzes the spectral oscillations due to self-interference of direct and reflected emission by a fluorophore located on a layered reflecting surface and yields a precise position determination (Fig. 5.1). Figure 5.2 illustrates the effect of self-interference on the emission spectrum of fluorescein immobilized at different heights on a silicon–silicon oxide wafer. This is compared with the smooth emission envelope of fluorescein immobilized on a glass slide where there is no self-interference. As shown, small height differences (10 nm in the figure) shift the fringes and change the period of oscillation, although the latter is less apparent.

It should be noted that in contrast to fluorescence interference-contrast microscopy [5], the axial position of the fluorophores in SSFM is determined from the spectral oscillations and not from the total intensity; therefore, variations in fluorophore density, emission intensity, and the excitation field strength do not affect the result.

5.3 Physical Model of SSFM

5.3.1 Classical Dipole Emission Model

An emitting fluorophore can be represented as a radiating dipole [7]. The emitter transition dipole, μ, the wave vector, **k**, and the electric field vector, **E**, lie in the same plane, which is the plane of polarization of the radiated light. In the far field, if the environmental factors remain constant, the amplitude of the electric field of a fluorescently emitted wave is proportional to the sine of the angle (α) between the dipole μ and the wave vector **k**. The emission of a classical dipole is, therefore, nonuniform, and has a donut shape (Fig. 5.3). The emission pattern of a dipole near a surface can be described by considering the intensity and polarization of both the direct and the reflected waves. Consider that light emitted from a dipole with polar tilt angle θ placed at

74   M.S. Ünlü et al.

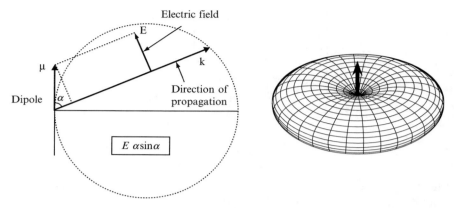

Fig. 5.3. *Left*: Intensity and polarization of electric field emitted by a classical dipole. *Right*: 3D emission pattern of a classical electric dipole.

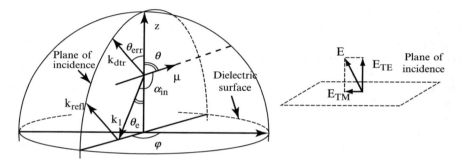

Fig. 5.4. *Left*: Direct and reflected waves emitted by a randomly oriented dipole. *Right*: Transverse electric and transverse magnetic components of the electric field of incident wave.

a distance d from a stack of mirrors is collected by a microscope objective, where the numerical aperture is defined by the collection cone of the objective (Fig. 5.4). Every emission direction within the collection cone can be described by θ_{em}, the tilt angle from the surface normal, and φ, the azimuthal angle to the dipole. The dipole radiates two coherent waves: one sent directly to the detector, and one incident on the mirror surface, which is then reflected so that it propagates parallel to the direct wave. As the distance between the dipole and the mirror is significantly smaller than the distance between the dipole and the microscope objective, these parallel waves arrive at the same spot on the detector.

The E_{TM} component of both the incident and direct waves lies in the plane of incidence, and E_{TE} is perpendicular to it. After reflection, the incident wave is modified by reflection coefficients R_{TE} and R_{TM} of the stack of mirrors. Also, there is an additional phase shift between the direct and the reflected waves because of the distance the incident light travels from the source to the stack of mirrors and back ($e^{i2kd\cos\theta_{em}}$). Light intensity collected by a microscope objective with maximum collection angle θ_{em}^{max} is integrated over $d\theta_{em}\sin\theta_{em}$. To account for fluorophores with different transition dipole orientations, the total collected light intensity is integrated over all possible dipole orientations. Usually, fluorophores in monolayers are randomly distributed in the horizontal plane, so their emission is integrated over φ; however, the range of polar tilt angles can sometimes be restricted.

For large enough separations between the fluorophores and the mirror, several oscillations of constructive and destructive interference appear within a small span of emitted wavelengths. This creates interference fringes in the emission spectrum, and the phase and contrast of the oscillations will depend on the location and orientation of dipoles above the mirror.

The emission of the dipole can be formulated as:

$$I = |E_{\text{TE dir}} + E_{\text{TE refl}}|^2 + |E_{\text{TM dir}} + E_{\text{TM refl}}|^2, \tag{5.1}$$

where

$$E_{\text{TE dir}} = E_{\text{TE inc}} \propto \cos\theta \sin\varphi, \tag{5.2}$$

$$E_{\text{TM dir/inc}} \propto \sqrt{1 - (\sin\theta_{em}\sin\theta\cos\varphi \pm \cos\theta_{em}\cos\theta)^2 - \sin^2\theta\sin^2\varphi}, \tag{5.3}$$

$$E_{\text{TE refl}} = E_{\text{TE inc}} R_{\text{TE}} e^{i2\phi}, \tag{5.4}$$

$$E_{\text{TM refl}} = E_{\text{TM inc}} R_{\text{TM}} e^{i2\phi}, \tag{5.5}$$

$$\phi = \frac{4\pi n}{\lambda} d\cos\theta_{em}, \tag{5.6}$$

and n is the index of refraction of the medium surrounding the dipole.

The total emission of a monolayer of random dipoles measured with an objective of maximum collection angle θ_{em}^{max} is:

$$I_{\text{total}} = \int_{\theta=0}^{\pi/2} \int_{\varphi=0}^{\pi/2} \int_{\theta_{em}=0}^{\theta_{em}^{max}} I(\theta,\varphi,\theta_{em}) d\theta d\varphi d\theta_{em} \sin\theta_{em}. \tag{5.7}$$

5.4 Acquisition and Data Processing

5.4.1 Microscope Setup

The measurements were performed with a system that combines an upright Leica DM/LM microscope and a Renishaw 100B micro-Raman spectrometer. The sample was positioned and scanned using a motorized submicrometer precision stage. An 1,800 grooves per millimeter grating was used with a spectral resolution of $2\,\text{cm}^{-1}$ at 500 nm. The light dispersed by the grating

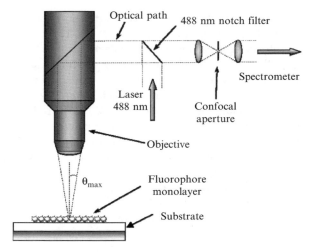

Fig. 5.5. Microscope setup for SSFM.

was imaged onto a thermoelectrically cooled charge-coupled device (CCD). For WL measurements, normal Koehler illumination with a halogen lamp was used, whereas for fluorescence measurements, the light source was an argon ion laser with a 488 nm line. The emission from the fluorophores was collected with a 5× (NA=0.12) objective and transmitted into the spectrometer through a notch filter, blocking the excitation light (Fig. 5.5).

5.4.2 Fitting Algorithm

Self-interference spectra are raw data composed of the envelope function represented by the emission profile of the free fluorophore, the oscillatory interference component, and high-frequency Gaussian noise introduced by the spectrophotometer. Both WL reflectivity and fluorescence self-interference spectra were fitted using a custom-built MATLAB application that separates the oscillatory component from the envelope function. This program automatically calculates the parameters of the system such as the thickness of thin films or position of the emitters above the mirror. The variation in the index of refraction of silicon oxide within the used wavelength span was taken into account in the fitting algorithm. The model also takes into account the complex reflectivity of the underlying stack of dielectrics and the orientation of the dipole moments of the emitters. Because the curve-fitting algorithm extracts the oscillatory term to determine the label height, the measurements are immune not only to potential quenching of the entire spectrum, but also to any spectral modifications or nonradiative transfer effects.

5.5 Experimental Results

5.5.1 Monolayers of Fluorophores on Silicon Oxide Surfaces: Fluorescein, Quantum Dots, Lipid Films

SSFM can be used to measure the vertical position of fluorophores separated from a silicon mirror by a transparent layer of silicon oxide. Monolayer surfaces of fluorescein, quantum dots [7], and lipid films [6] are investigated in the experiments below.

Fluorescein

To demonstrate detection of the axial position of a fluorescein monolayer from the silicon mirror, a sample was prepared with a spacer layer of silicon oxide etched to different heights, and a monolayer of fluorescein was immobilized on surface via isothiocyanate-aminosilane chemistry. The surface was then scanned in a grid pattern with fluorescence emission spectra taken every 200 μm, and the spectra were processed to yield the axial position of the fluorophores at each spot. Figure 5.6 displays the height data as a 3D gray-scale image, where it is apparent that nanometer axial resolution is obtained. Note that this is *not* the fluorescence intensity, which is relatively uniform.

Quantum Dots

Quantum dots (QDs) are a new area of research with potential applications in many fields. The excitation spectrum of most QDs is unusually broad. However, the emission properties of a QD depend on its size: smaller dots emit light with shorter wavelengths. Samples of QDs that are uniform in size have

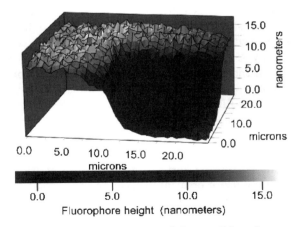

Fig. 5.6. Surface profile reconstructed from self-interference spectra.

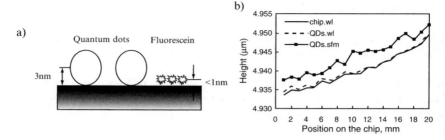

Fig. 5.7. (a) Average position of fluorescein and quantum dots above a Si-SiO$_2$ surface. (b) WL and SSFM measurements of a Si-SiO$_2$ chip before and after quantum dot deposition. QDs are sparse enough that their effect on determining the spacer layer thickness is insignificant. SSFM measurements show the source of emission is about 3 nm above the surface.

sharp emission peaks; for a broader distribution of sizes, the spectra widen as well. On the one hand, as a fluorescent label for biological research, quantum dots hold a number of advantages over conventionally used organic dyes as they are photostable, can withstand many excitation-emission cycles, and can provide a whole range of nonoverlapping spectra. On the other hand, QDs are less commonly used in staining biological specimens as they cannot be as easily functionalized with reactive groups for selective attachment. Chemical methods of adapting QDs to new environments are being developed by many research groups; in a recent achievement, QDs were used as fluorescent labels inside living cells [10].

To demonstrate visualization of a monolayer of QDs immobilized on a surface, a silicon oxide chip was silanized with aminopropyltriethoxysilane (APTES). CdSe quantum dots capped with ZnS were treated with mercaptoacetic acid to render them hydrophilic and negatively charged at neutral pH. The QDs were then electrostatically attached to the aminated surface [11]. SSFM measurements show that the average position of the emitters is elevated above the surface by about 3 nm (Fig. 5.7), which reflects the thickness of the QDs themselves. When compared with a monolayer of fluorescein, the best-fitting curve for the QD spectrum corresponds to a random orientation of the transition dipoles, whereas, in the case of fluorescein, the molecules seem to be more concentrated in the horizontal position.

Lipid Films

Deposition of lipid bilayers on surfaces has attracted considerable attention because of the possibility of creating biomembrane-based biosensors and for studying the fundamental properties of biomembranes. Artificial lipid films can be used as substitutes for biomembranes in investigating the structural and functional properties of various transmembrane properties in conditions similar to those found in cells.

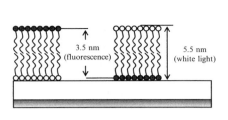
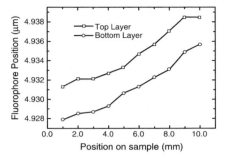

Fig. 5.8. *Left*: Structure of the lipid bilayer film. *Right*: Axial fluorophore positions measured across the surface.

The interaction between the lipid films and various membrane-bound components can be probed by diagnostic tools such as infrared spectroscopy, neutron reflectivity, surface-plasmon resonance (SPR), and atomic force microscopy (AFM). The finest resolution of lipid surface features can be achieved by AFM, but AFM is unsuitable for probing structures that are too delicate and cannot resolve buried components.

SSFM offers a unique opportunity to look inside the lipid membrane and detect the position of a fluorescently labelled component. To determine the position of a fluorescent label attached to the head groups of either the top or the bottom leaflet of a lipid bilayer separated by only about 4 nm, a layer of DPPE containing 2% DHPE-fluorescein was deposited by the Langmuir-Blodgett technique, followed by a layer of DPPE without the dye. The fluorescence emission spectra as well as WL reflectivity measurements were taken at the same locations on the chip. The chip was cleaned, followed by deposition of a new bilayer of lipids with the fluorescein-containing leaflet on top, and the measurements were repeated. The results are summarized in Fig. 5.8. The WL reflectivity measurements before and after lipid deposition were the same for both experiments (not shown). The fluorescence emission spectra show the average position of the fluorescent dye attached to either the top or the bottom head groups very accurately, fractions of a nanometer, inside the lipid bilayer.

5.5.2 Conformation of Surface-Immobilized DNA

DNA array technology, offering highly paralleled detection capability, has become a widespread tool in biological research with applications in expression screening, sequencing, and drug discovery. One of the defining characteristics of a DNA array is the availability of the single-stranded probes for hybridization with the target. The conformation of DNA molecules in an array may significantly affect the efficiency of hybridization; immobilized molecules located farther away from the solid support are closer to the solution state, and are more accessible to dissolved analytes. Recently, advances have been made to characterize surface-bound DNA probes, using optical or contact methods such as ellipsometry, optical reflectivity [12,13], neutron reflectivity [14], X-ray photoelectron spectroscopy [15], FRET [16,17], SPR [18,19], and AFM [20,21]. However, most experimental techniques characterize the DNA layer as a single entity, parameterizing its thickness or density. The advantage of SSFM is that unlike these other methods, it has the capability of examining the specific positions of internal elements of the DNA chain.

To study the conformation of single-stranded DNA (ssDNA) and double-stranded DNA (dsDNA) on glass surfaces with SSFM, 50 and 21-nt oligonucleotides were used [8]. In all studies, the first strand of the DNA was covalently bound to the surface at its 5′ end. Experiments were performed with fluorescein markers bound to either the first strand at its distal 3′ end, or the second strand at its 3′ or 5′ end.

WL Reflectivity Measurements

As schematically shown in Fig. 5.9, oligonucleotides carrying a 5′-amino tag are covalently bound to an aminated surface via a homobifunctional crosslinker. Using WL reflectivity, progressive growth of the surface-bound thin films during DNA immobilization steps is determined. The thickness of the silane layer is 0.8–1.0 nm, which roughly corresponds to a monolayer; phenylene isothiocyanate adds another 0.5–0.6 nm. Immobilization of DNA leads to a further increase in the optical thickness. The average film thickness is determined by WL reflection spectroscopy assuming an index of refraction of 1.46 for DNA as was measured for dense layers [22]. The optical film thickness for ssDNA of 21- and 50-nt is measured as 1.0–1.5 nm and 2.0–2.5 nm, respectively. Although the precision of absolute thickness will depend on the index, WL reflectivity provides an accurate relative measure of additional mass on the surface and can thus monitor the efficiency of hybridization. As shown in

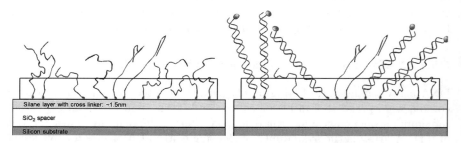

Fig. 5.9. The WL interference measurements of 50mers immobilized on the surface before (*left*) and after (*right*) hybridization. The height of the transparent layers (2 and 3 nm for left and right, respectively) correspond to average optical thickness of the DNA layer.

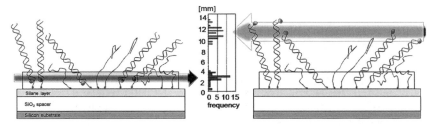

Fig. 5.10. The fluorescence interference measurements for the 50-bp dsDNA labeled at the proximal (*left*) and distal end (*right*).

Fig. 5.9 (right), adding complementary second strands to 50-mers results in an increase in the film thickness by ∼1.0 nm corresponding to a hybridization efficiency of ∼50%.

Determination of the Position of a Fluorescent Label

The optical thicknesses of the layers obtained by WL reflectivity measurements can be used together with the SSFM measurements to determine the position of the fluorophores relative to the surface. Figure 5.10 illustrates the schematics of the experiment to study the elevation of dsDNA fragments. The maximum elevation of the label is limited by the length of the double helix, which is ∼7 nm for 21-bp and ∼17 nm for 50-bp fragments. As dsDNA has a persistence length of ∼50 nm [23], short fragments in the experiments can be viewed as rigid rods on hinges. A rigid rod hinged to the surface would have an average height of the distal label with an average tilt angle of 60° from normal, assuming free rotation of one half of the length. Factors such as steric hindrance or surface interactions may limit this free rotation. Using SSFM, we measured the elevation of the label on top of DNA double helices to be 5.5 nm for 21-bp fragments and 10.5 nm for 50-bp fragments, which represents tilt angles from the normal of 40° and 50°, respectively. These values are a measure of the average distribution of heights within the microscope focal spot. If the label on the second strand is at the 3′ end, its location in the double helix will be at the bottom, close to the surface. Although we did not expect to see a significant variation, the proximal, 3′-end label on 50-bp DNA is elevated by ∼2–3 nm compared with only 0.5–1 nm in the case of the 21-bp fragment. This is because a 50-mer is long enough to have stable partial hybridization that could free the proximal end and yield an elevated label position, and also because of steric hindrance.

We also studied the conformation of ssDNA by measuring the height of fluorescent tags attached to the free end of surfacebound DNA oligonucleotides. Unlike dsDNA, ssDNA is flexible and little is known about the shape or size of ssDNAs on the surface. AFM measurements suggest that ssDNA immobilized on a surface exists in a globular conformation [21].

However, there are reports stating that because of steric hindrance from nearby molecules, ssDNA may change its conformation from a random coil to more extended forms [22,23]. The fluorescent label attached to the distal end of surface-bound single-stranded 21-mer is found to be close to the surface, within 1 nm. In a similar situation, 50-mers show much higher location of the label: 5.5 nm above the surface as illustrated, pointing out to a considerably more extended conformation compared with 21-mers, which may be due to steric effects from closely located grafts in the DNA layer. The surface density of immobilized ssDNA measured using a radiolabel is $\sim$35 fmol/mm^2 for both 21- and 50-mers, which translates into 11 nm distances between adjacent molecules or a 5.5 nm radius of free space around each. As the length of a fully extended 50-mer is 27.5 nm, it is enough to interact with its neighbors, at least intermittently. When a second, unlabeled strand is hybridized, the label at the distal end of the newly formed duplex extends out as well. Unlike the dsDNA in Fig. 5.10, the DNA layer now consists of two species: the dsDNA and the unhybridized strands, both of which are carrying a fluorescent marker but at different heights. The average position of the marker should be somewhere in between depending on the degree of hybridization. The binding efficiency is estimated by comparing the average height of the fluorescent layer above the surface for the two cases, and found to be between 30% and 50%. This rough estimate is close to the result obtained by WL reflectivity: 50% hybridization for both 21 and 50-mers, demonstrating self-consistency. As a further check, we have performed density measurements with radiolabeled DNA and found it consistent with 50% hybridization. Because of intrinsic limitations such as substrate-related quenching of radiative emission, we do not consider the radiolabeling for absolute determination of DNA densities, but rather only for relative estimation of DNA densities.

We also studied how the conformation of a surface-bound 50-nt labeled oligonucleotide changes when it is annealed with a 21-mer complementary to either its top or bottom part. When a 21-mer complementary to the top section was annealed, the position of the distal end increased from 5.5 to 6.5 nm. However, when an ssDNA–dsDNA construct has the doublestranded part proximal to the surface, the position of the label is lower than that of an unhybridized oligo, decreasing in average height from 5.5 to 3.5 nm. These data suggest that the distal bound 21-mer construct is nearly vertical, and the proximal bound construct has formed a rotation point allowing the flexible distal end to approach closer to the surface. The selected sequence of the oligonucleotides rules out the possibility of intramolecular DNA structures.

5.6 SSFM in 4Pi Configuration

SSFM is a powerful tool in determining the position of fluorescent labels and giving insight about the conformation of biological structures bound to surfaces. However, despite its high precision axial position determination capability, its lateral resolution is limited by several micrometers as low NA objectives have to be used to ensure spectral fringe visibility. The large spacer layer thickness results in a focal mismatch between the direct and the reflected image of the sample when high NA objectives are used. Also, with high NA objectives, the rapid change of fields for large angles due to the interference of direct and reflected signals smears out the spectral interference fringes and lowers the signal contrast. One remedy for increasing lateral resolution by using high NA objectives while maintaining the nanometer level axial localization capability is to use a second objective instead of a mirror and collect the emission from both sides (Fig. 5.11a). As the spherical wavefronts of the emission are collected by a symmetrical configuration, path length mismatch for higher collection angles for two interference pathways is eliminated unlike using one objective and a reflecting substrate. The path lengths of the two interference arms are adjusted by one of the mirrors mounted on a piezo controller, such that they are different by several tens of micrometers to induce modulations in the spectrum of the emitter. This configuration is similar to

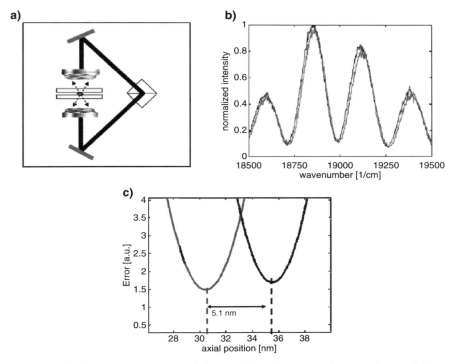

Fig. 5.11. (a) A 4Pi microscope. (b) Normalized spectra of a monolayer of Alexa Fluor 488 measured using 4Pi microscope at two axial positions that are 5 nm apart. (c) Minimization of error in fitting spectral responses of two positions.

4Pi confocal fluorescence microscopy introduced by Stefan Hell [1]. However, instead of using the 4Pi microscope for 3D image scanning, the emitted signal is collected by two high NA objectives and coupled to a spectroscopy system to localize monolayers with nanometer accuracy. Scanning of the tight collection spot is performed in lateral dimensions and axial profile is extracted form the spectral fringes similar to SSFM.

In Fig. 5.11b, typical spectral response of a 4Pi SSFM system is shown. The two distinct spectra are from the same spot of a monolayer that was moved by 5 nm in axial direction between the data acquisition. The fluorescent emitter is a monolayer of Alexa Fluor 488 dye deposited on a glass cover slip and placed in the common foci of the two high NA oil immersion objectives of the 4Pi microscope. The sample was excited by a 488 nm continuous wave laser using the two objectives. The spectral response clearly shows the shift in spectral fringes because of the change in the axial position of the monolayer of fluorophores. A simple fitting algorithm that utilizes a geometrical model and only includes the position of monolayer with respect to the geometrical focus and the external path length difference can be used to determine the relative difference in the axial positions of monolayers. As seen from Fig. 5.11c, the axial position values that minimize the error in fitting the spectral responses of two positions of the monolayer were actually 5.1 nm apart, which agrees well with the 5 nm physical change in the axial position of the layer.

5.7 Conclusions

We have developed a new technique, SSFM, which transforms the variation in emission intensity for different path lengths used in fluorescence interferometry to a variation in the intensity for different wavelengths in emission, encoding the high-resolution information in the emission spectrum. Using SSFM, we have demonstrated analysis of systems that include monolayers of fluorescein, quantum dots, and lipid films. More importantly, we have demonstrated conformation of surface-immobilized DNA by estimating the shape of coiled ssDNA, the average tilt of dsDNA of different lengths, and the amount of hybridization. We have also shown that localization of axial position of fluorescent structures with high precision is possible without sacrificing the lateral resolution using SSFM in 4Pi configuration.

These data provide important proofs of concept for the capabilities of novel optical methods for analyzing the molecular disposition of DNA and protein on surfaces. The determination of DNA conformations on surfaces and hybridization behavior provide information required to move DNA interfacial applications forward and thus impact emerging clinical and biotechnological fields.

Acknowledgments

The authors acknowledge Prof. İrşadi Aksun of Koc University for his contributions to modeling of fluorophore emission on layered surfaces and Dr. Stephen Ippolito for his help in setting up the microscope, the National Science Foundation (Grant DBI0138425), Air Force Office of Scientific Research (Grant MURI F-49620-03-1-0379), and National Institutes of Health-National Institute of Biomedical Imaging and BioEngineering (Grant 5R01 EB00 756-03) for their support. The views and conclusions contained in this document are those of the authors and should not be interpreted as representing the US Government.

References

1. S. Hell, E.H.K Stelzer, JOSA A **9**(12), 2159 (1992)
2. B. Bailey, D.L. Farkas, D.L. Taylor, F. Lanni, Nature **366**, 44 (1993)
3. P.T.C. So, C.Y. Dong, B.R. Masters, K.M. Berland, Ann. Rev. Biomed. Eng. **2**, 399 (2000)
4. M. Dyba, S.W. Hell, Phys. Rev. Lett. **88**, 163901 (2002)
5. A. Lambacher, P. Fromherz, Appl. Phys. A **63**, 207 (1996)
6. A.K. Swan, L. Moiseev, C.R. Cantor, B. Davis, S.B. Ippolito, W.C. Karl, B.B. Goldberg, M.S. Unlu, IEEE J. Select. Top. Quantum Electron. **9**, 294 (1996)
7. L. Moiseev, C.R. Cantor, I. Aksun, M. Dogan, B.B. Goldberg, A.K. Swan, M.S. Unlu, J. Appl. Phys. **96**, 5311 (2004)
8. L. Moiseev, M.S. Ünlü, A.K. Swan, B.B. Goldberg, C.R. Cantor, Proc. Natl. Acad. Sci. **103**, 2623 (2006)
9. R. Moller, A. Csaki, J.M. Kohler, W. Fritzsche, Nucleic Acids Res. **28**, e91 (2000)
10. X. Wu, H. Liu, J. Liu, K.H. Haley, J.A. Treadway, J.P. Larson, N. Ge, F. Peale, M.P. Bruchez, Nat. Biotech. **21**(1), 41 (2003)
11. W.C.W. Chan, D.J. Maxwell, X.H. Gao, R.E. Bailey, M.Y. Han, S.M. Nie, Curr. Opin. Biotech. **13**(1), 40 (2001)
12. L.A. Chrisey, G.U. Lee, C.E. O'Ferrall, Nucleic Acids Res. **24**, 3031 (1996)
13. D.E. Gray, S.C. Case-Green, T.S. Fell, P.J. Dobson, E.M. Southern, Langmuir **13**, 2833 (1997)
14. R. Levicky, T.M. Herne, M.J. Tarlov, S.K. Satija, J. Am. Chem. Soc. **120**, 9787 (1998)
15. T.M. Herne, M.J. Tarlov, J. Am. Chem. Soc. **119**, 8916 (1997)
16. M.T. Charreyre, O. Tcherkasskaya, M.A. Winnik, Langmuir **13**, 3103 (1997)
17. M.C. Murphy, I. Rasnik, W. Cheng, T.M. Lohman, T. Ha, Biophys. J. **86**, 2530 (2004)
18. K.A. Peterlinz, R.M. Georgiadis, T.M. Herne, M.J. Tarlov, J. Am. Chem. Soc. **119**, 3401 (1997)
19. L.K. Wolf, Y. Gao, R.M. Georgiadis, Langmuir **20**, 3357 (2004)
20. S.O. Kelley, J.K. Barton, N.M. Jackson, L.D. McPherson, A.B. Potter, E.M. Spain, M.J. Allen, M.G. Hill, Langmuir **14**, 6781 (1998)

21. L.S. Shlyakhtenko, A.A. Gall, J.J. Weimer, D.D. Hawn, Y.L. Lyubchenko, Biophys. J. **77**, 568 (1999)
22. Q. Du, M. Vologodskaia, H. Kuhn, M. Frank-Kamenetskii, A. Volodogski, Biophys. J. **88**, 4137 (2005)
23. S. Elhadj, G. Singh, R.F. Saraf, Langmuir **20**, 5539 (2004)

6

Applications of Optical Resonance to Biological Sensing and Imaging: II. Resonant Cavity Biosensors

M.S. Ünlü, E. Özkumur, D.A. Bergstein, A. Yalçin, M.F. Ruane, and B.B. Goldberg

6.1 Multianalyte Sensing

Interrogating binding interactions between proteins, segments of DNA or RNA, and biospecific small molecules is critically important for a great number of applications in biological research and medicine. Microarray technology has emerged over the past decade to address applications that seek to measure thousands or even millions of binding interactions at once. A microarray consists of a solid support, or substrate, with multiplicity of spots on its top surface each containing a different type of fixed capturing molecule. A sample solution containing unknown target molecules, or analytes, is introduced to the microarray surface typically via a small fluid chamber or flow cell. The amount of target material bound to any feature after washing the array gives an indication of the affinity between the target and capturing agent at that spot. There are a number of important applications in biological research that benefit from microarray throughput such as gene expression profiling or antigen–antibody interaction monitoring. Aside from research applications, microarrays may play a crucial role in a new era of medical diagnostics. Biomedical research is continuing to point toward molecular biomarkers that can be used to help doctors diagnose diseases sooner, with greater accuracy, and provide information that helps doctors personalize treatment plans.

Present microarray technology requires that the target molecules in the sample solution be labeled or attached with a fluorescent dye molecule for the purpose of determining how much target material has bound to each microarray feature using a fluorescence detector. The need to label the target molecules to visualize the results is a significant shortcoming of present technology and has been described previously [1–4]. In general, fluorescent labeling may suffer from bleaching of the labels, quenching effects from the surface, or low contrast between the label and the autofluorescence of the microarray substrate. In addition, proteins in particular may suffer from altered binding properties once they are labeled. For these reasons, label-free detection for both DNA and protein is preferred, particularly for use in medical diagnostics where variation and cost must be minimized.

A vast number of label-free detection techniques have surfaced over the last decade or two to address the problems of label-based detection [5–16]. The important figures of merit for these techniques are sensitivity to bound targets, throughput (number of features evaluated in a single test) and the ability to scale throughput, and cost per test. Sensitivity is often quantified by mass per area (pg mm^{-2}) or average height (pm) where 1 pm corresponds to roughly 1 pg mm^{-2} for protein or DNA [17–25].

6.2 Resonant Cavity Imaging Biosensor

6.2.1 Detection Principle

Two partially reflecting substrates are positioned such that their reflecting surfaces face each other and form the optical cavity (Fig. 6.1) [26]. The tunable wavelength laser light is collimated and incident from the back of one of the reflectors. The wavelength of the laser is swept in time and at specific wavelengths the resonant condition of the cavity is met, the light resonates inside the cavity and couples out. Beyond the cavity, the transmitted light is imaged on a camera, so that the resonant response at each location of the cavity is recorded by a corresponding pixel on the camera. The probe biomolecules are patterned on one of the reflector surfaces. When target biomolecules bind to their specific capturing agents, the local resonant response shifts in

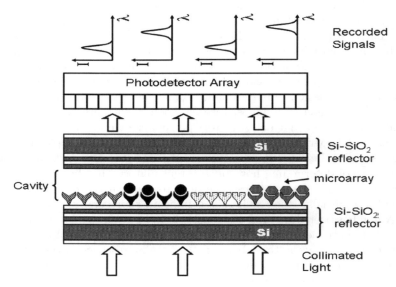

Fig. 6.1. Resonant cavity imaging biosensor (RCIB) setup.

wavelength. The presence of biomolecules bound to the surface is fundamentally detected by a small perturbation of the electric field inside the cavity. The effect of a slight phase delay is amplified by the resonant behavior of light within the cavity. The cavity enhancement can provide a significant sensitivity improvement over interference techniques that do not benefit from a high finesse optical cavity. The reflectors can be constructed with alternating dielectric layers of silicon and SiO_2 and may end with a layer of SiO_2 that allows the surface immobilization chemistries that are developed for glass slides to be used. The use of a parallel optical cavity, tunable wavelength laser, and a digital camera allows high throughput. Aside from the alignment and stability of the reflectors, and the internal operation of the tunable laser, there are no moving parts in the setup. Additionally, the reflectors described here may be made very inexpensive and hence disposable. In the present implementation, the cavity is formed with reflectors fabricated from alternating layers of Si and SiO_2, constituting a Bragg grating (Fig. 6.2) [27], and the wavelength of light is varied between 1,510 and 1,515 nm. The electric field distribution in the cavity forms a standing wave pattern with a minimum at the Si surface of the reflectors. It is beneficial to place the sensing surface on an approximately quarter wavelength thick layer of similar low-index material, namely SiO_2. Adding this extra oxide layer brings the binding surface some distance into the cavity where the field is maximum (Fig. 6.3). When it is placed at the field maximum, the sensing layer interacts with the highest number of photons and slight thickness changes will have a maximum effect on the field distribution, thereby shifting the resonant wavelength a maximum amount. If the sensing surface is positioned at a field minimum, small changes on the surface will have no effect on the resonant cavity and will therefore be undetectable. Adding an extra layer of SiO_2 also has the previously mentioned benefit of providing

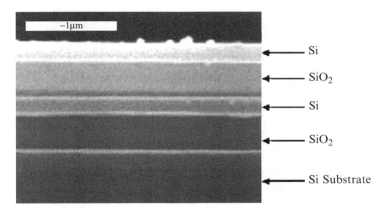

Fig. 6.2. The Bragg structures that are used as the reflectors in RCIB setup with alternating layers of $1/4$ wavelength thick Si and $3/4$ wavelength thick SiO_2. The image is taken with SEM.

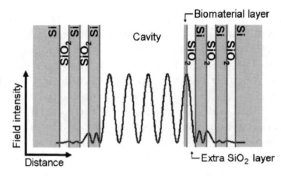

Fig. 6.3. Electric field pattern in the optical cavity.

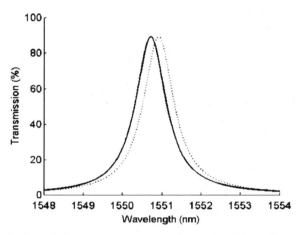

Fig. 6.4. Simulation of the transmittance of the cavity. The reflectors are 40.0 μm apart from each other with air in between. A final SiO_2 layer is included and shift corresponds to 5 nm step on the oxide.

a chemically compatible surface for microarray fabrication. Simulation of the cavity transmittance vs. wavelength with changing top surface oxide thickness is plotted in Fig. 6.4. Simulation was done in Matlab using the matrix method for calculating transmittance through layered media [28].

6.2.2 Experimental Setup, Data Acquisition, and Processing

An RCIB setup has been constructed. The illumination source is an IR tunable laser with a fiber-coupled output. Light is coupled into the system with a SMF-28 fiber and collimated to 10 mm FWHM. All the optics are antireflection coated to avoid interference effects. The laser power is set to 15 μW, and the wavelength is continuously swept from 1,510 to 1,515 nm in 10 s during data collection. The reflectors, which are 15 mm by 15 mm, are placed 40 μm apart from each other. One reflector is mounted on a tip-tilt stage to bring the reflectors parallel. The second reflector is mounted on a translational stage to adjust the reflector separation. The transmission from the cavity is imaged on an indium–gallium–arsenide (InGaAs) sensor array with 5 × magnification and an aperture setting the NA to 0.1. With these settings, $12 \times 12\,\mu\text{m}^2$ areas are imaged on the $60 \times 60\,\mu\text{m}^2$ pixels of the sensor array, which has 128×128 pixels. The laser wavelength is swept at a rate of $0.5\,\text{nm}\,\text{s}^{-1}$ while the camera captures images at a rate of 30 fps. Images are digitized to 12 bits (4,096 gray-levels) and transferred to a PC.

These data are transferred to Matlab and fit using several different methods. There is a trade-off between the accuracy and the speed with which the data can be fit. A more accurate (noise-resilient) curve fitting method is to fit the data to resonant line shape given by (6.1), which models a simple resonant cavity where

$$T = \frac{n_\text{i}}{n_\text{f}} \left| \frac{t^2 e^{jkd}}{1 - r^2 e^{-2jkd}} \right|^2, \qquad (6.1)$$

n_i and n_f are the refractive indices in and beyond the cavity, t and r are the transmission and reflection coefficients, respectively, of the reflectors, k is the wavenumber inside the cavity, and d is the reflector separation (Fig. 6.5). Further image processing subtracts the curvature inherent to the reflectors. Finally, a surface profile of small height variations on the SiO_2 surface can be obtained.

6.2.3 Experimental Results

The system sensitivity was tested by imaging etched features on the oxide surface (Fig. 6.6). The reflector that has a SiO_2 layer on top is used as the

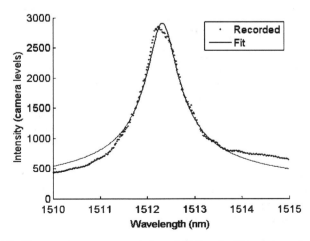

Fig. 6.5. Resonant curve recorded and fitting to this data using (6.1).

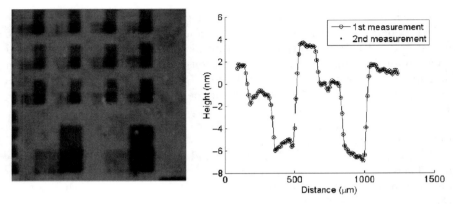

Fig. 6.6. *Left*: RCIB scan of the overlapping boxes. *Right*: Line cut of the image in two consecutive measurements.

sensing surface. Standard photolithography and wet (BOE) etching were used to pattern these features on the surface. Two overlapping rectangles, with depths of 3 and 6 nm were etched resulting in four regions with different oxide thickness; not etched, 3 nm deep, 6 nm deep, and 9 nm deep etched. These values are consistent with the AFM measurements of the same sample. System repeatability was tested by making consecutive measurements of the same sample. Averaging boxes measuring 5 pixels by 5 pixels, or an area of $50 \times 50\,\mu m^2$, the RMS deviation in consecutive measurements was 0.01 nm (Fig. 6.6).

6.2.4 Spectral Reflectivity Imaging Biosensor

The phase delay added by extra material can also be imaged using a simpler approach. We propose a surface profilometry technique to be used as a label-free microarray imaging device. Spectral reflectivity imaging biosensor (SRIB) uses wavelength dependent reflectivity of a silicon substrate with thick thermally grown SiO_2 (5–10 μm), to accurately find the film thickness at tens of thousands of different spots, and thus image the surface profile. This technique, having a much lower finesse, is expected to be less sensitive than RCIB, but should be more robust and offer higher throughput. A collimated laser beam that can be tuned from 764 to 784 nm is reflected from a pellicle beam splitter and is incident on the sample surface (Fig. 6.7). At any fixed wavelength, reflection from the sample is imaged on a CCD camera (camera pixel size is 13.7 μm and the magnification of the imaging system is 2.3). Then the wavelength is stepped, and another image is taken. Repeating this through the sweeping range of the laser, one can form a spectral reflectivity curve for every $36\,\mu m^2$ area on the sample in the field of view. The separate

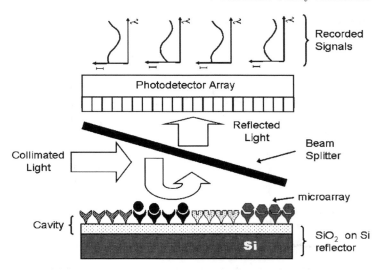

Fig. 6.7. Spectral reflectance imaging biosensor diagram.

reflections from oxide and silicon surfaces form an interference pattern and an exact oxide thickness corresponding to each pixel is found by curve fitting the recorded data.

Reflection of a single layer can be well estimated by Airy's formulas [28]. The reflection coefficient for an incoming field is given by:

$$r = \frac{r_{12} + r_{23}e^{-2i\phi}}{1 + r_{12}r_{23}e^{-2i\phi}} \qquad (6.2)$$

$$\phi = \frac{2\pi d}{\lambda} n_{SiO_2} \cos\theta, \qquad (6.3)$$

where r_{12} and r_{23} are the Fresnel reflection coefficients from the SiO_2 surface and SiO_2-Si interface, respectively, and ϕ is the optical phase difference between the two reflections. Here, d is the SiO_2 thickness, n_{SiO_2} is the refractive index of SiO_2, λ is the wavelength of the incident light, and θ is the incidence angle to the SiO_2–Si interface, which will be 0° for perpendicular incidence. Curve fitting tools are used to fit this equation to the recorded data at each pixel and to find the d. The setup was tested with etch marks of overlapping rectangles that was mentioned in the previous section. Silicon wafers with 6.1 μm of thermally grown SiO_2 were etched by photolithographic techniques. Three different depths of etch steps were created on the sample, which were 4, 12, and 16 nm, and their depths were confirmed with AFM scans. The sample was then imaged by SRIB, and data were processed with curve fitting. A line-cut from the image can be seen in Fig. 6.8. These etch marks were imaged with an accuracy of 0.5 nm.

94   M.S. Ünlü et al.

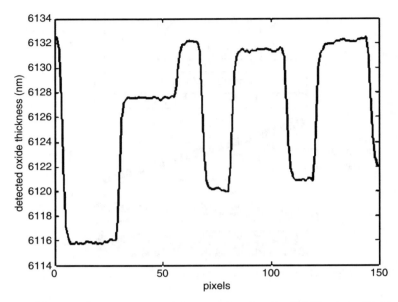

Fig. 6.8. Line-cut from a processed image taken by the SRIB setup. Etch marks at three different heights are seen: 4, 12, and 16 nm. The thickness of the oxide was ∼6,130 nm, initially.

Although the present data show that the system can image 250 spots with subnanometer sensitivity, balancing the laser intensity fluctuations and using a better imaging system is expected to improve sensitivity to 0.1 nm with more than 10,000 spots.

6.3 Optical Sensing of Biomolecules Using Microring Resonators

6.3.1 Basics on Microring Resonators

Integrated devices featuring resonant microcavities with high quality factors (Q) such as toroids, disks, rings, and spheres have been used as add-drop filters, optical switches, and in laser applications [29–33]. These devices have recently become popular for research in biochemical sensing [34–38] as the demand for highly sensitive and compact devices to detect biomolecules increased. Light is confined within the microring cavities by total internal reflections resulting in high Q resonant modes. When the refractive index of the cladding or the outside medium changes (e.g., due to binding of molecules), a new guiding condition is obtained for the mode, causing a shift in the resonant wavelength.

Microring resonators provide sensitivities comparable to surface plasmon resonance because of their high Q (∼12,000 in this study) [16,39], and they are robust and can be mass-manufactured using well-established silicon-integrated circuit fabrication techniques. However, they lack the very high-throughput potential that is demonstrated with RCIB.

The sensitivity of the system presented here is quantified through measurement of resonance shifts induced by a change in the refractive index of the medium (a bulk change). Additionally, sensing of biomolecules is demonstrated through observation of the change in resonant condition caused by molecular binding of the well-documented avidin–biotin couple on the surface.

6.3.2 Setup and Data Acquisition

The microring resonators used in this experiment are 60 μm in radius, and vertically coupled to waveguides (Fig. 6.9). The effective index of the cavity is measured as $n_{\text{eff}} \approx 1.5$ in deionized water (DI-H_2O) ambience, and the penetration depth for the guided mode at 1,550 nm in this environment is calculated as ∼360 nm. A fiber-coupled tunable wavelength IR laser is used as the light source. The cavity can support both TE and TM modes and with the use of a polarizer one of the modes is selected. An optical splitter placed after the polarizer separates the signal into two paths: one arm is chopped in free space at a frequency of 220 Hz and then coupled to the input waveguide of the microring resonator. The output waveguide is coupled to a photodetector as signal input. The second arm acts as the reference input to the photodetector for balanced detection. Common-mode noise cancellation

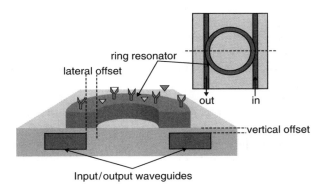

Fig. 6.9. Microring resonator is vertically coupled to input/output waveguides. Lateral and vertical offsets of the resonator allow control over the coupling coefficient. Receptor molecules are immobilized on the resonator surface, and target molecules are released on the surface in a solution during flow. The *top view* of the resonator is shown in the inset.

(mainly laser intensity fluctuations) is achieved through balanced detection, improving the signal-to-noise ratio, thus the overall system sensitivity. A flow-cell is used for controlled solution flow over the microring surface, and real time signal during solution flow is acquired and analyzed with LabVIEW. Direct measurement of a shift in resonance through repeated spectral scans is time consuming. Alternatively, the resonance shift can be measured indirectly by collecting data at a single wavelength and observing the intensity in real time, and the change in intensity can then be mapped to a shift in resonance. To obtain the maximum change in intensity due to a change in effective index, the wavelength at which the measurements are performed is selected to correspond to the point of maximum slope of the resonance curve. The high intensity stability necessary for these measurements is achieved by eliminating the noise components of laser output through balanced detection.

6.3.3 Data Analysis and Discussion

The bulk experiments to determine the sensitivity of the system were conducted by flow of a solution of known refractive index with respect to the refractive index of DI-H_2O that is used as a calibration reference. In Fig. 6.10, the measured shift in resonance with respect to a change in the refractive index of the medium (corresponding to various concentrations of phosphate buffered saline (PBS) solution) is plotted. Notice that the relation is highly linear. Using standard deviation δ (pm) of signal levels during PBS flow and the slope (m) of the plot in Fig. 6.10, the limit of detection (LOD) for a change in refractive index can be approximated as $LOD(n) = 3\delta \, m^{-1}$. We conclude that $LOD(n) = 1.8 \times 10^{-5}$ refractive index units (RIU) for our setup. The experiment to demonstrate detection of biomolecules consists of two parts: binding of biotinylated lectins to an avidin covered surface, and breaking the avidin–biotin bonds for surface regeneration. For binding, 3.5 µM biotinylated lectin solution is flowed over the surface. Once the signal level reached by the binding of biotin molecules to avidin covered surface is stabilized, DI-H_2O is flowed to wash away the unbound molecules. For surface regeneration, the surface is first washed with a chemical pH-7 buffer that breaks the avidin–biotin bonds, and then with DI-H_2O. The real-time intensity recording of binding/regeneration processes is shown in Fig. 6.11. The actual amount of change in the signal is measured between levels of DI-H_2O flow before and after the introduction of biotinylated lectin solution. At these levels, the outside medium is identical, so the change in intensity is caused only due to binding of molecules. The signal level at the end of the regeneration phase is close to the initial level indicating that partial recovery of surface is achieved.

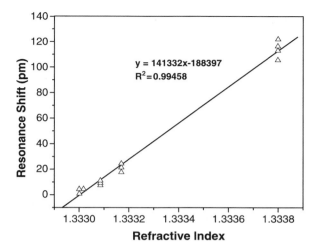

Fig. 6.10. Shift of resonance is linearly related to change in refractive index. The slope of the plotted curve is used to determine the limit of detection for refractive index changes.

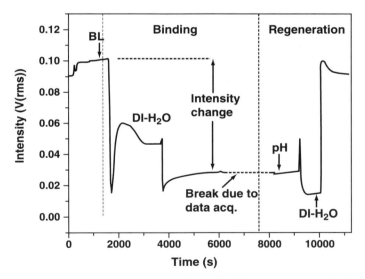

Fig. 6.11. The detected intensity in a real time binding experiment is plotted. The plot illustrates both binding and regeneration phases (BL: biotinylated lectin solution, DI-H_2O: deionized water, pH: pH-7 buffer). Binding efficacy is only measured between the signal levels when DI-H_2O is flowed across the sensor.

6.4 Conclusions

We have presented three optical techniques aimed at molecular sensing that employ optical resonance. A planar optical resonant cavity is used in the case of RCIB, where the high finesse of the optical cavity can provide high sensitivity, while the planar configuration tied with a camera and tunable laser can provide for simultaneous detection from many parallel locations. The technique was demonstrated using nanometer scale etches in a SiO_2 surface that model microarray features. We have also presented a second technique that observes the wavelength-dependent reflectivity from a silicon substrate with a thermally grown oxide layer to detect molecules binding to the top oxide surface. In this case, the Fresnel reflection measured can be considered to be the result of a low finesse optical cavity formed by the oxide layer together with the bound molecules. The benefit of this technique compared with the high finesse RCIB technique is that the monolithic structure avoids the difficulty maintaining the reflector alignment necessary in the latter technique. In the final technique presented, microring resonators employ a very strong resonant coupling to achieve very high sensitivity at the cost of added complexity and a larger minimum feature size.

Acknowledgments

The authors acknowledge Boston University Office of Technology Development for their support, the National Institute of General and Medical Sciences (NIH R21 GM 074872-02), and the Army Research Laboratory (ARL) for their support under the ARL Cooperative Agreement Number DAAD17-99-2-0070 and under the AMCAC-RTP Cooperative Agreement DAAD19-00-2-0004. The authors also thank Dr. Brent Little and the Little Optics Division of Nomadics Inc. for providing the microring resonator devices. The views and conclusions contained in this document are those of the authors and should not be interpreted as representing the Laboratory or the US Government.

References

1. S. Niu, R.F. Saraf, Smart Mater. Struct. **11**, 778 (2002)
2. R. Nadon, J. Shoemaker, Trends Genet.**18**, 265 (2002)
3. M.A. Cooper, Nat. Rev. Drug Discovery **1**, 515 (2002)
4. M. Schena, *Microarray Biochip Technology,* (Eaton Publishing, Natick, 2000)
5. J.M. Brockman, B.P. Nelson, R.M. Corn, Annu. Rev. Phys. Chem. **51**, 41 (2000)
6. K. Usui-Aoki, K. Shimada, M. Nagano, M. Kawai, H. Koga, Proteomics **5**, 2396 (2004)
7. C. Worth, B.B. Goldberg, M.F. Ruane, M.S. Ünlü, IEEE J. Select. Top. Quantum Electron. **7**, 874 (2001)
8. W. Lukosz, Biosensors Bioelectron. **6**, 215 (1991)
9. H. Nygren, T. Sandström, M. Stenberg, J. Immunol. Meth. **59**, 145 (1983)

10. T. Sandström, M. Stenberg, H. Nygren, Appl. Opt. **24**, 472 (1985)
11. J. Piehler, A. Brecht, G. Gauglitz, Anal. Chem. **68**, 139 (1996)
12. P.I. Nikitin, M.V. Valeiko, B.G. Gorshkov, Sensors Actuators B **90**, 46 (2003)
13. J.P. Landry, X.D. Zhu, J.P. Gregg, Opt. Lett. **29**, 581 (2004)
14. J. Lu, C.M. Strohsahl, B.L. Miller, L.J. Rotherberg, Anal. Chem. **76**, 4416 (2004)
15. S. Chan, Y. Li, L.J. Rothberg, B.L. Miller, P.M. Fauchet, Mater. Sci. Eng. C **15**, 277 (2001)
16. A. Yalcin, K.C. Popat, J.C. Aldridge, T.A. Desai, J. Hryniewicz, N. Chbouki, B.E. Little, O. King, V. Van, S. Chu, D. Gill, M.F. Anthes-Washburn, M.S. Ünlü, B.B. Goldberg, IEEE J. Select. Top. Quantum Electron. **12**, 148 (2006)
17. L. Moiseev, M.S. Ünlü, A.K. Swan, B.B. Goldberg, C.R. Cantor, Proc. Natl. Acad. Sci. **103**, 2623 (2006)
18. L.K. Wolf, Y. Gao, R.M. Georgiadis, Langmuir **20**, 3357 (2004)
19. L. Moiseev, C.R. Cantor, A.K. Swan, B.B. Goldberg, M.S. Ünlü, Proc. Int. Soc. Opt. Eng. Nanobiophotonics Biomed. Appl. **5331**, (2004)
20. J.J. Ramsden, Opt. Biosens. J. Mol. Recogn. **10**, 109 (1997)
21. J.A. DeFeijter, J. Benjamins, F.A. Veer, Biopolymers **17**, 1759 (1978)
22. S.N. Timasheff, in *Handbook of Biochemistry and Molecular Biology*, vol. 3, ed. by G. Fasman (CRC Press, Cleveland, 1977), p. 372
23. P. Schaaf, P. Dejardin, A. Schmitt, Langmuir **3**, 1131 (1987)
24. T.M. Davis, W.D. Wilson, Anal. Biochem. **284**, 348 (2000)
25. S. Nui, A. Label-free, Ph.D. dissertation, Virginia Polytechnic Institute and State University, 2004
26. D. Bergstein, M.F. Ruane, M.S. Ünlü, in *International Semiconductor Device Research Symposium, Maryland, USA*, 7–9 December 2005
27. M.K. Emsley, O.I. Dosunmu, M.S. Ünlü, IEEE J. Select. Top. Quantum Electron. **8**, 948 (2002)
28. P. Yeh, *Optical Waves in Layered Media* (Wiley, New York, 1988)
29. B.E. Little, S.T. Chu, W. Pan, D. Ripin, T. Kaneko, Y. Kokubun, E.P. Ippen, IEEE Photon. Technol. Lett. **11**, 215 (1999)
30. M. Cai, G. Hunziker, K. Vahala, IEEE Photon. Technol. Lett. **11**, 686 (1999)
31. V. Van, T.A. Ibrahim, K. Ritter, P.P. Absil, F.G. Johnson, R. Grover, J. Goldhar, P.T. Ho, IEEE Photon. Technol. Lett. **14**, 74 (2002)
32. L. Yang, D.K. Armani, K.J. Vahala, Appl. Phys. Lett. **83**, 825 (2003)
33. B. Liu, A. Shakouri, J.E. Bowers, Appl. Phys. Lett. **79**, 3561 (2001)
34. K.J. Vahala, Nature **424**, 839 (2003)
35. R.W. Boyd, J.E. Heebner, Appl. Opt. **40**, 5742 (2001)
36. S. Arnold, M. Khoshsima, I. Teraoka, S. Holler, F. Vollmer, Opt. Lett. **28**, 272 (2003)
37. F. Vollmer, D. Braun, A. Libchaber, M. Khoshsima, I. Teraoka, S. Arnold, Appl. Phys. Lett. **80**, 4057 (2002)
38. J. Hryniewicz, N. Chbouki, B.E. Little, O. King, V. Van, S. Chu, D. Gill, in *Biophotonics/Optical Interconnects and VLSI Photonics/WBM Microcavities*, 2004 Digest of the LEOS Summer Topical Meetings, 33 (2004)
39. Bardin, I. Kasik, A. Trouillet, V. Matejec, H. Gagnaire, M. Chomat, Appl. Opt. **41**, 2514 (2002)

7

Biodetection Using Silicon Photonic Crystal Microcavities

P.M. Fauchet, B.L. Miller, L.A. DeLouise, M.R. Lee, and H. Ouyang

7.1 Photonic Crystals: A Short Introduction

7.1.1 Electromagnetic Theory

Lord Rayleigh was the first to study electromagnetic wave propagation in one-dimensional (1D) periodic media and to identify the angle-dependent, narrow band in which light propagation is prohibited. However, it was not until a full century later, when Yablonovitch [1] and John [2] in 1987 combined Maxwell's equations with solid-state physics theorems to introduce the concept of photonic bandgaps in two and three dimensions. Many subsequent developments in fabrication, theory, and applications (e.g., fiber optics, integrated optics, and negative refraction materials) have since followed.

As shown in Fig. 7.1, photonic crystals (PhCs) are periodically structured media [3] in one, two, or three dimensions. PhCs can be designed to produce photonic bandgaps. Light with photon energies or frequencies that fall inside this bandgap cannot propagate through the PhC. The periodicity in length scale is proportional to the wavelength of light inside the bandgap. PhCs are the electromagnetic analog of crystalline atomic lattices, in which interference in the electron wavefunction produces the forbidden bands. Hence, the study of PhCs is also governed by the Bloch-Floquet theorem.

To study the propagation of light, Maxwell's equations are solved as an eigenproblem. We assume that light propagates in a mixed dielectric medium with no free charges or currents. Maxwell's equations are then given by:

$$\nabla \boldsymbol{H}(\boldsymbol{r},t) = 0 \quad \nabla \times \boldsymbol{E}(\boldsymbol{r},t) + \frac{1}{c}\frac{\partial \boldsymbol{H}(\boldsymbol{r},t)}{\partial t} = 0$$
$$\nabla \varepsilon(\boldsymbol{r})\boldsymbol{E}(\boldsymbol{r},t) = 0 \quad \nabla \times \boldsymbol{H}(\boldsymbol{r},t) - \frac{\varepsilon(\boldsymbol{r})}{c}\frac{\partial \boldsymbol{E}(\boldsymbol{r},t)}{\partial t} = 0 \ , \quad (7.1)$$

where $\varepsilon(r) = \varepsilon r + a_i$ is the dielectric function, a_i is the primitive lattice vector in all three dimensions, and c is the speed of light in vacuum. By combining

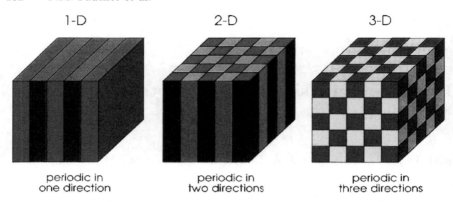

Fig. 7.1. 1D, 2D, and 3D photonic crystals. The periodicity is comparable to the wavelength of light. After Joannopoulos, [3]

the source-free Faraday's and Ampere's laws at a fixed frequency ω, one can obtain a harmonic mode solution:

$$\begin{aligned}\boldsymbol{H}(\boldsymbol{r},t) &= \boldsymbol{H}(\boldsymbol{r})e^{i\omega t} \\ \boldsymbol{E}(\boldsymbol{r},t) &= \boldsymbol{E}(\boldsymbol{r})e^{i\omega t}.\end{aligned} \qquad (7.2)$$

The two curl equations can then be written as:

$$\begin{aligned}\nabla \times \boldsymbol{E}(\boldsymbol{r}) + \frac{i\omega}{c}\boldsymbol{H}(\boldsymbol{r}) &= 0 \\ \nabla \times \boldsymbol{H}(\boldsymbol{r}) - \frac{i\omega}{c}\varepsilon(\boldsymbol{r})\boldsymbol{E}(\boldsymbol{r}) &= 0 ,\end{aligned} \qquad (7.3)$$

leading to the following equation:

$$\nabla \times \left(\frac{1}{\varepsilon(\boldsymbol{r})}\nabla \times \boldsymbol{H}(\boldsymbol{r})\right) = \left(\frac{\omega}{c}\right)^2 \boldsymbol{H}(\boldsymbol{r}). \qquad (7.4)$$

By casting (7.4) as a periodic eigenproblem in analogy with Schrodinger's equation, the solution can be obtained from the Bloch-Floquet theorem:

$$\boldsymbol{H}(\boldsymbol{r}) = \boldsymbol{H}_{n,\boldsymbol{k}}(\boldsymbol{r})e^{i\boldsymbol{k}\cdot\boldsymbol{x}}, \qquad (7.5)$$

with eigenvalues $\omega_n(\boldsymbol{k})$ that satisfy:

$$(\boldsymbol{\nabla} + i\boldsymbol{k}) \times \left(\frac{1}{\varepsilon(\boldsymbol{r})}(\boldsymbol{\nabla} + i\boldsymbol{k}) \times \boldsymbol{H}_{n,\boldsymbol{k}}(\boldsymbol{r})\right) = \left(\frac{\omega_n(\boldsymbol{k})}{c}\right)^2 \boldsymbol{H}_{n,\boldsymbol{k}}(\boldsymbol{r}). \qquad (7.6)$$

Equation (7.6) yields a different Hermitian eigenproblem over the primitive cell at each Bloch wavevector k. These eigenvalues $\omega_n(k)$ are continuous functions of k and can be plotted in a usual dispersion diagram. As the eigensolutions are periodic functions of k, a numerical computational approach, such as the plane-wave expansion method, is often used to calculate the PhC bandgap diagram in reciprocal space.

7.1.2 One-Dimensional and Two-Dimensional PhC

Because Maxwell's equations are scale-invariant, it is convenient to use normalized frequencies ω or wavelengths (a/λ), where a is the lattice constant and λ is the wavelength of light in vacuum. Thus, the same solutions can be applied to any wavelength simply by choosing the appropriate a. Figure 7.2 shows the dispersion curve in a 1D PhC if light propagation is restricted along the optical axis z. If the incident wave travels off-axis, the symmetry in the xy plane is destroyed, and the degeneracy of the bands is lifted. This angle of incidence dependent bandgap imposes practical constraints. Indeed, the take advantage of the 1D bandgap, the light beam must be well collimated and aligned along the z-direction.

Let us now consider 2D PhC structures. In these structures, the electromagnetic field can be decomposed into two polarizations. In the transverse electric (TE) mode, the **E** field is in plane, and the **H** field is perpendicular to the plane. For the transverse magnetic (TM) mode, the situation is reversed. There are two common topologies for making 2D PhCs (Fig. 7.3): square lattice with high index rods surrounded by a low index medium and hexagonal lattice with low-index holes embedded in high index medium. The rod structure is better suited for TM light, and the holes are better suited for TE light. It is also desirable to have bandgaps that overlap in frequency in each symmetry direction of the crystal. Thus, geometries with a periodicity

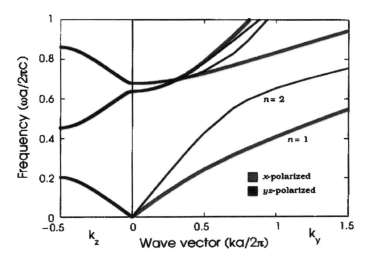

Fig. 7.2. Dispersion curve for a 1D PhC, for light propagating on the z-axis only (*left*) and for light propagating off axis (*right*). On the right, the thick curves labeled $n = 1$ correspond to the x-polarized band and the thin curves labeled $n = 2$ represent the yz-polarized band. When light propagates on axis, the bands are degenerate. After Joannopoulos, Ref. 3

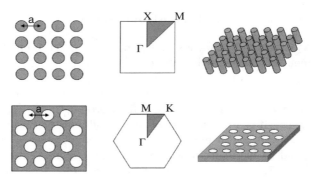

Fig. 7.3. From *left* to *right*: Schematic of 2D PhC topography, Brillouin zone and 3D view of a PhC slab. The *top row* describes a square lattice of dielectric rods in air; the *bottom row* corresponds to a triangular lattice of air holes in a dielectric substrate

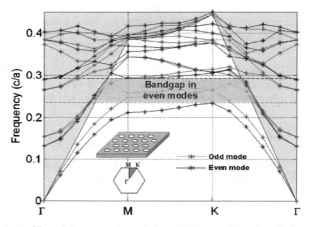

Fig. 7.4. Projected band diagram for a finite-thickness slab of air holes in a dielectric slab. The *shaded region* indicates light cone; only modes below the light cone can propagate and the rest of the modes suffer from severe radiation loss. The "bandgap" below the light cone exists only for the even (TE-like) modes

that is nearly the same in different directions are preferred, and in the case of 2D PhCs, this is an hexagonal lattice.

It is impossible to fabricate a 2D PhC with an infinite slab thickness. An alternative solution is to use thin PhC slabs. The major feature that distinguishes the two systems is the light cone, which is a continuum of states indicated by a shaded region in Fig. 7.4. The modes in the light cone suffer from severe radiation loss, while states that lie beneath the light cone are guided modes. These guided modes are confined to the slab because of the high refractive index contrast between the slab material (e.g., Si) and the substrates (e.g., air or SiO_2). The bandgap is then determined by the range

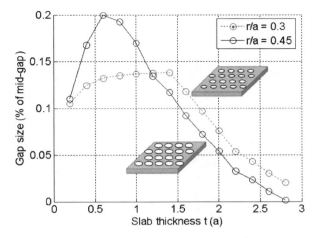

Fig. 7.5. Gap size (as a fraction of the mid-gap frequency) versus Si slab ($n = 3.45$) thickness for PhC slabs with air hole radius to periodicity ratio r/a of 0.3 and 0.45, respectively. The optimum slab thickness is different for these two cases because of the difference in slab effective refractive index

of frequencies for which no guided modes exist. For a slab surrounded by identical top and bottom layers, two categories of slab modes [even (TE-like) and odd (TM-like)] are then defined according to the reflection symmetry with respect to the horizontal plane of midthickness of the slab.

The slab thickness plays an important role in determining whether a PhC slab has a bandgap. For example, if the slab is too thin (i.e., less than half a wavelength), the mode cannot be confined within the slab. As shown in Fig. 7.5, the photonic bandgap width depends on the slab thickness, and the optimal slab thickness varies with different r/a – or different effective slab refractive index. However, the bandgap width is not be the only consideration in selecting the slab thickness. With the microcavities to be discussed later, longer photon decay times or higher quality factors Q can be achieved by using slightly thicker slabs.

7.1.3 Microcavities: Breaking the Periodicity

By intentionally introducing a defect in the PhC, localized electromagnetic states can arise inside the photonic bandgap. These localized modes are the optical analog of the donor or acceptor states produced inside the bandgap of semiconductor crystals. Figure 7.6 illustrates the electric field distribution of a fundamental mode in a PhC microcavity. Instead of staying inside the high refractive index Si, the electric field concentrates in lower refractive index air holes. This property makes it possible for the E-field to interact with the analyte infiltrated inside the air holes.

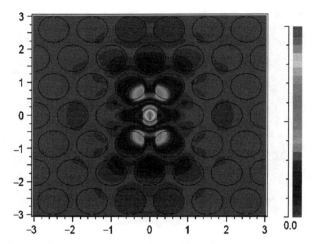

Fig. 7.6. Field confinement at the symmetry-breaking defect in a 2D PhC microcavity. The electric field is mostly confined in defect region

7.1.4 Computational Algorithms

There are two standard computational approaches for studying PhCs: frequency-domain and time-domain methods. The plane-wave expansion algorithm belong to the frequency-domain group. It does a direct computation of the eigenstates and eigenvalues of Maxwell's equations, and is thus better suited for calculating band diagrams. By solving for the eigenvalues, one can precisely predict eigenstates without missing very closely spaced modes, and each field is computed at a definite frequency. One of the disadvantages using this method is that all of the lowest eigenstates have to be computed before reaching the states of interest near the bandgap. It is especially problematic in calculating defect states, in which the lower bands are "folded" many times (depending on the array size) in the Brillouin zone. Moreover, frequency-domain approaches require a supercell with periodic boundaries. Hence, a large array size has to be chosen when computing a single-point-defect PhC microcavity because the interactions between repeated supercells must be minimized.

In contrast, the FDTD (finite-domain-time-difference) method, which is a time-domain approach, computes the fields at each computational cycle at a definite time. This approach is well-suited to compute field distributions and resonance decay-times (quality factor Q). By finding transmission peaks in the Fourier transform of the system response to the input, resonant modes can be identified. One advantage of the FDTD methods is that by sending a pulse into the structure, the response can be obtained at all the frequencies at the same time. Ideally, FDTD methods can be used to calculate structures with arbitrary geometries. As the frequency resolution of the Fourier spectrum is inversely related to the simulation time, a long runtime is required to obtain an acceptable resolution, and the disadvantage in terms of computation time can be severe.

7.2 One-Dimensional PhC Biosensors

7.2.1 Preparation and Selected Properties of Porous Silicon

One-dimensional PhC biosensors have been made using porous silicon (PSi). PSi is obtained by the electrochemical etching of a crystalline silicon wafer in a hydrofluoric acid (HF) solution [4]. The properties of PSi, including the thickness of the porous layer, its porosity, the average pore diameter, and the pore nanomorphology, can be controlled precisely. As a result, the optical properties of PSi can be tuned widely, which makes it a very flexible material for optical biosensors. The internal surface of PSi is very large, ranging from a few to hundreds of square meters per gram, which makes PSi very suitable for capturing biological targets.

The dissolution of silicon requires the presence of fluorine ions (F^-) and holes (h^+). The pore initiation and growth mechanisms are qualitatively understood. Pore growth can be explained by several models [5–7]. If the silicon/electrolyte interface becomes rough shortly after etching starts, the surface fluctuations of the Si/electrolyte interface either grow (PSi formation) or disappear (electropolishing). In forward bias for p-type substrates, the holes can still reach the Si/electrolyte interface as the electric field lines are focused at the tip of the pores. Thus, holes preferentially reach the Si/electrolyte interface deep in the pores, where etching can proceed rapidly (Fig. 7.7). In contrast, no holes reach the end of the Si rods, effectively stopping the etching there. In addition to this electrostatic effect, the random walk of the holes toward the Si/electrolyte interface makes it more likely that they reach it at or near the pore's tip, also resulting in preferential etching at the pore's tip. When an n-type substrate is used, porous silicon formation takes place in reverse bias. Another important mechanism becomes predominant if the Si rods are narrow enough (typically much less than 10 nm). In this size regime, the electronic states start to be affected by quantum confinement. When the motion of carriers is restricted in one or more dimensions, the hole states in the valence band are pushed to lower energy by quantum confinement, which produces a potential barrier to hole transport from the wafer to the Si rods. The holes can no longer drift or diffuse into the Si rods and further etching stops except at the pore's tip.

When the current density decreases, the number of holes at the pore tips drops, which produces smaller pores. Thus, the porosity (defined as the percentage of void space in the material) can be precisely controlled by the etching current density. Figure 7.8 shows the dependence of the porosity on current

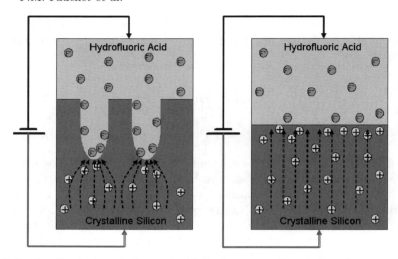

Fig. 7.7. The dissolution of silicon when holes are injected from the substrate toward the Si/electrolyte interface. On the *left*, porous silicon formation takes place when the hole flux is relatively small. On the *right*, electropolishing takes place when the hole flux is large

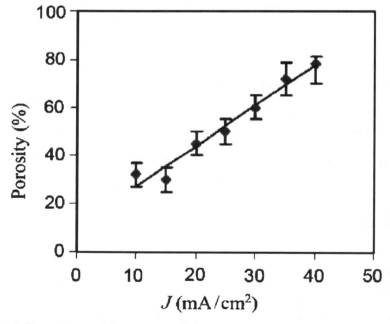

Fig. 7.8. Dependence of the porosity of the current density for highly-doped n-type silicon

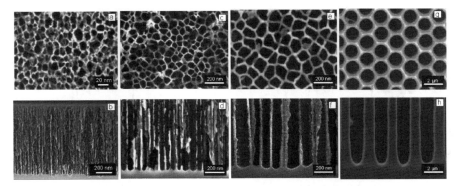

Fig. 7.9. *Top* view and cross-sectional SEM images of mesoporous silicon with an average pore diameter of approximately 20 nm, formed in a highly doped p-type silicon substrate (0.01 ohm-cm), using an electrolyte with 15% HF in ethanol. (**c, d**): SEM images of 60 nm macropores formed in a very highly doped n-type silicon substrate (0.001 ohm-cm) using an electrolyte with 6% HF. (**e, f**): SEM images of 120 nm macropores formed in highly doped n-type silicon substrate (0.01 ohm-cm), using an electrolyte with 6% HF. (**g, h**): SEM images of 1.5 μm macropores etched from low doped p-type silicon (20 ohm-cm), using an HF/dimethylformamide electrolyte

density for highly doped n-type (0.01 ohm-cm) silicon [8]. The pore morphology is also affected by the choice of doping type and concentration, as illustrated in Fig. 7.9 [8]. The top view SEM images show PSi samples with different pore diameters ranging from mesopores (pore size between 10 and 50 nm) to macropores (pore size >50 nm). The bottom-row figures are cross-sectional SEM images of the same samples. The mesopores formed in p+ silicon substrates have very branchy pore walls (Fig. 7.9a, b). The macropores formed in n-type wafers (Fig. 7.9c–h) have much smoother pore walls and larger pore sizes. Note that most mesoporous silicon samples and certainly all microporous silicon samples (pore sizes <10 nm) exhibit strong luminescence in the visible to near infrared region [9]. This results from quantum confinement of electrons and holes in nanometer-sized quantum structures, which increases the bandgap [4, 10, 11] and enhances the radiative recombination rate [12].

7.2.2 Sensing Principle

PSi is a good host material for label-free optical biosensing applications because its optical properties (photoluminescence and reflectance) are highly sensitive in the presence of chemical and biological species inside the pores [13]. PSi optical biosensors with a variety of configurations such as single layer Fabry-Perot cavities [14], Bragg mirrors [15], rugate filters [16], and microcavities [17–19] have been experimentally demonstrated for the detection of toxins [20], DNA [17], bacteria [21], and proteins [8, 14, 22]. The capture of

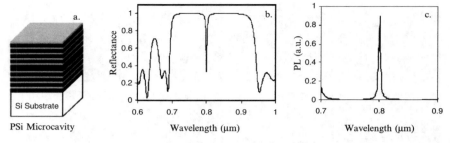

Fig. 7.10. (a) A 1D microcavity with a symmetry-breaking defect layer sandwiched between two Bragg reflectors; (b) The reflectance spectrum of such a PSi microcavity displays a sharp dip near the middle of the high-reflectance bandgap; (c) the photoluminescence of bulk PSi consists of a broad band ranging from <700 to ≃900 m, whereas the PL spectrum of the microcavity consists of a narrow peak centered on the reflectance dip

biological or chemical molecules inside the pores increases the effective refractive index of the PSi structures, thus resulting in a red shift in the photoluminescence or reflectance spectra of 1D PSi PhCs.

A PSi microcavity is a 1D photonic band gap structure that contains a defect (symmetry breaking) layer sandwiched between two Bragg mirrors [23]. Each Bragg mirror is a periodic stack of layers with two different porosities and quarter wavelength optical thickness. Figure 7.10 shows a PSi microcavity and its typical reflectance and photoluminescence (PL) spectra. Depending on the thickness of the defect layer, the reflectance spectrum contains one or several sharp resonance dips in the bandgap. The reflectance spectrum modulates the measured PL spectrum and produces one or several very narrow PL peaks. The position of the reflectance dip or photoluminescence peak is determined by the optical thickness (refractive index times the physical thickness) of each layer in the structure. A slight change of the refractive index inside the pores causes a shift of the spectrum.

The effective dielectric constant of PSi is related to its porosity by the Bruggeman effective medium model [24]:

$$(1-P)\frac{\varepsilon_{si} - \varepsilon_{PSi}}{\varepsilon_{si} + 2\varepsilon_{PSi}} + P\frac{\varepsilon_{void} - \varepsilon_{PSi}}{\varepsilon_{void} + 2\varepsilon_{PSi}} = 0, \quad (7.7)$$

where P is the porosity, ε_{PSi} the effective dielectric constant of porous silicon, ε_{si} the dielectric constant of silicon, and ε_{void} the dielectric constant of the medium inside the pores. Equation (7.7) shows that the effective refractive index of PSi ($n_{eff}^2 = \varepsilon_{PSi}$) increases as the porosity decreases and as ε_{void} increases. In Fig. 7.11, n_{eff} is plotted as a function of porosity [10]. In sensing applications, ε_{void} increases due to the binding of targets to the internal surface of the pores. Thus, the overall effective dielectric constant of the porous structure ε_{PSi} increases, which causes a red shift in the reflectance dips.

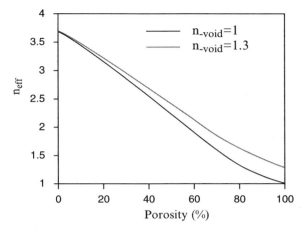

Fig. 7.11. Effective refractive index of porous silicon as a function of porosity for two values of the index of refraction of the pores

Figure 7.11 shows that, for a given increase of ε_{void}, the effective refractive index change is larger for higher porosity layers.

To selectively detect the targets of interest, the internal surface of PSi needs to be functionalized. Highly selective elements, such as DNA segments and antibodies, can be immobilized on the internal surface of the pores as the bioreceptors or probe molecules. When the sensors are exposed to the target, the probe molecules selectively capture the target molecules. The molecular recognition events are then converted into optical signals via the increase of the refractive index.

7.2.3 One-Dimensional Biosensor Design and Performance

The figure of merit describing the sensitivity of affinity sensors is $\Delta\lambda/\Delta n$, where $\Delta\lambda$ is the wavelength shift and Δn is the change of the ambient refractive index. For a PSi microcavity sensor with an average porosity of 75%, $\Delta\lambda/\Delta n$ is ~ 550 nm [25], which is much larger than for sensing platforms that rely on the interaction between the evanescent tail of the field and the analyte [26]. For a system able to detect a wavelength shift of 0.1 nm, the minimum detectable refractive index change of the pores is 2×10^{-4}.

The sensitivity of PSi microcavity sensors can be improved by increasing its quality factor Q ($Q = \lambda/\Delta\lambda$, where λ is the resonance center wavelength and $\Delta\lambda$ is the full width half maximum of the resonance dip). For higher Q-values, the spectral features are sharper and smaller shifts can be detected. The Q-factor can be increased by increasing the contrast in the porosity of the different layers. However, a higher contrast in porosity is usually produced by a higher contrast in pore opening, which may not be favorable for biosensing applications when easy infiltration of the target throughout the entire multilayer structure is required. For a given porosity contrast, the Q-factor can also be increased by increasing the number of periods of the Bragg mirror. In practice, the number of periods cannot be increased arbitrarily for two reasons. First, uniform infiltration of the molecules becomes more difficult in thicker devices. Second, maintaining a constant HF concentration at the tip of very deep pores is difficult [27], which may lead to an undesired porosity or refractive index gradient.

PSi-based PhC sensors are not perturbed by large, unwanted bioparticles. When the PSi sensor is exposed to a complex biological mixture, only the molecules that are smaller than the pores can be infiltrated into the sensor. Furthermore, an increase of refractive index on the top of the microcavity due to the presence of large, unwanted objects only causes changes to the side lobes in the reflectivity spectrum, not to the resonance dip [28]. Thus, PSi microcavities are more reliable than planar sensing platforms, where the nonspecific binding of large size objects present in a "dirty" environment may produce a false-positive signal.

The pore size also affects the sensitivity of PSi biosensors because the targets do not completely fill the pores but instead are attached to the pore walls. For a PSi layer of a given porosity, the internal surface area decreases as the pore size increases. The effective refractive index change of a layer with larger pores is thus smaller as the percentage of the pore volume occupied by the biological species is smaller. The spectra for microcavities with fixed porosities (e.g., 80% for the high porosity layer and 70% for the low porosity layer) but different pore diameters (ranging from 20 to 180 nm) have been calculated. For a given coating thickness (with $n_{\text{layer}} = 1.42$, a typical value for biomolecules), the resonance red shift decreases as the pore size increases, as shown in Fig. 7.12 [8]. For a 0.3 nm thick coating layer, a microcavity with 40 nm pores produces a red shift of 10 nm, while a microcavity with 100 nm pores produces a red shift of only 3 nm. Thus, to optimize the sensitivity, the pore size should be as small as possible while still allowing for easy infiltration of the biological material. Note that the amount of red shift can be used to precisely measure the amount of material captured inside the pores [29].

7.2.4 Fabrication of One-Dimensional PhC Biosensors

The porosity can be controlled by the etching current density. Once a layer has been etched, further etching using a different current density does not affect it, as explained in Sect. 2.1. Stacks of layers with different refractive indices can thus be formed by switching the current density during etching [30]. Figure 7.13 shows a cross-sectional SEM image of a multilayer structure etched, using a periodic current density pulse train. The current density determines the porosity and the pulse duration determines the thickness. Figure 7.14 shows top view and cross-sectional SEM images, and reflectance spectra for two different types of microcavities with different pore sizes. The

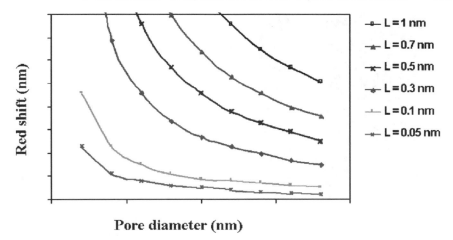

Fig. 7.12. Calculated red shift of the reflectance dip vs. pore diameter for layers of fixed porosities (70 and 80%) and various thicknesses L of the coating on the pore walls

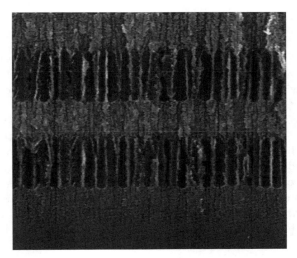

Fig. 7.13. Three low-porosity layers and two high-porosity layers etched in silicon using a periodic current density. Note the continuity of the pores and the sharpness of the interfaces between high and low porosity layers. The period is approximately 1 μm

mesoporous silicon microcavity with pore diameters from 10 to 50 nm was fabricated on a p+ wafer (0.01 ohm-cm) with 15% HF in ethanol. The macroporous silicon microcavity with pore diameters from 80 to 150 nm was etched in an n+ wafer (0.01 ohm-cm) with 5.5% HF in DI water [10].

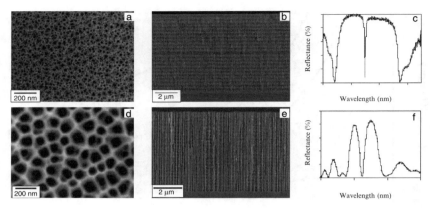

Fig. 7.14. *Top* view and cross-sectional SEM, and reflectance spectra of PSi microcavities made of mesoporous Si (*top*) and macroporous silicon (*bottom*)

7.3 Selected Biosensing Results

7.3.1 DNA Detection

Mesoporous silicon microcavities were used for the detection of small DNA segments (22 base pairs) [17, 31]. Multiple photoluminescence peaks as narrow as 3 nm of full width half maximum (FWHM) were measured from a microcavity with a thick defect layer. The first step in the sensor preparation involved the silanization of the thermally oxidized porous silicon sample with 3-glycidoxypropyltrimethoxy silane, which was hydrolyzed to a reactive silanol by using DI water (pH$\sim$4). The thermal oxidation treatment not only produced a silica-like surface, but also improved the stability of the luminescence. After silanization, the probe DNA, which has an amine group attached to the 3' end of the sequence, was immobilized onto the pore surface via diffusion. The amine group attacks the epoxy ring of the silane, thereby opening it up and forming a bond. Finally the DNA attached porous silicon sensor was exposed to the complementary strand of DNA (cDNA). By comparing the photoluminescence spectrum of the sensor before and after the DNA hybridization, the DNA/cDNA binding was detected, as shown in Fig. 7.15 [31]. The detection limit of the sensor was determined to be in the picomolar range.

7.3.2 Bacteria Detection

Gram(-) bacteria are responsible for a large numbers of deaths resulting from infection. To detect Gram(-) bacteria, we selected a target molecule present in this bacterial subclass and not in Gram(+) bacteria. Lipopolysacharide (LPS) is a primary constituent of the outer cellular membrane of Gram(-) bacteria [32]. LPS is composed of three parts: a variable polysaccharide chain,

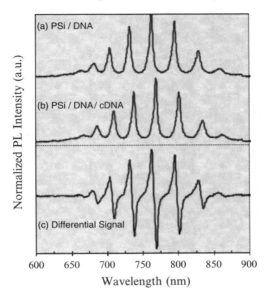

Fig. 7.15. Photoluminescence spectra of a mesoporous silicon microcavity with a thick defect layer supporting six clearly resolved reflectance dips (PL peaks). (**a**) After functionalization of the pore surface with probe DNA segments; (**b**) after capture of complementary DNA segments; (**c**) differential spectrum

a core sugar, and lipid A. As lipid A is common to all LPS subtypes, it is a natural target. An organic receptor, tetratryptophan *ter*-cyclo pentane (TWTCP) was designed and synthesized as the probe molecule. TWTCP specifically binds to diphosphoryl lipid A in water with a dissociation constant of 592 nM [59].

After the microcavity sensor was exposed to 3-glycidoxypropyltrimethoxy silane, a mixture of TWTCP and glycine methyl ester was applied to the sensor. Glycine methyl ester was used as a blocker molecule to avoid the reaction of the four amino groups of the tetratryptophan receptor with the epoxide-terminated PSi surface. As shown in Fig. 7.16, upon exposure of lysed Gram(-) cells (*Escherichia coli*) to the immobilized TWTCP sensor, a 4 nm red shift of the photoluminescence spectrum of the microcavity was detected [21]. However, when the sensor was exposed to a solution of lysed Gram(+) cells (*Bacillus subtilis*), no shift of the spectrum was observed. These results were confirmed with all Gram(-) and Gram(+) bacteria tested.

7.3.3 Protein Detection

A protein biosensor based on macroporous silicon microcavity was demonstrated with the streptavidin-biotin couple [8]. Biotin is a small molecule, while streptavidin is relatively large (67 kDa), making its infiltration difficult into

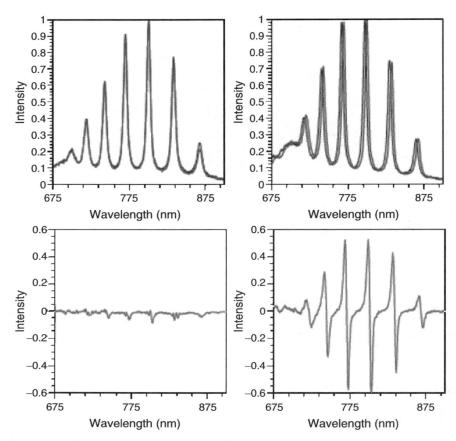

Fig. 7.16. Photoluminescence spectra of a mesoporous silicon biosensor similar to the one shown in Fig. 7.15 before and after exposure to bacterial cell lysates (*top row*) and differential spectra (*bottom row*). The spectra on the *left* have been obtained for bacteria that are Gram(+) and the spectra on the *right* for bacteria that are Gram(-)

mesopores but easy into macropores. Furthermore, each streptavidin tetramer has four equivalent sites for biotin (two on each side of the complex), which makes it a useful molecular linker.

To create a biotin-functionalized sensor for the capture of streptavidin, microcavities were first thermally oxidized, then silanized with aminopropyltriethoxysilane to create amino groups on the internal surface. The probe molecules, Sulfo-NHS-LC-LC-biotin, were then immobilized inside the pores. As shown in Fig. 7.17a, the red shift of the resonance increased with increasing biotin concentration. A 10 nm red shift corresponds to a nearly complete (95%) biotin surface coverage.

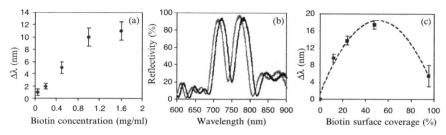

Fig. 7.17. (a) Increase of the reflectance dip red shift when more probe molecules (biotin) are added; the shift eventually saturates when all the pore walls are covered by biotin. (b) After exposure to target molecules (streptavidin), the microcavity spectrum shifts to the red. (c) The measured red shift upon capture of the targets depends on the coverage of the pore walls by the probe molecules

To study how the biotin surface coverage affects the binding to streptavidin, sensors derivatized with different biotin concentrations were exposed to the same amount of the target: 50 ml of streptavidin with a concentration of 1 mg ml^{-1}. After exposure to streptavidin, a red shift was detected, which is attributed to the specific binding of streptavidin to the biotin-derivatized macroporous microcavity (Fig. 7.17b). For the control samples that were silanized but did not contain biotin, no shift was detected after exposure to streptavidin, which indicates that nonspecific absorption of streptavidin inside the microcavity is nonexistent or negligible.

The red shift caused by streptavidin binding to biotin is a function of the biotin surface coverage. Figure 7.17c shows that there is an optimum biotin surface coverage that maximizes the red shift, and therefore the capture of streptavidin. This behavior is in good agreement with observations by other researchers using different sensing techniques [34]. The spacing between each neighboring biotin needs to be large enough so that biotin can reach the pocket-like binding site of streptavidin. The roll-off of the red shift showing in Fig. 7.17c is due to the fact that when the biotin surface density becomes very high and each biotin is closely surrounded by its neighbors, biotin cannot protrude deep enough into the binding site of streptavidin, thus decreasing the probability of capturing streptavidin. Other proteins have also been detected using 1D PhC microcavities made of PSi, including intimin, a protein secreted by pathogenic *E. coli* [22].

7.3.4 IgG Detection

Immunoglobulin (IgG) is the most common type of antibody synthesized in response to a foreign substance (antigen). Antibodies have a specific molecular structure capable of recognizing a complementary molecular structure on the antigen, such as proteins, polysaccharides, and nucleic acids. From the X-ray crystal structure, the longest dimension of IgG is approximately 17 nm [35].

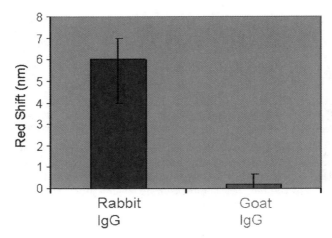

Fig. 7.18. A PSi microcavity biosensor functionalized for the capture of one type of IgG can discriminate against another type of IgG with a contrast ratio $\gg 10\times$

The detection of rabbit IgG (150 kDa) was investigated through multiple layers of biomolecular interactions in a macroporous silicon microcavity sensor. The silanized sensor was first derivatized with biotin, which can selectively capture streptavidin. Exposure of biotinylated goat anti-rabbit IgG to the sensor results in its attachment to the surface through the binding between biotin and streptavidin. The sensor used goat anti-rabbit IgG as the probe molecule to selectively capture rabbit IgG. A red shift of the spectrum can be detected when each layer of molecules is added to the sensor. As shown in Fig. 7.18 when the sensor was exposed to 50 ml of a solution containing rabbit IgG at a 1 mg ml^{-1} concentration, a 6 nm red shift was detected [25]. When the sensor was exposed to 50 ml goat IgG (1 mg ml^{-1}), which does not bind to the goat anti-rabbit IgG, the red shift was negligible (<0.5 nm).

7.4 Two-Dimensional PhC Biosensors

7.4.1 Sample Preparation and Measurement

Silicon-on-insulator (SOI) wafers were used to fabricate 2D PhC microcavity biosensors [36]. The Si slab thickness was approximately 400 nm. Rigorous 3D simulations using the FDTD method were performed prior to the fabrication. The incident light was coupled along the Γ-M direction because the in-plane leakage of the resonance mode is mainly in the Γ-M direction and hence the coupling efficiency is higher along the Γ-K direction. The ridge waveguide was tapered down from 2 to 0.6 μm with an external surface cross-section of 2 μm × 400 nm, which predominantly select fundamental TE mode as the propagation mode.

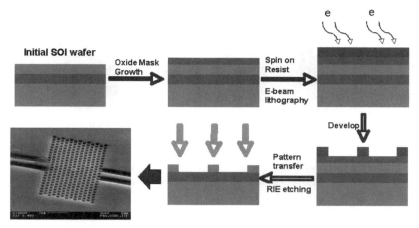

Fig. 7.19. Fabrication flow starting from a silicon on insulator (SOI) wafer and ending with the 2D PhC microcavity inserted in a silicon wire waveguide

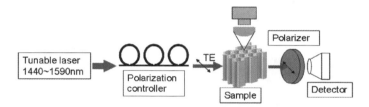

Fig. 7.20. Experimental setup for measuring the transmission of a 2D PhC microcavity. A microscope objective is used to observe the alignment of all the components

The fabrication procedure is depicted in Fig. 7.19. After the SOI wafer was cleaned and an oxide mask grown, negative (PMMA) and positive (FOX) resists were used for pattern transformation. After electron-beam lithography and resist development, reactive ion etching was performed to transfer the pattern onto the Si slab. The ridge waveguide facets were then polished to improve the coupling efficiency.

Figure 7.20 illustrates the measurement setup. An HP8168F laser, tunable from 1,440 to 1,590 nm, was used as the light source. The TE polarized laser beam was focused onto the input tapered ridge waveguide (from 2 μm to 0.6 μm) using a tapered lensed fiber (with a spot size of 2 μm at focus), and the transmitted signal was measured by a photodetector. To optimize the intensity of the TE polarized-light, a polarization controller was inserted between the laser and the input fiber.

7.4.2 Sensing Principle

The sensing principle is very similar to the one used in the 1D PhC biosensors. In contrast to chemical sensors [37], in biological sensors, the biomolecule

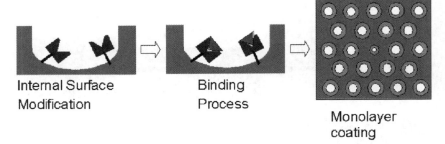

Fig. 7.21. After functionalization of the air hole surfaces with probe molecules (*left*), targets are captured by the biosensor (*middle*), which produces a coating on the air hole walls (*right*)

recognition strongly depends on the surface chemistry, thus, instead of filling up the pores, the molecules prefer to form a monolayer coating at the pore (air hole) wall As shown in Fig. 7.21, highly selective probes (e.g., DNA, antibody) can be immobilized on the internal surface of the air holes and form a monolayer that can capture specific target molecules (e.g., matching DNA, proteins). Hence, the coating causes a refractive index change only in the vicinity of the pore wall.

7.4.3 Selected Biosensing Results

Protein Detection

To test the device performance, glutaraldehyde-bovine serum albumin (BSA) coupling was used as the model system. The air hole is ∼30 times larger than the BSA hydrodynamic diameter [38], so that the infiltration proceeds easily, and both glutaraldehyde and BSA can form a uniform monolayer coated on the pore walls.

To prepare the surface for capturing biomolecules, the device was first thermally oxidized at 800°C to form a silica-like interface for binding of amino groups. To functionalize the sensor and capture glutaraldehyde, the microcavity internal surface was silanized. Subsequently, the target, BSA, was immobilized covalently on the sidewall. The experiment protocol is as follows: (1) clean the sensor surface with DI water and dry it under nitrogen flow. Store the sensor in moist ambient; (2) drop 5 μl of 2% aminopropyltrimethoxysilane on the sensor for 20 min; (3) rinse the sensor with DI water and dry with nitrogen flow, and then bake at 100°C for 10 min; (4) drop 5 μl of 2.5% glutaraldehyde in Hepes buffer on the sensor for 30 min; soak in buffer for 10 min to dilute the unreacted agents and dry it under nitrogen flow; (5) drop 5 μl of 1% BSA on the sensor for 30 min; then soak in buffer for 20 min and dry it with nitrogen.

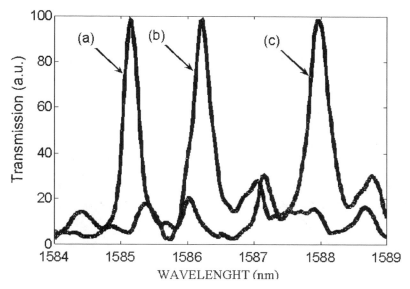

Fig. 7.22. Transmission near the 2D PhC microcavity resonance for the functionalized sensor (**a**), after exposure to the probe, glutaraldehyde (**b**), and subsequent exposure to the target, bovine serum albumin

Figure 7.22 illustrates the normalized transmission spectra measured at three different binding stages. The PhC microcavity consisted of a triangular array of cylindrical air pores in a 400 nm-thick silicon (Si) slab. The lattice constant a was 465 nm, and the pore diameter r was $0.3a$. A defect was introduced by reducing the center pore diameter to $0.18a$, leading to a transmission resonance in the bandgap close to 1.58 μm. Curve (a) shows the initial transmission after the oxidation and silanization. Curve (b) was measured after exposure to glutaraldehyde. A resonance red shift of 1.1 nm is observed. Curve (c) shows a red shift of 1.7 nm after BSA binding corresponding to a total shift of 2.8 nm with respect to the initial spectrum. To verify the experimental results, a plane-wave expansion calculation with 32 grid points per supercell was performed. The simulation of the resonance red shift assumes that the refractive index of the dehydrated proteins is 1.45. This value agrees with an ellipsometry measurement [29] and with the literature values [39].

Figure 7.23 plots the predicted resonance red shift vs. coating thickness. Using ellipsometry, the thickness of the dehydrated glutaraldehyde monolayer was measured as 7 ± 1 Å. We then calculated that glutaraldehyde should introduce a resonance red shift of 0.98 ± 0.2 nm, which is in good agreement with the experimental data. The thickness of a dehydrated BSA layer measured with the same method is approximately 15 ± 5 Å, which should introduce an additional red shift of 2.7 ± 1 nm or a total red shift of 3.8 ± 1 nm. The experimental data are again in good agreement with the model.

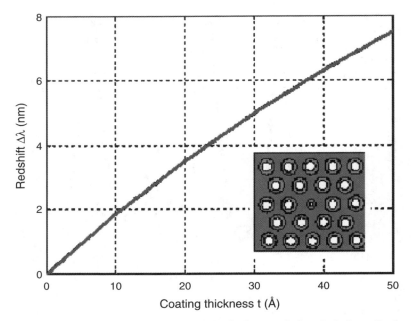

Fig. 7.23. Calculated red shift vs. coating thickness of the air hole walls for the device structure tested in Fig. 7.22

The biomolecules can also attach to the sidewall, the bottom, or the top of the device. 3D simulations (Fig. 7.24) using FDTD methods were carried to simulate the molecule-position dependence. Within the molecule size range of interests (i.e. <50 nm), the coating on the pore wall contributes the most to the spectral shift, whereas the presence of biomolecules elsewhere produces a much smaller shift. This is not surprising, as the mode is strongly confined in the Si slab and on the air hole sidewalls, the interaction of the mode and the analyte inside the device is much stronger than on top of the device.

Toward Single Molecule Detection

As the electric field is strongly localized in the defect region, what would happen if the biomolecules were present only in the defect region? [40] The sensitivity $\Delta\lambda/\Delta t$ decreases by a factor of 4, but the total amount of analyte that is detectable drops by nearly two orders of magnitude, from 2.5 pg to 0.05 pg. To demonstrate the potential for single bioparticle detection, a microcavity with a large defect (diameter ∼900 nm) was fabricated. Latex spheres (density of 1.05 g cm^{-3}, refractive index $n = 1.59$) with a diameter of approximately 370 nm were dropped on the PhC. As shown in Fig. 7.25,

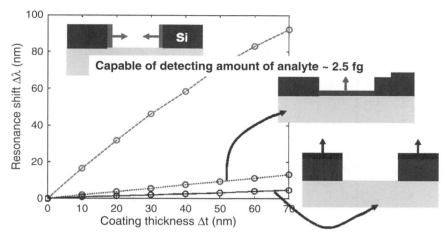

Fig. 7.24. Calculated resonance red shift vs coating thickness, for a biomolecule layer on the side walls of the air holes, at the bottom of the air holes, or on top of the device

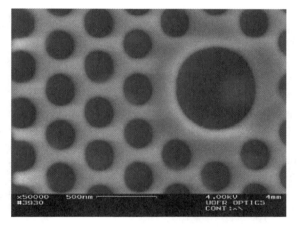

Fig. 7.25. Close up of near the central defect of a 2D PhC microcavity in which one latex microsphere has been captured

one latex microsphere was inside the defect region and the rest of the spheres remained on top of the device. The normalized transmission spectra of the PhC microcavity shown in Fig. 7.26 demonstrate that the device can detect a single bioparticle. For a device with a quality factor Q of ∼2,000, a spectral shift of 1 nm can be measured very easily. Simulations show that biomolecules smaller than 50 nm can be detected.

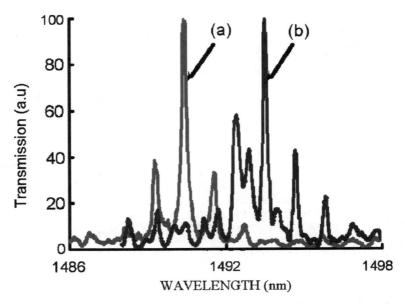

Fig. 7.26. Transmission near the resonance of the 2D PhC microcavity when the defect air hole contains one latex microsphere (curve b) or no microsphere (curve a). The microsphere's diameter was 370 nm

7.5 Conclusions

PhC microcavities made of silicon are a platform of choice for biosensing. After proper functionalization, both 1D and 2D devices have excellent performance in terms of sensitivity and selectivity. It is important to note that a complete biosensor system requires other components (such as prefiltration and concentration modules [41]) that have not been described here but can also be made in silicon and integrated with the PhC microcavity.

Acknowledgments

The work described in this chapter was supported by the US National Science Foundation through grant BES 04279191. Fabrication of the 2D PhC microcavities was performed in part at the Cornell NanoScale Facility, which is supported by NSF through grant ECS 03-35765. The technical contributions of Dr. Selena Chan, Dr. Marc Christophersen, and Dr. Chris Striemer are gratefully acknowledged.

References

1. E. Yablonovitch, Phys. Rev. Lett., **58**, 2059 (1987)

2. S. John, Phys. Rev. Lett., **58**, 2486 (1987)
3. J.D. Joannopoulos, R.D. Meade, J.N. Winn, in *Photonic Crystals: Molding the Flow of Light*, (Princeton Press, 1995)
4. P. M. Fauchet, *in Encyclopedia of Applied Physics*, Update 2, (Wiley-VCH Verlag, 1999), pp. 249–272
5. R.L. Smith, S.D. Collins, J. Appl. Phys. **71**, R1 (1992)
6. A.G. Cullis, L.T. Canham, P.D.J. Calcott, J. Appl. Phys. **82**, 909 (1997)
7. V. Lehmann, *Electrochemistry of Silicon*, (Wiley-VCH, 2002)
8. H. Ouyang, M. Christophersen, R. Viard, B.L. Miller, P.M. Fauchet, Adv. Funct. Mater. **15**, 1851 (2005)
9. P.M. Fauchet, in *Semiconductors and Semimetals*, vol. 49, ed. by D.J. Lockwood, (Academic Press, San Diego, 1998), pp. 206–252
10. L.T. Canham, Appl. Phys. Lett. **57**, 1046 (1990)
11. V. Lehmann, U. Gosele, Appl. Phys. Lett. **58**, 856 (1991)
12. M.S. Hybertsen, Phys. Rev. Lett. **72**, 1514 (1994)
13. M.P. Stewart, J.M. Buriak, Adv. Mater. **12**, 859 (2000)
14. K.-P.S. Dancil, D.P. Greiner, M.J. Sailor, J. Am. Chem. Soc. **121**, 7925 (1999)
15. P.A. Snow, E.K. Squire, P.S.J. Russell, L.T. Canham, J. Appl. Phys. **86**, 1781 (1999)
16. F. Cunin, T.A. Schmedake, J.R. Link, Y.Y. Li, J. Koh, S.N. Bhatia, M.J. Sailor, Nat. Mat. **1**, 39 (2002)
17. S. Chan, Y. Li, L.J. Rothberg, B.L. Miller, P.M. Fauchet, Mat. Sci. Eng. C **15**, 277 (2001)
18. L. Pavesi, La Rivista del Nuovo Cimento **20**, 1 (1997)
19. H. Ouyang, M. Christophersen, P.M. Fauchet, Phys. Stat. Sol. (a) **202**, 1396 (2005)
20. H. Sohn, S. Letant, M.J. Sailor, W.C. Trogler, J. Am. Chem. Soc. **122**, 5399 (2000)
21. S. Chan, S.R. Horner, P.M. Fauchet, B.L. Miller, J. Am. Chem. Soc. **123**, 11797 (2001)
22. H. Ouyang, L.A. DeLouise, B.L. Miller, P.M. Fauchet, Anal. Chem. **79**, 1502 (2007)
23. J.E. Lugo, H.A. Lopez, S. Chan, P.M. Fauchet, J. Appl. Phys. **91**, 4966 (2002)
24. W. Theiβ, Surf. Sci. Rep. **29**, 91 (1997)
25. H. Ouyang, M. Lee, B.L. Miller, P.M. Fauchet, Proc. SPIE **5926**, 59260J (2005)
26. J. Scheuer, W.M.G. Green, G.A. DeRose, A. Yariv, IEEE J. Sel. Top. Quant. Electron. **11**, 476 (2005)
27. M. Thonissen, M.G. Berger, S. Billat, R. Ares-Fischer, M. Kruger, H. Luth, W. Theiss, S. Hillbrich, P. Grosse, G. Lerondel, U. Frotscher, Thin Solid Films **297**, 92 (1997)
28. H. Ouyang, L.A. DeLouise, M. Christophersen, B.L. Miller, P.M. Fauchet, Proc. SPIE **5511**, 71 (2004)
29. H. Ouyang, C.C. Striemer, P.M. Fauchet, Appl. Phys. Lett. **88**, 163108 (006)
30. S. Chan, P.M. Fauchet, Appl. Phys. Lett. **75**, 274 (1999)
31. S. Chan, P.M. Fauchet, Y. Li, L.J. Rothberg, B.L. Miller, Phys. Stat. Sol. (a) **182**, 541 (2000)
32. L.S. Young, W.J. Martin, R.D. Meyer, R.J. Weinstein, E.T. Anderson, Ann. Inern. Med. **86**, 456 (1997)
33. R.D. Hubbard, S.R. Horner, B.L. Miller, J. Am. Chem. Soc. **123**, 5811 (2001)

34. L.S. Jung, K.E. Nelson, P.S. Stayton, C.T. Campbell, Langmuir **16**, 94221 (2000)
35. E.O. Saphire, P.W.H.I. Parren, R. Pantophlet, M.B. Zwick, G.M. Morris, P.M. Rudd, R.A. Dwek, R.L. Stanfield, D.R. Burton, I.A. Wilson, Science **293**, 1155 (2001)
36. M. Lee, P.M. Fauchet, Opt. Exp. **15**, 4530 (2007)
37. E. Chow, A. Grot, L.W. Mirkarimi, M. Sigalas, G. Girolami, Opt. Lett. **29**, 1093 (2004).
38. I.D. Kuntz, W. Kauzmann, Adv. Protein Chem., **28**, 239 (1974)
39. M. Malmsten, J. Colloid. Interf. Sci., **166**, 333 (1994)
40. M.R. Lee, P.M. Fauchet, Paper presented at the 4th IEEE LEOS International Conference on Group IV Photonics, Tokyo, Japan, September 2007
41. C.C. Striemer, T.R. Gaborksi, J.L. McGrath, P.M. Fauchet, Nature **445**, 749 (2007)

8

Optical Coherence Tomography with Applications in Cancer Imaging

S.A. Boppart

8.1 Introduction

Optical coherence tomography (OCT) is a rapidly emerging optical imaging technique for a wide range of biological, medical, and material investigations [1, 2]. OCT was initially developed in the early 1990s, and has provided researchers with a novel means by which biological specimens and nonbiological samples can be visualized. A primary advantage of OCT is the ability to image tissue microstructure in situ at micron-scale image resolution, without the need for excision of a specimen for tissue processing. The optical ranging in OCT is analogous to ultrasound B-mode imaging, except that OCT uses low-coherence light rather than high-frequency sound. Cross-sectional OCT images can be generated, as is commonly done in ultrasound, or en face OCT sections can be acquired, as in confocal and multiphoton microscopy. The OCT imaging principle involves optical ranging, where the optical reflection of light from a low-coherence optical source is spatially localized using interferometry. The OCT image that is assembled is a gray-scale or false-color multidimensional spatial representation of backscattered light intensity. The signal intensity represented within an OCT image represents the differential backscattering contrast between different tissue types on a micron scale. OCT performs imaging using light; therefore, it has a one to two order-of-magnitude higher spatial resolution than ultrasound. Because the optical imaging beam can be transmitted readily through air, OCT beam-delivery systems do not require contact with the specimen or sample, as do ultrasound probes. Spectroscopic characterization of tissue and cellular structures is also possible within the optical spectrum of the light source.

Ophthalmology was the first medical application area for OCT, where high-resolution tomographic imaging of the anterior eye and retina is possible [3]. The transparency of the eye at visible and near-infrared wavelengths, and its accessibility using optical instruments and techniques, has enabled extensive research and clinical applications for diagnostic OCT imaging. The diagnosis of many retinal diseases is possible because OCT can provide images of retinal pathology with micron-scale resolution.

OCT has also been applied in a wide range of nontransparent tissues [4,5]. In nontransparent tissues such as skin, muscle, and other soft tissues, the imaging depth is limited by optical attenuation due to scattering and absorption. Optical scattering decreases with increasing wavelength. Therefore, while ophthalmic OCT imaging has primarily been performed at 800 nm wavelengths, OCT imaging in nontransparent tissues has been typically performed with wavelengths of 1.0–1.3 µm. Imaging depths up to 2–3 mm can be achieved using a system detection sensitivity of 100–110 dB. In early exploration imaging studies, OCT has been performed in virtually every organ system to investigate applications in cardiology [6–8], gastroenterology [9,10], urology [11, 12], neurosurgery [13], and dentistry [14], to name a few. Using short coherence length, short-pulsed light sources, high-resolution OCT has been demonstrated with axial resolutions of less than 5 µm [15–17]. High-speed real-time image acquisition rates have also been achieved, which enables volumetric imaging [18, 19]. New imaging modes in OCT have been demonstrated, such as Doppler OCT imaging of blood flow [20–22] and birefringence imaging to investigate laser intervention [23–25]. OCT beam delivery systems including transverse imaging catheter/endoscopes and forward imaging devices have enabled OCT imaging of internal structures [26–28], and most recently, catheter-based OCT has been used to perform in vivo imaging in animal models and human patients [29–33].

8.2 Principles of Operation

OCT performs optical ranging in tissue, using high spatial resolution and high dynamic range detection of backscattered light as a function of optical delay. While the speed of ultrasound waves in tissue is relatively slow and detectable using electronics, the velocity of light is extremely high. Therefore, the time delays of the reflected light cannot be measured directly and interferometric detection techniques must be employed. Low-coherence interferometry or optical coherence domain reflectometry are techniques that are used in OCT. First developed in the telecommunications field for measuring optical reflections from faults or splices in optical fibers [34], low-coherence interferometry can also be used for localizing the optical reflections in biological tissue. Subsequently, the first applications in biological samples included one-dimensional optical ranging in the eye to determine the location of different ocular structures [35, 36].

Time delays of reflected light-off of tissue boundaries are typically measured using a Michelson-type interferometer (Fig. 8.1). Other interferometer designs, such as a Mach–Zehnder interferometer, have been implemented to optimize the delivery of the OCT beam and the collection of the reflected signals [37, 38]. Light reflected from the specimen or sample is interfered with

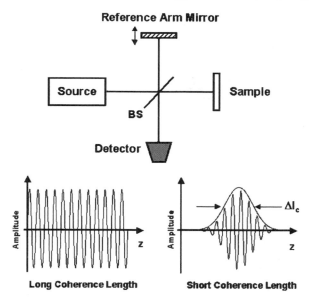

Fig. 8.1. Schematic illustrating the concept of low coherence interferometry. Using a short coherence length light source and a Michelson-type interferometer, interference fringes are observed only when the path lengths of the two interferometer arms are matched to within the coherence length of the light source in a time-domain OCT system. The full-width half-maximum of the envelope of the autocorrelation function is equal to the coherence length (Δl_c) and axial resolution in OCT. This envelope also represents the axial point-spread-function of an OCT system

light reflected from a reference path of known path length, which spatially determines the location of the reflection in depth. Interference of the light between the two arms of the interferometer can occur only when the optical path lengths of the two arms match within the coherence length (axial resolution) of the optical source. If the reference arm optical path length is scanned, as in a time-domain OCT system, different delays of backscattered light from within the sample are measured, and a single column of depth-dependent data is collected. The interference signal is detected at the output port of the interferometer, digitized, and stored on a computer. Following a depth (z) scan, the incident beam is scanned in the transverse direction (x) and multiple axial measurements are performed. A two-dimensional data array is generated, which represents the optical backscattering through a cross-sectional plane in the specimen (Fig. 8.2). Similarly, the OCT beam can be translated in the third (y) dimension, and a series of two-dimensional cross-sectional images can be collected to form a three-dimensional volume. The logarithm of the backscatter intensity is then mapped to a false-color or gray-scale and displayed as an OCT image. Typically, the interferometer in an OCT instrument can be implemented using fiber optic couplers, and beam-scanning can be

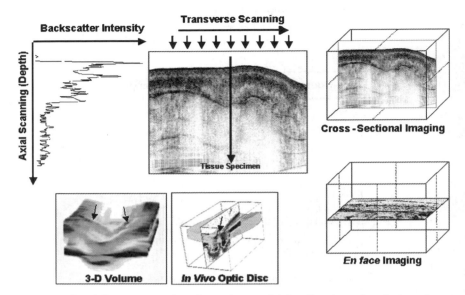

Fig. 8.2. An OCT image is based on the spatial localization of variations in optical backscatter from within a specimen. For depth-priority scanning, images are acquired by performing axial measurements of optical backscatter at different transverse positions on the specimen and displaying the resulting two-dimensional data set as a gray-scale or false-color image. OCT images can also be acquired with transverse-priority scanning to collect an en face image similar to optical sectioning in confocal or multiphoton microscopy. Multiple planes in either mode can be assembled for three-dimensional OCT

performed with small mechanical galvanometers or another beam-delivery instrument to yield a compact and robust system (Fig. 8.3). The axial resolution in OCT images is determined by the coherence length of the light source, in contrast to the tight focus and confocal region in confocal microscopy. The axial point spread function of the OCT measurement is defined by the signal detected at the output of the interferometer (detector arm), and is the electric-field autocorrelation of the source. The coherence length of the light is the spatial width of the field autocorrelation, and the envelope of the field autocorrelation is equivalent to the inverse Fourier transform of the optical source power spectrum. The width of the autocorrelation function (axial resolution), therefore, is inversely proportional to the width of the optical power spectrum. If the optical source has a Gaussian spectral distribution, then the free-space axial resolution Δz is given by

$$\Delta z = \frac{2 \ln 2}{\pi} \frac{\lambda^2}{\Delta \lambda}, \tag{8.1}$$

where Δz and $\Delta \lambda$ are the full widths at half maximum of the autocorrelation function and power spectrum, respectively, and λ is the central wavelength of

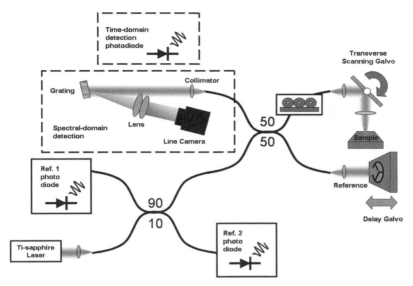

Fig. 8.3. Schematic representation of an OCT system implemented using fiber optics. The Michelson interferometer is implemented using a fiber-optic coupler. Light from the low-coherence source is split and sent to a sample arm with a beam delivery instrument and a reference arm with an optical path-length scanner. Reflections from the arms are combined and the output of the interferometer is detected with either a photodiode or a linear CCD array in a spectrometer. Components for both a time-domain and a spectral-domain OCT system are shown

the source. The dependence of the axial resolution on the bandwidth of the optical source is plotted in Fig. 8.4. To achieve high axial resolution (approaching 1 μm), therefore, requires extremely broad bandwidth optical sources. The curves plotted in Fig. 8.4 are for three commonly used wavelengths, 800, 1,300, and 1,500 nm. From (8.1), higher resolutions can be achieved with shorter wavelengths. However, shorter wavelengths are more highly scattered in biological tissue, and frequently result in less imaging penetration.

Evident from this discussion, the axial and transverse resolutions in OCT are not directly related, as in conventional microscopy. However, the transverse resolution in an OCT imaging system is determined by the focused spot size, according to the principles of Gaussian optics, and is given by

$$\Delta x = \frac{4\lambda}{\pi} \frac{f}{d}, \qquad (8.2)$$

where d is the beam diameter on the objective lens and f is the focal length of the objective lens. Large numerical aperture optics can be used to focus the beam to a small spot size and provide high transverse resolution.

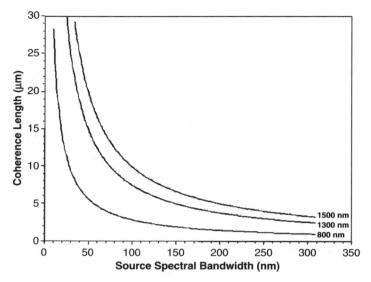

Fig. 8.4. Dependence of coherence length (axial resolution) on optical source bandwidth. Curves are plotted for 800, 1,300, and 1,500 nm, three common wavelengths used in OCT. High axial imaging resolution is achieved with broad spectral bandwidths and shorter wavelengths. Shorter wavelengths, however, are more highly absorbed in biological tissue, decreasing imaging penetration depth

The transverse resolution is also related to the depth of focus or the confocal parameter $2z_R$ (two times the Raleigh range).

$$2z_R = \frac{\pi \Delta x^2}{2\lambda}. \tag{8.3}$$

Increasing the transverse resolution (decreasing the spot size at the focus) subsequently results in a reduced depth of field. For OCT imaging, the confocal parameter or depth of focus is typically chosen to match the desired depth of imaging. High transverse resolutions are often required and may be utilized in OCT. However, the short depth of field requires additional optical or mechanical techniques to spatially track the focus in depth along with the axial OCT scanning.

Relatively small incident powers of 1–10 mW are commonly required for OCT imaging. Typically, the real-time ($\sim$30 frames per second) acquisition can be achieved at a signal-to-noise ratio of $\sim$100 dB with 5–10 mW of incident optical power. Recent advances in spectral-domain OCT (SD-OCT) [39, 40] and swept-source OCT (SS-OCT) [41, 42] have enabled extremely fast axial scanning and real-time volumetric imaging at micron-scale resolution, while maintaining sufficiently high SNR for imaging larger volumes of tissue.

8.3 Optical Sources for Optical Coherence Tomography

OCT imaging systems have primarily used superluminescent diodes (SLDs) as low coherence light sources [43]; however, new sources are becoming widely available. SLDs are attractive because they are compact, have high efficiency, low noise, and are commercially available at a range of wavelengths including 800 nm, 1.3 μm, and 1.5 μm. Output powers, however, are typically less than a few milliwatts, which frequently limits the use of these sources to slow acquisition rates in order to preserve signal-to-noise ratio. The available bandwidths are relatively narrow, permitting imaging with 10–15 μm resolution. Because of the need for broader bandwidth sources, manufacturers have been developing SLD sources that have bandwidths of 50–100 nm (3–10 μm resolution, depending on wavelength).

Advances in short-pulse, solid-state, laser technology make these sources attractive for OCT imaging in research applications. Femtosecond solid-state lasers can generate tunable, low-coherence light at powers sufficient to permit high-speed OCT imaging [15–17]. The titanium:sapphire (Ti:Al_2O_3) laser from 0.7 to 1.1 μm has been a commonly used source, not only for OCT, but most often for multiphoton microscopy. Imaging resolutions of less than 1 μm have been demonstrated at 800 nm using a Ti:Al_2O_3 laser source [17]. To demonstrate the improvement in resolution afforded by a titanium:sapphire laser, a comparison between the power spectra and autocorrelation functions of a 800 nm center wavelength SLD and a short pulse ($\approx$5.5 fs) titanium:sapphire laser is shown in Fig. 8.5. An order-of-magnitude improvement in axial resolution is noted for the short pulse titanium:sapphire laser source [16]. In an effort to develop more compact and convenient sources compared to a titanium:sapphire laser, superluminescent fiber sources have been investigated [44,45] as well as pumped optical fibers that can produce extremely broad supercontinuum [46–48]. Since the titanium:sapphire (Ti:Al_2O_3) laser technology is routinely used in multiphoton microscopy applications for its high peak intensities to enable multiphoton absorption and subsequent emission of fluorescence from exogenous fluorescent contrast agents [49], the simultaneous generation of OCT and multiphoton microscopy images has been demonstrated, providing complementary image data using a single optical source [50].

8.4 Fourier-Domain Optical Coherence Tomography

OCT has most-commonly been performed by rapidly scanning the reference arm path-length to acquire a single scan in depth. This mode has been termed time-domain OCT (TD-OCT). In this mode, faster image acquisition rates were therefore dependent on the rate at which this optical path-length could be varied, and rapid-scanning optical delay lines were previously developed for

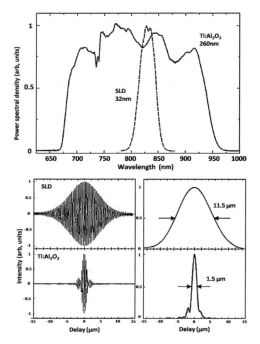

Fig. 8.5. Comparison of optical output spectra (*top*) with interference signals (*lower left*) and envelopes (*lower right*) for a Kerr-lens modelocked titanium:sapphire (Ti:Al$_2$O$_3$) laser vs. a superluminescent diode (SLD). The broad optical bandwidth of the titanium:sapphire laser (260 nm) permits a free-space axial resolution of 1.5 μm. In comparison, the superluminescent diode with a 32 nm spectral bandwidth permits an axial resolution of 11.5 μm. Figure reprinted with permission [16]

this purpose. In the mid 1990s, it was realized, and in recent years, demonstrated that alternatives to this time-domain mode of scanning were techniques that simultaneously collected light from all depths into the tissue as the beam was positioned at each transverse position. The light sources for these techniques are either a broadband source in which the collected light is detected by a linear CCD array in a spectrometer (spectral-domain OCT, SD-OCT) [39, 40], or a narrow-band source which is rapidly swept over a range of frequencies and the collected light is detected by a photodiode (swept-source OCT, SS-OCT) [41, 42]. For both of these techniques, depth-dependent scattering information is obtained by taking the inverse Fourier transform of this acquired spectral data. Recently, these techniques have become preferred for the improvement in acquisition rate and sensitivity that they provide over the time-domain OCT systems.

Recent use of the SS-OCT has demonstrated real-time acquisition rates. The primary advantages to using a swept-source is that all the optical power is contained within a narrow wavelength range at any one point in time, and then is swept to a different wavelength. Wide tuning ranges make high imaging resolution possible and at the rate of hundreds of thousands of axial scans per second, fast volumetric imaging can be performed. The development of fast swept sources has utilized innovative methods, including a rotating polygonal mirror array [41] or Fourier-domain mode-locking [42].

High-speed SD-OCT has been demonstrated in the living human eye and in other real-time applications. Because the acquisition rate is no longer dependent on a mechanically scanned reference arm, significantly faster rates are achievable, up to typical rates of 30,000 axial scans per second, depending on the read-out rates of the linear CCD arrays in the spectrometer. The computational complexity has been increased for both SD-OCT and SS-OCT because the inverse Fourier transform of the acquired data is required for each axial scan line. Because dedicated digital-signal-processing (DSP) chips are readily available, it appears that the computational power necessary for this application is feasible and available to the research investigator. Spectral-domain OCT also has the advantages of improved phase stability, because no mechanical scanning is performed, and higher signal-to-noise ratio, because individual wavelengths are now detected by separate elements in a linear detector array. Linear detector arrays, especially indium–gallium–arsenide arrays used for detecting the commonly used 1,300 nm OCT wavelength, are expensive, shifting the costs of rapid scanning optical delay lines for time-domain systems to the detector array for SD-OCT.

8.5 Beam Delivery Instruments for Optical Coherence Tomography

The OCT imaging technology is modular in design. This is most evident in the optical instruments through which the OCT beam can be delivered to the tissue. Because OCT is fiber-optic based, single optical fibers can be used to deliver the OCT beam and collect the reflected light, thereby making the beam delivery system potentially very small, on the order of the size of an optical fiber (125 μm diameter) itself. The OCT technology can also be readily integrated into existing optical instruments such as research and surgical microscopes [51, 52], ophthalmic slit-lamp biomicroscopes [3], catheters [26], endoscopes [29], laparoscopes [27], needles [28], and hand-held imaging probes [27] (Fig. 8.6). Imaging penetration is determined by the optical absorption and scattering properties of the tissue or specimen. The imaging penetration for OCT ranges from tens of millimeters for transparent tissues such as the eye to less than 3 mm in highly scattering tissues such as skin. To image highly scattering tissues deep within the body, novel beam-delivery instruments have been developed to relay the OCT beam to the site of the tissue to be imaged. An OCT catheter has been developed for insertion into biological lumens such as the gastrointestinal tract [29]. Used in conjunction

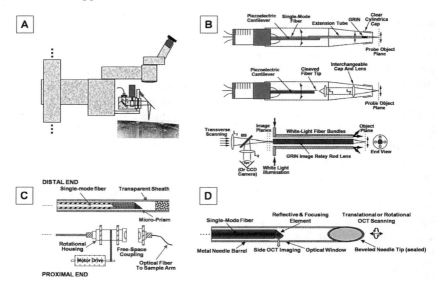

Fig. 8.6. Beam delivery instruments. The OCT beam can be delivered through a number of new or modified optical instruments including (**a**) surgical and research microscopes, (**b**) hand-held probes and laparoscopes, (**c**) fiber-optic catheters, and (**d**) optical needle-probes

with endoscopy, the 1 mm diameter catheter can be inserted through the working channel of the endoscope for simultaneous OCT and video imaging [31]. Similar catheters have been used to image plaques within the living human coronary arteries [33]. Minimally invasive surgical procedures utilize laparoscopes, which are long, thin, rigid optical instruments to permit video-based imaging within the abdominal cavity. Laparoscopic OCT imaging has been demonstrated by passing the OCT beam through the optical elements of a laparoscope [27,32]. Deep solid-tissue imaging is possible with the use of fiber-needle probes [28]. Small (400 µm diameter) needles housing a single optical fiber and micro-optic elements have been inserted into solid tissues and rotated to acquire OCT images. Recently, microfabricated micro-electro-optical-mechanical systems (MEOMS) technology has been used to miniaturize the OCT beam scan mechanism [53].

8.6 Spectroscopic Optical Coherence Tomography

Spatially distributed spectroscopic information can be extracted from the tissue specimen using spectroscopic OCT (SOCT) algorithms and techniques [54–57] (Fig. 8.7). In structural OCT imaging, the amplitude of the envelope of the field autocorrelation is acquired and used to construct an image based on the magnitude of the optical backscatter at each position. Spectral information

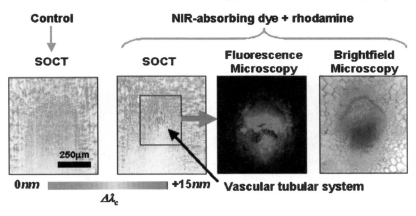

Fig. 8.7. Spectroscopic OCT imaging. Spatially resolved spectroscopic information can be extracted from the detected OCT signal. By digitizing the interferogram fringes and transforming the data using a Fourier-transform, the spectrum of the returned light can be determined. When compared to the original laser spectrum, spectral differences can be identified, which are related to the absorption and scattering of the incident light. Spectroscopic OCT images of control and experimental botanical specimens where a highly absorbing and fluorescent chemical dye has accumulated in the vascular system. Corresponding fluorescence and bright-field microscopy images are shown. Figure modified with permission from [58]

can be obtained by digitizing the full interference signal and applying digital signal processing algorithms to transform the data from the time (spatial) domain to the frequency (spectral) domain. Transformation algorithms include the Morlet wavelet and the short-time Fourier transform with attention to reduction of windowing artifacts. While spectroscopic data can be extracted from each point within the specimen, there is an inherent trade-off between high spatial resolution and high spectral resolution and novel algorithms are being investigated to optimize this data. Once the spectral data is obtained at a point in the tissue, the spectral center of mass can be calculated and compared with the original spectrum from the laser source. Spectral shifts to longer or shorter wavelengths, from the original center of mass, are displayed on a 2D image using a multidimensional hue-saturation-luminance (HSL) color space. At localized regions within the tissue, a color is assigned with a hue that varies according to the direction of the spectral shift (longer or shorter wavelength) and a saturation that corresponds to the magnitude of that shift. The luminance is held constant.

Longer wavelengths of light are scattered less in turbid media. In a homogeneously scattering sample, one can observe using spectroscopic OCT that shorter wavelengths are scattered near the surface and a smooth color-shift occurs with increasing depth, as longer wavelengths are scattered. In more heterogeneous samples, such as tissue, scattering objects such as cells and sub-cellular organelles produce variations in the spectroscopic OCT data.

Spectroscopic OCT images can indicate changes in the spectroscopic properties of the tissue; however, further investigation is needed to determine how the spectroscopic variations relate to the biological structures and how this information can be used for diagnostic purposes.

8.7 Applications to Cancer Imaging

The noninvasive, noncontact, high-resolution, real-time imaging capabilities of OCT and its many modes of operation have enabled a wide range of new applications in biology, medicine, surgery, and materials investigations. In this section, applications in tumor cell biology and cancer imaging are presented. OCT has successfully made a transition from being a laboratory-based experimental technology to one that is useful clinically [59]. In the coming years, results from long-term controlled clinical trials will further demonstrate the usefulness of this technology.

8.7.1 Cellular Imaging for Tumor Cell Biology

Although previous studies have demonstrated in vivo OCT imaging of tissue morphology, most have imaged tissue at ~10–15 µm resolutions, which does not allow differentiation of cellular structure. The ability of OCT to identify the mitotic activity, the nuclear-to-cytoplasmic ratio, and the migration of cells has the potential to not only impact the fields of cell, tumor, and developmental biology, but also impact medical and surgical disciplines for the early diagnostics of disease such as cancer.

High-resolution in vivo cellular and subcellular imaging has been demonstrated in the *Xenopus laevis* (African frog) tadpole (Fig. 8.8) [16,60]. Many of the cells in this common developmental biology animal model are rapidly dividing and migrating during the early growth stages of the tadpole, providing an opportunity to image dynamic cellular processes. Cell dynamics can also be tracked in three-dimensional volumes of high-resolution OCT data as they migrate through an engineered tissue scaffold along a chemoattractant-induced gradient (Fig. 8.9) [61]. From this 3D data set, time-dependent cell position was color-coded. The ability of OCT to characterize cellular dynamics such as mitosis and migration is relevant for cancer diagnostics and for the investigation of tumor metastasis in humans.

Combining the coherence gating of OCT with high numerical aperture microscope objectives enables high axial- and transverse-resolution imaging deep within highly scattering specimens. This optical configuration has been called optical coherence microscopy (OCM), and can improve the optical sectioning capability of confocal microscopy. For several years, investigators have recognized that OCT and MPM can utilize a single laser source for multimodality imaging. Recently, a microscope that integrates OCM and multiphoton (MPM) fluorescence imaging has been used to image cells in 3D engineered

8 OCT with Applications in Cancer Imaging 139

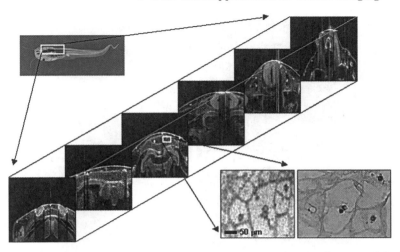

Fig. 8.8. In vivo morphological and cellular imaging in a developmental biology animal model. Photograph and cross-sectional OCT images of a *Xenopus laevis* (African frog) tadpole specimen showing developing tissue morphology, as well as individual cells and corresponding histology

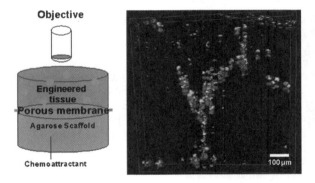

Fig. 8.9. Tracking cell migration. A population of macrophage cells is tracked in three dimensions over the course of 3 h using OCT. Macrophage cell migration was induced with a chemoattractant at one end of the 3D scaffold, separated by a semi-permeable membrane. The time-dependent positions of the cells are color-coded white (time = 0 h) and grey (time = 3 h). Modified figure used with permission from [61]

tissue scaffolds [62, 63]. OCT and MPM provide complementary image data. OCT can image deep through transparent and highly scattering structures to reveal the 3D structural information. OCT, however, cannot detect the presence of a fluorescing particle. In a complementary manner, MPM can localize fluorescent probes in three-dimensional space and provide insight into cell function. MPM can detect the fluorescence, but not the microstructure

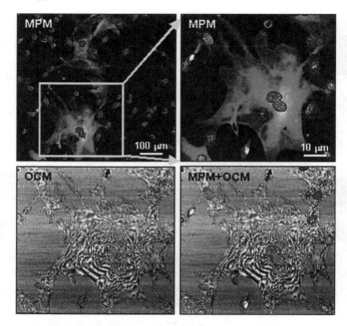

Fig. 8.10. Integrated multimodality microscopy using optical coherence microscopy (OCM) and multiphoton microscopy (MPM). Coregistered image data provides insight into structure (OCM) and function (MPM) of cells in 2D and 3D culture conditions, under the influence of external mechanical stimuli, and following pharmacological interventions. Fibroblasts with green-fluorescent-protein (GFP) transfected to be expressed with vinculin, a cell surface adhesion protein, are shown. A nuclear fluorescent dye has also been used to identify the location of the nuclei. OCM images based on scattering show the spatial distribution of the cell and extracellular components. Figure modified and used with permission from [63]

nor the location of the fluorescence relative to the microstructure. Hence, the development of an integrated microscope capable of OCT and MPM uniquely enables the simultaneous acquisition of microstructural data and the localization of fluorescent probes for precisely coregistered structural and functional imaging (Fig. 8.10).

8.7.2 Translational Breast Cancer Imaging

OCT has been used to differentiate between the morphological structure of normal and neoplastic tissue for a wide-range of tumors [13, 64–67]. The use of OCT to identify tumors and tumor margins in situ will represent a significant advancement for medical or image-guided surgical applications. Beam delivery systems such as a compact and portable hand-held surgical probe or a modified surgical microscope permits OCT imaging within the surgical field, while the main OCT instrument can be remotely located in the surgical suite. The use of OCT in image-guided surgical procedures represents a paradigm shift for the surgical oncology community. While the surgical oncologist typically resects macroscopic solid tumors, consideration must still be given to the microscopic, cellular extent of the disease. To address this, surgeons typically resect with large margins in an effort to remove any occult cells or nests of tumor. Resected tissue is sent to the surgical pathology lab where margins are examined following histological processing to ensure that they are clean. During this time, a fully staffed operating room and anesthetized patient must often wait for the decision on these margins. Therefore, moving the high-resolution microscopic imaging of tumor margins from the surgical pathology lab into the operating room would represent a substantial reduction in costs, including time, costs, and patient health. The use of OCT for examining tumor margins has many advantages, as well as introducing new challenges. Large surface areas (and three-dimensional volumes) of tissue must be examined in real-time at micron resolutions. Improvements in data management, acquisition rates, automated tissue identification, and zooming capabilities all must be addressed.

An example of the use of OCT for identifying tumors at varying stages is shown in Fig. 8.11. Using a well-characterized carcinogen-induced rat mammary tumor model that mimics the progression and histological findings of human ductal carcinoma of the breast, OCT images were acquired at varying time-points and compared to corresponding histology [67]. In good agreement, the OCT images identify not only late-stage morphological changes and the clear tumor margin, but also early ductal changes and evidence of abnormal cells located away from the primary tumor.

A portable OCT system has been constructed for intraoperative imaging of surgical margins following lumpectomy procedures for the surgical treatment of breast cancer (Fig. 8.12). Different OCT biopsy needles have also been constructed for imaging into solid tumors or for guiding needle-biopsy procedures in breast cancer [28] (Fig. 8.12). Ongoing intraoperative and image-guided procedure studies are underway to determine the sensitivity and specificity of OCT compared with the gold-standard histopathological analysis.

8.8 Optical Coherence Tomography Contrast Agents

When imaging biological tissues, it is often desirable to enhance the signals measured from specific structures. Contrast agents that produce a specific image signature have been utilized in virtually every imaging modalities, including ultrasound, computed tomography, magnetic resonance imaging, and optical microscopy, among many others. There are multiple optical properties that are amenable for generating contrast in optical images, including OCT images (Fig. 8.13). Recently, new engineered contrast agents and molecular contrast techniques specifically designed for OCT have been developed and

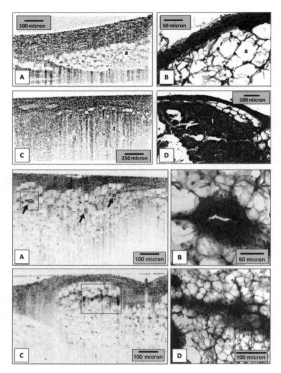

Fig. 8.11. Cancer imaging using OCT. OCT can identify neoplastic changes at varying stages of tumor growth. Top image set: Morphological OCT and histological imaging of (**a,b**) normal and (**c,d**) late-stage carcinogen-induced mammary tumor in a rat model. The tumor (t) mass is evident compared to the low-backscattering adipose (a) cells. Bottom image set: Early-stage ductal changes detected in this model using OCT and confirmed with histology. Ducts are imaged in cross-section (**a,b**) and through a longitudinal section (**c,d**). Figure modified and used with permission from [67]

characterized [68,69]. In contrast to fluorescent probes, which are commonly used as contrast agents in fluorescence, confocal, and multiphoton microscopy, new classes of optical contrast agents suitable for OCT must be based on mechanisms other than the detection of incoherently emitted fluorescence because OCT detects only coherent light. Agents that alter the local scattering or absorption properties are used (Fig. 8.14), such as oil-filled microspheres that have scattering nanoparticles of melanin, gold, carbon, or iron-oxide either embedded in their protein shell or encapsulated in their liquid-filled core [70]. Fluorescence-based microscopy techniques have the advantage of very low background signal in the absence of autofluorescence. This advantage can also be obtained in OCT by using novel dynamic contrast agents that are physically modulated in space using an external magnetic or electric field [71,72]. While

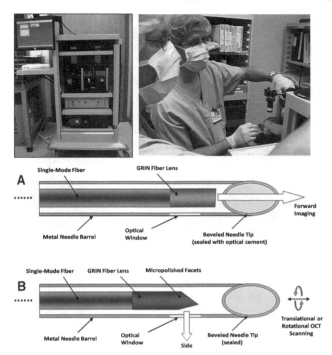

Fig. 8.12. Portable OCT system for real-time imaging during surgical operations and procedures. The compact system utilizes a superluminescent diode with a center wavelength at 1,300 nm wavelength and spectral-domain detection. (**a**) Forward-directed and (**b**) side-directed fiber-optic OCT needle probes can be used to collect data from within solid tissue masses or during needle-biopsy procedures

the resolutions of OCT may be insufficient to resolve individual nanometer or micron-sized scattering agents, the use of modulating particles enables sub-resolution detection of particles moving in and out of the OCT beam, or by modulating not only the agent itself, but also the local tissue environment, producing both changes in the amplitude of the reflected light (for structural OCT) and the phase of the reflected light (for phase-resolved spectroscopic or Doppler OCT). Because OCT detects changes in optical scattering, the sensitivity of detecting small relative changes in scattering is less than detecting relatively large changes in local absorption, particularly if absorption is due to an exogenously administered contrast agent. There is an increasing interest in near-infrared contrast agents in fluorescence-based microscopy techniques because longer excitation and emission wavelengths can propagate deeper into and further out of scattering tissue. Spectroscopic OCT techniques, which can detect spatially resolved spectral changes, can be used to identify the presence of a near-infrared absorbing dye if the absorption spectra of the dye overlaps with the optical source spectrum [58].

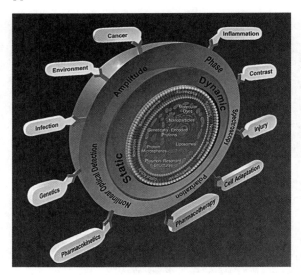

Fig. 8.13. Metaphorical diagram of an optical contrast agent illustrating the multiple optical substrates, properties, and applications that can be utilized for optical probes and contrast agents. There are many properties and principles to exploit, in addition to the more common fluorescence and bioluminescence properties. Figure used with permission from [68]

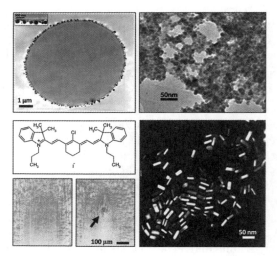

Fig. 8.14. Contrast agents for OCT. As with every other imaging modality, contrast agents enhance the diagnostic ability of the modality and permit site-specific detection of features. A wide array of novel contrast agents have recently been developed including (*top left*) microspheres with scattering nanoparticles embedded in the core or shell, (*top right*) magnetically-modulated agents of free or encapsulated iron-oxide, (*bottom left*) absorbing near-infrared dyes detected with spectroscopic OCT, and (*bottom right*) plasmon-resonant nanorods as scattering and absorbing agents

Finally, plasmon-resonant gold nanoparticles in the form of spheres, shells, cages, and rods have been utilized as strong as wavelength-specific absorbers and scatterers [73–75]. By changing the size and structure of these nanoparticles, the wavelength-dependent absorption and scattering can be tuned throughout the near-infrared and visible wavelengths. These nanoparticles have been used as contrast agents for OCT, and also proposed as multifunctional therapeutic agents because their strong absorption properties can be used to induce local hyperthermia in cells and tissues.

All of these agents are expected to be highly biocompatible and composed of materials that have been previously found to be suitable for in vivo use. These agents, with protein, iron-oxide, gold, or biocompatible molecular surfaces, may be functionalized with antibodies or molecules to target them to specific molecules, cells, or tissue types and thus provide additional selectivity that can enhance the utility of OCT as an emerging diagnostic technique.

8.9 Molecular Imaging using Optical Coherence Tomography

The advantages of OCT, compared to other imaging techniques, are numerous. In particular, OCT can provide imaging resolutions that approach those of conventional histopathology and imaging can be performed in situ. Despite its advantages, a serious drawback to OCT is that the linear scattering properties of pathological tissue probed by OCT are often morphologically and/or optically similar to the scattering properties of normal tissue. For example, although morphological differences between normal and neoplastic tissue may be obvious at later tumor stages, it is frequently difficult to optically detect early-stage tumors. To improve the ability of discriminating tissue types in this scenario, molecular imaging techniques are being developed.

Molecular imaging involves the generation of images or maps of tissue that contain molecular-specific information. The use of exogenous contrast agents in OCT (described above) is one example of molecular imaging, when such agents are functionalized to target and label specific molecular structures on cells or in tissue, such as cell-surface receptors that are over-expressed in tumors. In many instances, it may be more desirable to detect the presence of endogenous molecules without the addition exogenous agents that could alter the biology or the viability of the tissue. Pump and probe techniques have been used to detect the presence of molecules that have transient absorption states in the tissue that are induced by an external pump beam [76]. Spectroscopic OCT has been used to probe for variations in the oxygenation state of tissue, recognizing the spectroscopic differences between oxy- and deoxy-hemoglobin [77]. Both of these methods, however, rely on a limited set of molecules or molecular features that can be detected.

A new method called Nonlinear Interferometric Vibrational Imaging (NIVI) has been developed to achieve molecular contrast in OCT imaging by exploiting optical nonlinearities [78,79]. The nonlinear processes, in particular, are coherent anti-Stokes Raman scattering (CARS) and second harmonic generation (SHG), but the general idea could easily be extended to other nonlinear processes such as third harmonic generation (THG), coherent Stokes Raman scattering (CSRS). For these, the nonlinear process generates an optical signal that remains coherent with the incident light. Therefore in NIVI, the origin of this coherent signal can be spatially resolved in three-dimensions using OCT-like interferometers. The heterodyne detection scheme common in OCT also imparts significant advantages in detecting these nonlinear signals, including increased sensitivity, full reconstruction of the magnitude and phase information of the sample Raman susceptibility [80], background rejection of noncoherent four-wave-mixing processes arising from molecules such as water [81], and in the case where reference molecules are placed in the reference arm of the interferometer, molecular specificity can be obtained. Figure 8.15 shows a schematic of the energy-level diagram involved in CARS, along with how a reference molecular species would be incorporated into

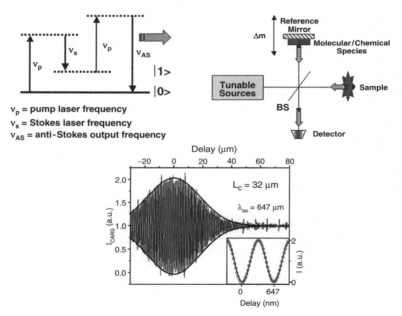

Fig. 8.15. Molecular OCT imaging. Nonlinear optical signatures in tissue are used to identify specific molecular bonds in a technique called Nonlinear Interferometric Vibrational Imaging (NIVI). An energy level diagram is shown (*top left*) for coherent anti-Stoke Raman scattering (CARS). By placing a reference molecular species in the reference arm of the OCT interferometer (*top right*), coherent CARS signals are generated in the reference molecular species, and from the sample to produce spatially resolved images of molecular composition. An interferogram from two independently generated CARS beams is shown (*bottom*) for the benzene molecule, demonstrating the familiar-looking OCT interferogram, but with molecular specificity

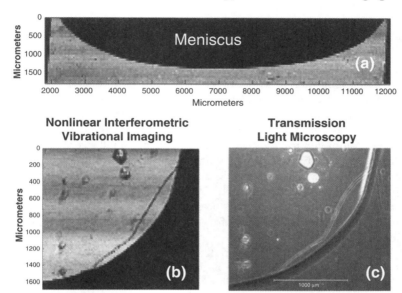

Fig. 8.16. Multidimensional Nonlinear Interferometric Vibrational Imaging (NIVI). NIVI, tuned to the C–H vibrational resonance, is performed on a cuvette containing acetone. Signal is generated only from the acetone and not from the cuvette or the particulates in the acetone. NIVI images are shown for the meniscus and bottom of the cuvette, and compared with a transmission light microscopy image of the cuvette bottom. Modified figures used with permission from [82]

the OCT interferometer. A representative interferogram generated from two independent samples of benzene is shown, demonstrating that a molecularly specific interferogram can be generated, and used to enhance molecular contrast and perform molecular imaging using NIVI [78]. This has been extended to multidimensional imaging (Fig. 8.16), and on-going research is investigating NIVI for 3D molecular imaging in tissues. While this method is still in development, the ability to image and map the three-dimensional spatial distribution of individual endogenous molecules such as DNA, RNA, proteins, or other diagnostic biomolecules [79] is likely to extend the diagnostic capabilities and clinical utility of OCT to the molecular level.

8.10 Conclusions

The capabilities of OCT offer a unique and informative means of imaging biological specimens and nonbiological samples. The noncontact nature of OCT and the use of low-power near-infrared radiation for imaging allow this technique to be noninvasive and safe. Conventional structural OCT imaging does not require the addition of fluorophores, dyes, or stains to provide contrast in images under most conditions. Instead, OCT relies on the inherent optical contrast generated from variations in optical scattering and index of refraction. These factors permit the repeated use of OCT for extended imaging. OCT permits the cross-sectional imaging or en face sectioning of tissue and samples, enabling in vivo structure to be visualized in opaque specimens, or in specimens too large for high-resolution confocal or light microscopy.

Imaging at cellular and subcellular resolutions with OCT is an important area of ongoing research. While many developmental biology animal models have been commonly used for their scientific value, small size, ease of care and handling, and readily visualized cellular features, cellular imaging in humans, particularly in situ, is a challenge because of the smaller cell sizes (10–20 µm) compared to larger undifferentiated cells in developing organisms. To our advantage, poorly differentiated cells present in many neoplastic tissues tend to be larger, increasing the likelihood for detection using OCT at current imaging resolutions.

Clinical applications of OCT are likely to continue to increase, particularly as more commercial OCT systems become available. With successful integration into clinical ophthalmic imaging, OCT applications in gastroenterology, cardiology, and oncology are likely to become more commonplace. Finally, novel means to perform molecular imaging using OCT will be pursued, using targeted OCT-specific contrast agents or advanced nonlinear optical methods for identifying and mapping the molecular and ultrastructural composition of biological tissue. OCT represents a multifunctional investigative tool, which not only complements many of the existing imaging technologies available today, but overtime, is also likely to become established as a major optical imaging modality.

Acknowledgments

I thank my students, post-doctoral fellows, and research personnel in the Biophotonics Imaging Laboratory at the Beckman Institute for their hard work and dedication in advancing the OCT technology. I appreciate the insight and collaborative contributions of Profs. Kenneth Suslick, Martin Gruebele, and Keith Singletary from the University of Illinois at Urbana-Champaign and Alexander Wei from Purdue University for using OCT to explore new applications. I also thank my colleagues in Biomedical Optics for their contributions to this work, and my clinical collaborators and colleagues at Carle Foundation Hospital and Carle Clinic Association, Urbana, Illinois, USA. Additional information can be obtained at http://biophotonics.uiuc.edu. Prof. Stephen Boppart's email address is boppart@uiuc.edu.

References

1. D. Huang, E.A. Swanson, C.P. Lin et al., Science **254**, 1178 (1991)
2. B.E. Bouma, G.J. Tearney (eds.), *Handbook of Optical Coherence Tomography* (Marcel Dekker, New York, 2001)
3. J.S. Schuman, C.A. Puliafito, J.G. Fujimoto, *Optical Coherence Tomography of Ocular Diseases*, 2nd edn. (Slack, Thorofare, NJ, 2004)
4. J.M. Schmitt, A. Knuttel, M. Yadlowsky et al., Phys. Med. Biol. **39**, 1705 (1994)
5. J.G. Fujimoto, C. Pitris, S.A. Boppart et al., Neoplasia **2**, 9 (2000)
6. M.E. Brezinski, G.J. Tearney, B.E. Bouma et al., Circulation **93**, 1206 (1996)
7. G.J. Tearney, M.E. Brezinski, S.A. Boppart et al., Circulation **94**, 3013 (1996)
8. I.E. Jang, B.E. Bouma, D.H. Kang et al., J. Am. Coll. Cardiol. **39**, 604 (2002)
9. G.J. Tearney, M.E. Brezinski, J.F. Southern et al., Am. J. Gastroenterol. **92**, 1800 (1997)
10. J.A. Izatt, M.D. Kulkarni, H.W. Wang et al., IEEE J. Sel. Top. Quant. Electron. **2**, 1017 (1996)
11. G.J. Tearney, M.E. Brezinski, J.F. Southern et al., J. Urol. **157**, 1915 (1997)
12. E.V. Zagaynova, O.S. Streltsova, N.D. Gladkova et al., J. Urol. **167**, 1492 (2002)
13. S.A. Boppart, M.E. Brezinski, C. Pitris et al., Neurosurgery **43**, 834 (1998)
14. B.W. Colston Jr., M.J. Everett, L.B. Da Silva et al., Appl. Opt. **37**, 3582 (1998)
15. B. Bouma, G.J. Tearney, S.A. Boppart et al., Opt. Lett. **20**, 1486 (1995)
16. W. Drexler, U. Morgner, F.X. Kartner et al., Opt. Lett. **24**, 1221 (1999)
17. B. Povazay, K. Bizheva, A. Unterhuber et al., Opt. Lett. **27**, 1800 (2002)
18. H. Ren, T. Sun, D.J. MacDonald et al., Opt. Lett. **31**, 927 (2006)
19. E.C.W. Lee, J.F. de Boer, M. Mujat et al., Opt. Exp. **14**, 4403 (2006)
20. Z. Chen, T.E. Milner, S. Srinivas et al., Opt. Lett. **22**, 1119 (1997)
21. J.A. Izatt, M.D. Kulkarni, S. Yazdanfar et al., Opt. Lett. **22**, 1439 (1997)
22. J.K. Barton, A.J. Welch, J.A. Izatt, Opt. Exp. **3**, 251 (1998)
23. J.F. de Boer, T.E. Milner, M.J.C. van Germert et al., Opt. Lett. **22**, 934 (1997)
24. M.J. Everett, K. Schoenenberger, B.W. Colston Jr. et al., Opt. Lett. **23**, 228 (1998)
25. K. Schoenenberger, B.W. Colston Jr., D.J. Maitland et al., Appl. Opt. **37**, 6026 (1998)
26. G.J. Tearney, S.A. Boppart, B.E. Bouma et al., Opt. Lett. **21**, 1 (1996)
27. S.A. Boppart, B.E. Bouma, C. Pitris et al., Opt. Lett. **22**, 1618 (1997)
28. A.M. Zysk, S.G. Adie, J.J. Armstrong et al., Opt. Lett. **32**, 385 (2007)
29. G.J. Tearney, M.E. Brezinski, B.E. Bouma et al., Science **276**, 2037 (1997)
30. A. Das, M.V. Sivak Jr., A. Chak et al., Gastrointest. Endosc. **54**, 219 (2001)
31. X.D. Li, S.A. Boppart, J. Van Dam et al., Endoscopy **32**, 921 (2001)
32. F.I. Feidchtein, G.V. Gelikonov, V.M. Gelikonov et al., Opt. Exp. **3**, 257 (1998)
33. H. Yabushita, B.E. Bouma, S.L. Houser et al., Circulation **106**, 1640 (2002)
34. K. Takada, I. Yokohama, K. Chida et al., Appl. Opt. **26**, 1603 (1987)
35. A.F. Fercher, K. Mengedoht, W. Werner, Opt. Lett. **13**, 186 (1988)
36. C.K. Hitzenberger, Invest. Ophthalmol. Vis. Sci. **32**, 616 (1991)
37. A.M. Rollins, J.A. Izatt, Opt. Lett. **24**, 1484 (1999)
38. B.E. Bouma, G.J. Tearney, Opt. Lett. **4**, 531 (1999)
39. M.A. Chroma, M.V. Sarunic, C. Yang et al., Opt. Exp. **11**, 2183 (2003)
40. B. Cense, N.A. Nassif, T.C. Chen et al., Opt. Exp. **12**, 2435 (2004)
41. S.H. Yun, G.J. Tearney, J.F. de Boer et al., Opt. Exp. **11**, 2953 (2003)

42. R. Huber, M. Wojtkowski, J.G. Fujimoto, Opt. Exp. **14**, 3225 (2006)
43. H. Okamoto, M. Wada, Y. Sakai et al., J. Lightwave Tech. **16**, 1881 (1998)
44. B.E. Bouma, L.E. Nelson, G.J. Tearney et al., J. Biomed. Opt. **3**, 76 (1998)
45. M. Bashkansky, M.D. Duncan, L. Goldberg et al., Opt. Exp. **3**, 305 (1998)
46. I. Hartl, X.D. Li, C. Chudoba et al., Opt. Lett. **26**, 608 (2001)
47. D.L. Marks, A.L. Oldenburg, J.J. Reynolds et al., Opt. Lett. **27**, 2010 (2002)
48. Y. Wang, Y. Zhao, J.S. Nelson et al., Opt. Lett. **28**, 182 (2003)
49. W. Denk, J.H. Strickler, W.W. Webb, Science **248**, 73 (1990)
50. E. Beaurepaire, L. Moreaux, F. Amblard et al., Opt. Lett. **24**, 969 (1999)
51. S.A. Boppart, G.J. Tearney, B.E. Bouma et al., Proc. Nat. Acad. Sci. USA **94**, 4256 (1997)
52. A.G. Podoleanu, J.A. Rogers, D.A. Jackson et al., Opt. Exp. **7**, 292 (2000)
53. Y. Pan, H. Xie, G.K. Fedder, Opt. Lett. **26**, 1966 (2001)
54. U. Morgner, W. Drexler, F.X. Kartner et al., Opt. Lett. **25**, 111 (2000)
55. R. Leitgeb, M. Wojtkowski, A. Kowalczyk et al., Opt. Lett. **25**, 820 (2000)
56. C. Xu, S.A. Boppart, Opt. Exp. **12**, 4790 (2004)
57. C. Xu, P.S. Carney, S.A. Boppart, Opt. Exp. **13**, 5450 (2005)
58. C. Xu, J. Ye, D.L. Marks et al., Opt. Lett. **29**, 1647 (2004)
59. A.M. Zysk, F.T. Nguyen, A.L. Oldenburg et al., J. Biomed. Opt. **12**(5), 051403 (2007)
60. S.A. Boppart, B.E. Bouma, C. Pitris et al., Nat. Med. **4**, 861 (1998)
61. W. Tan, A.L. Oldenburg, J.J. Norman et al., Opt. Exp. **14**, 7159 (2006)
62. C. Vinegoni, T.S. Ralston, W. Tan et al., Appl. Phys. Lett. **88**, 053901 (2006)
63. C. Xu, C. Vinegoni, T.S. Ralston et al., Opt. Lett. **31**, 1079 (2006)
64. A.M. Sergeev, V.M. Gelikonov, G.V. Gelikonov et al., Opt. Exp. **1**, 432 (1997)
65. S.A. Boppart, A.K. Goodman, C. Pitris et al., Br. J. Obstet. Gynaecol. **106**, 1071 (1999)
66. Y.T. Pan, T.Q. Xie, C.W. Du et al., Opt. Lett. **28**, 2485 (2003)
67. S.A. Boppart, W. Luo, D.L. Marks et al., Breast Cancer Res. Treat. **84**, 85 (2004)
68. S.A. Boppart, A.L. Oldenburg, C. Xu et al., J. Biomed. Opt. **10**, 041208 (2005)
69. C. Yang, Photochem. Photobiol. **81**, 215 (2005)
70. T.M. Lee, A.L. Oldenburg, S. Sitafalwalla et al., Opt. Lett. **28**, 1546 (2003)
71. A.L. Oldenburg, J.R. Gunther, S.A. Boppart, Opt. Lett. **30**, 747 (2005)
72. A.L. Oldenburg, F.J. Toublan, K.S. Suslick et al., Opt. Exp. **13**, 6597 (2005)
73. A.L. Oldenburg, M.N. Hansen, D.A. Zweifel et al., Opt. Exp. **14**, 6724 (2006)
74. C. Loo, A. Lin, L. Hirsch et al., Technol. Cancer Res. Treat. **3**, 33 (2004)
75. J. Chen, F. Saeki, B.J. Wiley et al., Nano Lett. **5**, 473 (2005)
76. K.D. Rao, M.A. Choma, S. Yazdanfar et al., Opt. Lett. **28**, 340 (2003)
77. D.J. Faber, E.G. Mik, M.C.G. Aalders et al., Opt. Lett. **28**, 1436 (2003)
78. C. Vinegoni, J.S. Bredfeldt, D.L. Marks et al., Opt. Exp. **12**, 331 (2004)
79. D.L. Marks, S.A. Boppart, Phys. Rev. Lett. **92**, 123905 (2004)
80. G.W. Jones, D.L. Marks, C. Vinegoni et al., Opt. Lett. **31**, 1543 (2006)
81. D.L. Marks, C. Vinegoni, J.S. Bredfeldt et al., Appl. Phys. Lett. **85**, 5787 (2004)
82. D.L. Marks, C. Vinegoni, J.S. Bredfeldt et al., Opt. Photon. News **12**, 23 (2005)

9

Coherent Laser Measurement Techniques for Medical Diagnostics

B. Kemper and G. von Bally

9.1 Introduction

Holography and speckle interferometry are well established tools for industrial nondestructive testing and quality control, which are in general performed by the detection of object displacements while applying static stress, temperature changes, shock waves, or by vibration monitoring [1–3]. Therefore, up to the present, various holography and speckle interferometry systems for macroscopic as well as microscopic applications have been developed. Also for biomedical applications holographic and speckle interferometric metrology opens up new perspectives for the visualization and the detection of displacements and movements. Here, for the early recognition of malignant cancers, for example, it is of interest to distinguish between different tissue elasticities [4–7]. Also, in the fields of Life Sciences and Biophotonics, there is a requirement for an optical instrumentation of timely and spatially high resolution analysis, measurement, and documentation in supracellular, cellular, and subcellular range. For such applications, the special requirements of tissues and cellular probes have to be taken into account and a lateral resolution in microscopic dimensions as well as very compact and flexible speckle interferometric arrangements are required. Additionally, for in vivo applications, a high repetition rate of the results for online monitoring is necessary. This can be obtained by the application of micro-optics, optical fiber technology, miniaturized (color) CCD sensors, and fast digital image processing systems. Furthermore, spatial phase shifting methods and suitable techniques for nondiffractive digital holographic reconstruction, which require in contrast to temporal techniques considerably less stability of the experimental setup and the investigated specimens, allow for a quantitative determination of the object wave phase. From this information the underlying displacement data as well as the object shape can be obtained under suitable geometric conditions [1,8]. In this contribution, in an overview, experimental results obtained with several holographic and (speckle) interferometric arrangements based on endoscope and microscope optics are presented to demonstrate applications and possibilities of holographic interferometric metrology on biological specimens and cellular samples. The results of these investigations show that it is possible to detect minimal-invasive elasticity differences of biological tissues, which may be used for an early tumor and cancer recognition. Furthermore, data obtained from long-term digital holographic investigations on toxin-induced reactions of adherent cancer cells demonstrate application prospects of digital holographic phase contrast microscopy in marker-free life cell imaging [9, 10].

9.2 Electronic Speckle Pattern Interferometry (ESPI)

9.2.1 Double Exposure Subtraction ESPI

Figure 9.1 shows schematically a mainly out-of-plane sensitive ESPI for the investigation of rough surfaces. Coherent laser light is divided into an object illumination wave O and a reference wave R. The image of the object surface that is covered by a speckle pattern due to the speckle effect [11] is imaged on an image recording device (e.g., a charge coupled device (CCD) sensor). The superimposition of O and R is performed by a second beam splitter BS. In double exposure subtraction ESPI, two speckle patterns are recorded subsequently on a CCD sensor superimposed by a reference wave. The intensities I and I' of the two object states are [1]

$$I(x,y) = I_O(x,y) + I_R(x,y) + 2\sqrt{I_O(x,y)I_R(x,y)}\cos(\psi(x,y))$$
$$I'(x,y) = I_O(x,y) + I_R(x,y) \qquad (9.1)$$
$$+ 2\sqrt{I_O(x,y)I_R(x,y)}\cos(\Delta\varphi(x,y) + \Delta\phi(x,y)).$$

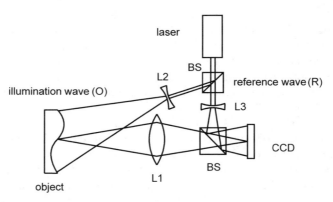

Fig. 9.1. ESPI setup. *BS* beam splitter; L1, L2, L3 lenses; *CCD* CCD sensor

In (9.1) I_O represents the intensity of the object wave, which can be assumed equal before and after the object deformation or movement between two recordings and I_R is the intensity of the reference wave. The term $\Delta\psi(x,y) = \phi_O(x,y) - \phi_R(x,y)$ represents the stochastic phase difference while $\Delta\phi(x,y)$ is the phase difference caused by the object deformation **d** and the sensitivity vector $\mathbf{S} = \mathbf{k}_Q + \mathbf{k}_B$, which takes into account the illumination direction $\mathbf{k}_Q$, the observation direction $\mathbf{k}_B$, and the light wavelength λ [1]:

$$\Delta\phi(x,y) = \frac{2\pi}{\lambda} \mathbf{S} \cdot \mathbf{d}. \qquad (9.2)$$

Subtraction of the two intensities (9.1) yields

$$(I - I')(x,y) = 4\sqrt{I_O(x,y)I_R(x,y)} \, \sin\left[\psi(x,y) + \frac{\Delta\phi(x,y)}{2}\right] \sin\frac{\Delta\phi(x,y)}{2}. \qquad (9.3)$$

To display the result in video repetition rate on a monitor the modulus $|I - I'|$ is calculated. By observation of $|I - I'|$ qualitative information about the object deformation is obtained.

9.2.2 Spatial Phase Shifting (SPS) ESPI

Investigations on biological specimen require a method for quantitative determination of the phase difference $\Delta\phi$ that works even under instable conditions. For this reason, spatial phase shifting (SPS) methods [12, 13] are applied for sign correct phase difference determination that require only a single interferogram to determine the phase distribution of one object state. Another advantage of this technique in comparison with temporal phase shifting procedures [14] is that no movable parts (e.g., Piezo translators) in the optical path and no additional electronic synchronization are necessary.

For the realization of SPS, the speckle image of the investigated object surface is superimposed in the image plane with spatial carrier fringes, e.g., with vertical or horizontal orientation. This is achieved by an adequately chosen, almost constant spatial phase gradient $\beta(x,y) = (\beta_x, \beta_y)$ that is generated by a tilt between object wave and reference wave.

The intensity of the interference pattern formed on the CCD sensor can be described as [14]

$$I(x_m, y_n) = I^0(x_m, y_n) \left[1 + \gamma(x_m, y_n) \operatorname{sinc}\left(\frac{\Phi}{2}\right) \cos(\phi(x_m, y_n) + m\beta_x + n\beta_y + C)\right]. \qquad (9.4)$$

In (9.4) $m = 1, ..., M$ and $n = 1, ..., N$ are the indices of the CCD pixel co-ordinates x_m, y_n. $I^0(x_m, y_n) = I_O^0(x_m, y_n) + I_R^0(x_m, y_n)$ is the sum of the intensities of object wave and reference wave. The parameter $\gamma(x_m, y_n) = 2\sqrt{I_O(x_m, y_n)I_R(x_m, y_n)}/I^0(x_m, y_n)$ denotes the modulation of the intensity patterns. The term $\text{sinc}(\Phi/2) = \sin(\Phi/2)/(\Phi/2)$ describes the intensity integration on a single CCD pixel in the angle range Φ. $\phi(x_m, y_n)$ is the distribution of the object wave phase and C an additional constant phase off-set.

For the calculation of the object wave phase from each recorded intensity pattern, in addition to Fourier transformation methods [15–17], a variable three-step algorithm [12] can be applied by taking into account three intensity values I_{k-1}, I_k, I_{k+1} (in horizontal or vertical orientation) of neighboring CCD pixels:

$$\phi_k + k\beta + C = \arctan\left(\frac{1-\cos\beta}{\sin\beta}\frac{I_{k-1}-I_{k+1}}{2I_k - I_{k-1} - I_{k+1}}\right) \quad \text{modulo} \quad 2\pi. \tag{9.5}$$

The algorithm requires for a correct phase evaluation that the mean speckle size is at least the size of three pixels of the digitized interferograms. For the parameter β, either β_x or β_y may be chosen. The adjustment of the phase gradients can be performed by analysis of the 2D frequency spectrum of the carrier fringe pattern in the interferograms by 2D digital fast Fourier transform (FFT) [13, 16].

The phase difference $\Delta\phi_k$ modulo 2π between two phase states ϕ_k, ϕ'_k of the object wave front is calculated:

$$\Delta\phi_k = (\phi'_k + k\beta + C) - (\phi_k + k\beta + C) = \phi'_k - \phi_k. \tag{9.6}$$

Figure 9.2 illustrates the evaluation process of spatial phase-shifted interferograms for displacement detection. Figure 9.2a,b shows the spatial phase-shifted interferograms I, I' obtained from two different displacement states of a tilted metal plate. In Fig. 9.2c,d the corresponding phase distributions ϕ, ϕ' (mod 2π), calculated by (9.5) from Fig. 9.2a,b, are depicted. The corresponding correlation fringe pattern calculated by (9.3) is depicted in Fig. 9.2e. The resulting phase difference distribution $\Delta\phi$ mod 2π obtained by subtraction of Fig. 9.2c,d modulo 2π as well as the according filtered phase difference distribution is shown in Fig. 9.2f,g. Figure 9.2h,i finally represents the unwrapped phase difference after removal of the 2π ambiguity and the related pseudo 3D representation of the data. From the data in Fig. 9.2h the underlying displacement of the plate is calculated by taking into consideration recording and imaging geometry of the experimental setup by (9.2).

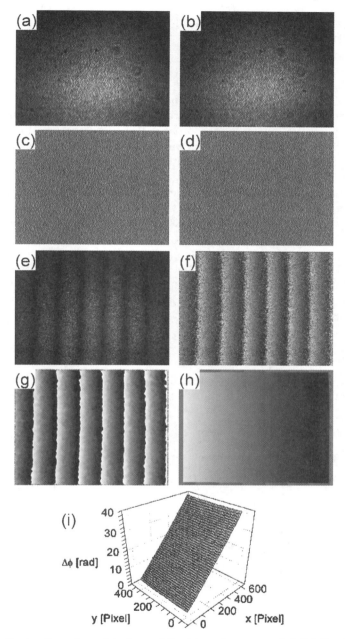

Fig. 9.2. Evaluation of spatial phase-shifted interferograms. (**a**),(**b**) Spatial phase-shifted interferograms I, I' of two displacement states of a tilted metal plate; (**c**),(**d**) Phase distributions ϕ, ϕ' (mod 2π) calculated by (9.5) from (**a**) and (**b**); (**e**) Correlation fringe pattern calculated by (9.3); (**f**),(**g**) Raw and filtered phase difference distribution mod 2π obtained by subtraction of (**c**) and (**d**) modulo 2π; (**h**),(**i**) Unwrapped phase difference and corresponding pseudo 3D representation of $\Delta\phi$

9.3 Endoscopic Electronic Speckle Pattern Interferometry (ESPI)

Endoscopy is a wide spread intracavity observation technique routinely used in minimal invasive diagnostics and industrial intracavity inspection. The combination of endoscopic imaging with speckle interferometric metrology allows the development of tools for a nondestructive quantitative detection of defects within body cavities, including the analysis of shape, structure, displacements, and vibrations of the object. In this way Electronic-Speckle-Pattern Interferometry (ESPI) opens up new perspectives for biomedical applications, especially in minimally invasive diagnostics in the medical field [18]. Contrary to earlier attempts in holographic endoscopy where single interferograms with Q-switched lasers were recorded [19], an on-line process analysis can be performed with a rate near video repetition frequency. This can be achieved by endoscopic ESPI systems with an external (proximal) interferometric arrangement using standard endoscope optics [4,20] or by an arrangement where the ESPI system is positioned in the endoscope tip [21] (distal arrangement). Such endoscope ESPI systems open up the possibility to replace the operator's tactile sense, which is lost in endoscopic surgery, by visual information ("endoscopic taction") [4,5,22].

9.3.1 Proximal Endoscopic ESPI

The setup for proximal endoscopic ESPI with the speckle interferometer positioned outside the cavity is illustrated in Fig. 9.3. The advantage of such

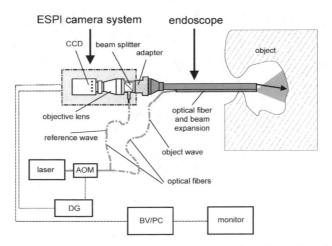

Fig. 9.3. Arrangement for proximal endoscopic ESPI. *CCD* CCD sensor, *AOM* acousto-optic modulator, *DG* double pulse generator, *BV/PC* digital image processing system

an arrangement is that standard CCD cameras can be used. Additionally, it is possible to form modular concepts of the ESPI system with flexible as well as rigid standard endoscopes for imaging. The light source is a cw laser (e.g., an argon-ion laser, $\lambda = 514.5$ nm or a frequency doubled Nd:YAG laser, $\lambda = 532$ nm). The regulation of the light intensity is performed by an acousto-optic modulator (AOM) using the first diffraction order. Behind the AOM, the laser beam is divided into the object illumination wave and the reference wave, with both coupled into single-mode optical fibers. Spatial phase shifting (SPS) (see Sect. 9.2.2) is applied to obtain quantitative information about displacements and motions. Therefore, an interference pattern between object wave and reference wave is generated by positioning the end of the reference wave fiber with an off-set to the optical axis. The interference patterns of the superimposed object wave and reference wave are recorded by a progressive-scan (full frame) camera with a CCD sensor. A fast digital image processing system (BV/PC) allows an image acquisition of the intensity patterns up to 25 Hz. Simultaneously, the corresponding correlation fringe patterns obtained by image subtraction are displayed online with a rate of 12.5 Hz on an external monitor. The exposure time and the delay between two exposures are variable in order to be suitable for the measurement conditions of biological specimen (for further details see [4, 13, 23]). Figure 9.4 shows on the left panel a mobile endoscope ESPI system based on an analog CCD camera and an image processing system with frame grabber interface. On the right panel of Fig. 9.4 a modular endoscope ESPI system with IEEE1394 ("FireWire") technology consisting of a laser unit, a speckle interferometric module, and a notebook-computer for image processing is depicted.

One of the current limitations of endoscopic imaging in the gastrointestinal tract is the absence of tactile perception, making it impossible to assess gastrointestinal wall elasticity during routine endoscopic examinations. For this reason it is of interest to distinguish between tissue elasticity for the

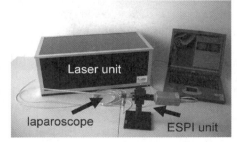

Fig. 9.4. (*Left*) Mobile endoscope ESPI system based on an analoge CCD camera and an image processing system with frame grabber interface. (*Right*) Modular endoscope ESPI system with IEEE1394 ("FireWire") technology consisting of a laser unit, a speckle interferometric module, and a notebook-computer for image processing

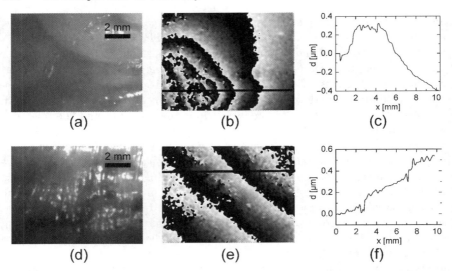

Fig. 9.5. In vitro investigations on a human intestinal specimen by stimulation with an endoscopic ultra sound tube. (**a**),(**d**) White light images of a tissue part without pathological findings and of the tumorous tissue; (**b**),(**c**) Phase difference distributions (mod 2π); (**c**),(**f**) Pseudo 3D plots of the displacement along the marked cross sections in (**b**) and (**e**)

early recognition of malignant cancers, for example. In vitro investigations on intestinal specimens with carcinoma tissue have been carried out. For the experiment, the tissue is stimulated by an endoscopic ultrasonic tube (single pulse, amplitude: $\approx 200\,\mu m$). Afterwards, during the relaxation process of the tissue, series of stroboscopic phase difference distributions are captured.

Figure 9.5 depicts characteristic results. Figure 9.5a,d shows the white light images of an investigated area of the specimen without pathological findings in comparison to a tumorous part of the tissue. In Fig. 9.5b,e exemplary results of phase difference distributions mod 2π obtained during the relaxation process of the tissue are depicted. The corresponding calculated surface deformation along the cross sections in Fig. 9.5b,e is plotted in Fig. 9.5c,f. The phase difference shows concentric fringes for the tissue part without pathological findings (Fig. 9.5b) and a parallel fringe distribution for the hardened tumorous tissue (Fig. 9.5e). Thus, the tissue part without pathological findings can be distinguished from the tumorous tissue of diminished elasticity.

9.3.2 Distal Endoscopic ESPI

To avoid the phase instabilities of image fiber bundles and the aberrations of standard rod lens endoscope imaging systems in endoscopic ESPI, a compact realization of the ESPI camera system in the endoscope tip (distal arrangement) is of particular advantage [21]. In addition, a smaller speckle size can be

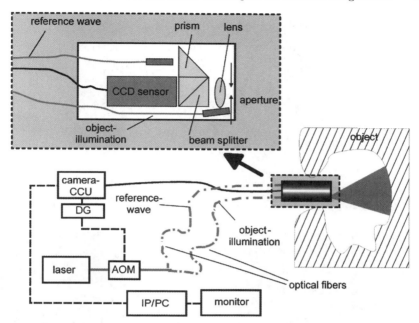

Fig. 9.6. Concept of a distal endoscope ESPI sensor. *CCD* CCD sensor (1/4 or 1/6 in.)

obtained by short optical path length distances in a compact interferometric alignment, thus achieving a higher lateral resolution up to microscopic dimensions. The concept of an out-of-plane sensitive distal ESPI sensor is illustrated in Fig. 9.6. The coherent light source is an amplitude modulated cw laser. The object illumination wave (O) as well as the reference wave (R) are guided by single-mode optical fibers. For the object illumination O as well as for the reference wave R additional microlens systems are applied that are fixed at the end of each optical fiber.

To achieve an arrangement suitable for endoscopic applications, the reference wave is redirected by 180° with a prism and a beam splitter. The interference patterns of the superimposed waves R and O are recorded by a one-chip color CCD sensor (e.g., 1/4 inch sensor or 1/6 inch sensor; pixel resolution, 752 × 582 pixels; maximum diameter, 8 or 4 mm). For the quantitative displacement detection, an aperture is used to regulate the speckle size to three CCD pixels for the application of spatial phase shifting techniques [9, 23]. Therefore, a spatial carrier fringe pattern between O and R is generated by positioning the end of the reference wave fiber with a lateral off-set out of the optical interferometer axis. Figure 9.7 depicts a photo of a distal endoscope ESPI sensor based on a 1/6 inch CCD sensor.

Figure 9.8 shows results of investigations with a distal endoscopic speckle interferometer based on 1/4 inch color CCD sensor. In Fig. 9.8a the white

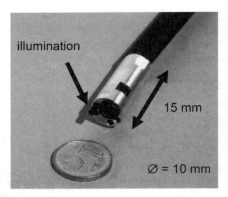

Fig. 9.7. Distal endoscope ESPI sensor based on a 1/6 in. CCD sensor

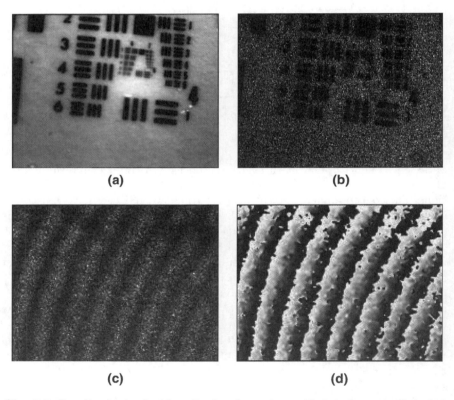

Fig. 9.8. Results obtained with a distal endoscopic speckle interferometer based on a 1/4 in. CCD sensor. (**a**) White light image of an USAF 1951 test chart; (**b**) Corresponding speckle image ($\lambda = 514.5$ nm), a lateral resolution of 8.7 µm is obtained; (**c**) Correlation fringe pattern of displacement measurements on a tilted white painted metal plate; (**d**) Corresponding filtered phase difference distributions

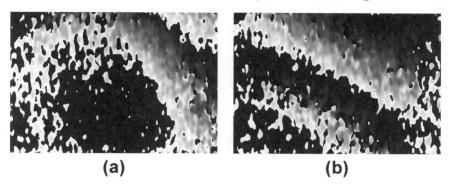

Fig. 9.9. Results of in vitro investigations on a human stomach: Filtered phase difference distributions of a "healthy" area (**a**) and a region containing an adeno carcinoma (**b**)

light image of a USAF 1951 test chart positioned in a distance of 5 mm in front of the sensor is depicted. Figure 9.8b shows the corresponding speckle image affected by illumination with coherent light ($\lambda = 514.5$ nm). Within the speckle image a lateral resolution of 8.7 μm is obtained. Figure 9.8c,d presents results from a recorded series of displacement measurements with a white painted metal plate, which was tilted between two recordings to demonstrate the fringe resolution and quality. Shown are the correlation fringe pattern (Fig. 9.8c) and the respective filtered wrapped phase difference distributions modulo 2π (Fig. 9.8d) converted to 256 gray levels. The maximum displacement resolution of the system in this configuration can be estimated from the noise of the phase difference map to $\lambda/7$. To demonstrate the performance of the distal ESPI sensor on biological specimens analogous to the experiments described in Sect. 9.3.1, investigations on tumorous human stomach gastric wall (in vitro) have been carried out.

In Fig. 9.9 the filtered wrapped phase difference distribution of a "healthy" part of the specimen is depicted (imaged area about $1\,\text{cm}^2$), which has been stimulated manually with a test needle. The results are concentric fringe patterns around the tip of the needle. In contrast, an area containing an adeno carcinoma shows a parallel fringe pattern (Fig. 9.9b), indicating that the tissue elasticity in this region is diminished, which is in agreement with the results from Sect. 9.3.1.

9.4 Microscopic (Speckle) Interferometry

In microscope ESPI, the motivation is to integrate the interferometric metrology into microscope systems to achieve an enhancement of these techniques by an additional high resolution detection and visualization of movements, displacements [24, 25], and refractive index changes. Figure 9.10 shows the

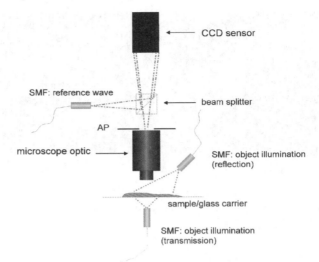

Fig. 9.10. Setup for a microscope (speckle) interferometer. Light source: frequency doubled cw Nd:YAG laser ($\lambda = 532$ nm); SMF: single-mode fibers for object illumination and reference wave; the illumination of the sample can be performed optional with incident light or in transmission; AP: aperture for regulation of the speckle size and suppression of scattered light

schematic setup for a microscopic (speckle) interferometer. The object is imaged by a microscope lens on a CCD sensor. The coherent object illumination is performed either in transmission or with incident light by single mode optical fibers. The speckle size (or in the case of smooth waves disturbing scattered light due to transparent specimens) is regulated by an additional aperture AP that is positioned behind the microscope lens. The detection of optical path length changes affected by micro changes of the sample can be performed by a spatial phase shifting method as described in Sect. 9.2.2. Therefore, the reference wave is generated by an optical single mode fiber placed in the plane of AP that is positioned with an off-set from the optical axis of the interferometer to produce a nearly constant phase gradient between object wave and reference wave.

Figure 9.11 shows results obtained by a microscope (speckle) interferometer setup in combination with a ×10 magnification microscope lens. In Fig. 9.11a the white light image of an USAF 1951 test chart is depicted. Figure 9.11b shows the corresponding speckle image effected by illumination with coherent laser light ($\lambda = 532$ nm) with a lateral resolution of 7.8 µm. In Fig. 9.11c–f results from displacement measurements on a dyed probe of an intestine carcinoma which was tilted in the experiment are depicted. Figure 9.11c shows the white light image of the investigated area. The correlation fringe pattern, the corresponding phase difference distribution, and the filtered phase difference fringes are presented in Fig. 9.11d–f.

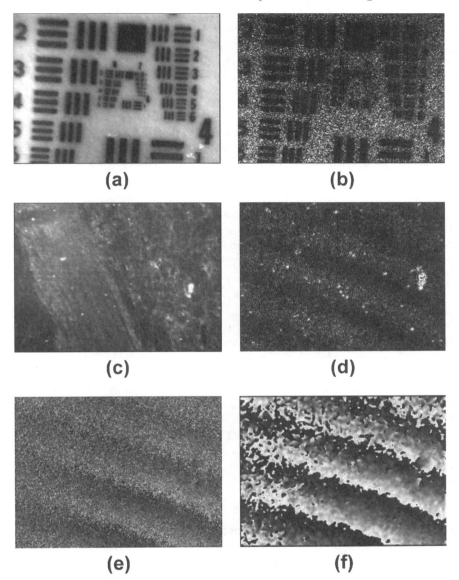

Fig. 9.11. Results of measurements obtained with a ×10 magnification microscope lens. (**a**) White light image of an USAF 1951 test chart; (**b**) Speckle image of (**a**) by illumination with coherent laser light ($\lambda = 532$ nm); (**c**) White light image of a dyed part of intestine carcinoma fixed on a glass slide; (**d**) Correlation fringe pattern of the tilted specimen in (**a**); (**e**) Corresponding phase difference distribution modulo 2π; (**f**) Filtered phase difference distribution

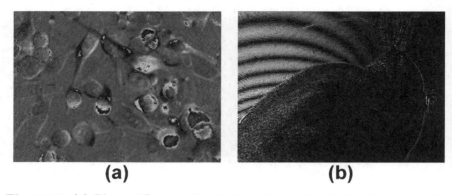

(a) (b)

Fig. 9.12. (a) Phase difference distribution of optical path length changes due to micro-changes and micro-movements of human liver tumor cells (HepG2) after $t = 75$ min (×40 magnification immersion microscope lens). (b) Correlation fringe pattern obtained with a ×10 magnification microscope lens during investigations of a living water flea. For both measurements the illumination of the specimen was performed in transmission, $\lambda = 532$ nm

The results demonstrate that displacements can be detected while the lateral resolution of the speckle image is in microscopic dimensions. Figure 9.12a shows results obtained by application of the interferometric arrangement in Fig. 9.10 in combination with a ×40 magnification immersion microscope lens for monitoring microchanges of human liver tumor cells (HepG2) in cell culture medium. The depicted phase difference distribution (illumination in transmission, $\lambda = 532$ nm) due to optical path length changes corresponds to micromovements and microchanges of the specimens after $t = 75$ min. Figure 9.12b shows a correlation fringe pattern obtained from the investigation of a living water flea utilizing a ×10 magnification microscope lens, which is an exemplary result of a recorded series of 50 images and demonstrates the feature of online monitoring.

9.5 Digital Holographic Microscopy

9.5.1 Principle and Measurement Setup

In Life Sciences and Biophotonics, a quantitative minimal invasive analysis of dynamic life processes with resolution in cellular and subcellular scale is of particular interest. In connection with microscopy, digital holography provides contact-less, marker-free, quantitative phase contrast imaging.

Digital holography is based on the classic holographic principle, with the difference that the hologram recording is performed by a digital sensor, e.g., a CCD or CMOS camera [26, 27]. The subsequent reconstruction of the holographic image that includes the information about the object wave is carried

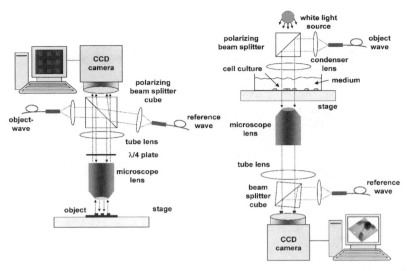

Fig. 9.13. Schematics for digital holographic microscopy using incident light (*left*) and inverse transmission arrangements (*right*) [28]

out numerically with a computer. Digital holographic microscopy (DHM) permits nondestructive, marker-free, and quantitative high resolution full-field phase contrast imaging of transparent samples such as living cells [9, 10].

Figure 9.13 depicts the schematics of two "off-axis" setups for digital holographic microscopy, which are particularly suitable for the integration into commercial microscopy systems. The coherent light of a laser (e.g., a frequency doubled Nd:YAG laser, $\lambda = 532$ nm) is divided into object illumination and reference wave, using singlemode optical fibers. The left panel of Fig. 9.13 shows an incident light illumination arrangement for the investigation of reflective objects. The set-up on the right panel of Fig. 9.13 is designed to investigate transparent objects such as living cell cultures. In both cases the coherent laser light for the illumination of the sample is coupled into the optical path of the microscope's condenser by a beam splitter. The reference wave is superimposed onto the light reflected or transmitted by the object by a second beam splitter with a slight tilt against the object wave front. Thus, "off-axis" holograms are generated and recorded by a CCD camera. After hologram acquisition, the data is transmitted by an IEEE1394 ("FireWire") interface to a digital image processing system; thus bypassing the need for cost intensive frame grabber cards with hardware-specific software. This approach provides the advantage that commercial standard microscope optics with high numerical aperture (e.g., water immersion and oil immersion) can be used in combination with an optimized (Koehler-like) illumination of the sample. Furthermore, such a modular integration of the additional optical components for digital holography does not restrict the conventional function of the microscopy systems [28].

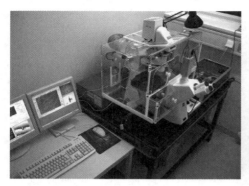 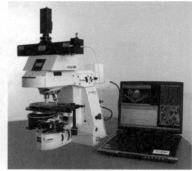

Fig. 9.14. (*Left*) Inverse microscope with P.A.L.M. MicroBeam and attached digital holographic microscopy module. The illumination of the sample is performed in transmission mode (built in cooperation with P.A.L.M. Microlaser Technologies GmbH, Bernried, Germany). An incubator allows temperature stabilized long-term investigations of livings cells. (*Right*) Upright fluorescence microscope with digital holographic microscopy module for illumination of the sample in transmission and reflection (built in cooperation with Carl Zeiss Jena GmbH, Germany) [28]

Figure 9.14 shows an inverse (left) and an upright (right) microscope with attached digital holographic microscopy modules. The typical hologram capturing time depends on the applied imaging device (here CCD cameras) and amounts within the range of 1 ms. The reconstruction rate of the digital holograms depends mainly on the capability of the applied image processing system and the size of the digitized holograms. With current computer systems (e.g., a Pentium IV 2.8 GHz) and the reconstruction algorithm described in Sect. 9.5.2 at a frame size of 512×512 pixels reconstruction rates up to $\sim$4 Hz can be achieved.

9.5.2 Nondiffractive Reconstruction

The reconstruction of the digitally recorded holograms is performed numerically with a computer. In general, Fresnel-transformation based digital holographic reconstruction methods generate not only the information contained in the object wave but in addition the intensity of the reference wave ("zero order") and a "twin image" [29]. Furthermore, the size of the reconstructed holographic image depends on the reconstruction distance to the hologram plane. Thus, in digital holographic microscopy, the reconstruction of the digitally captured holograms is performed by the application of a nondiffractive reconstruction method (NDRM) [30–34]. The intensity distribution $I_{\text{HP}}(x, y, z_0)$ in the hologram plane HP, located at $z = z_0$, is formed by the interference of the object wave $O(x, y, z = z_0)$ and the reference wave $R(x, y, z = z_0)$:

$$I_{\text{HP}}(x,y,z_0) = \text{O}(x,y,z_0)\text{O}^*(x,y,z_0) + \text{R}(x,y,z_0)\text{R}^*(x,y,z_0)$$
$$+ \text{O}(x,y,z_0)\text{R}^*(x,y,z_0) + \text{R}(x,y,z_0)\text{O}^*(x,y,z_0)$$
$$= I_\text{O}(x,y,z_0) + I_\text{R}(x,y,z_0)$$
$$+ 2\sqrt{I_\text{O}(x,y,z_0)I_\text{R}(x,y,z_0)}\cos\Delta\varphi_{\text{HP}}(x,y,z_0), \quad (9.7)$$

with $I_\text{O} = \text{OO}^* = |\text{O}|^2$ and $I_\text{O} = \text{RR}^* = |\text{R}|^2$ (* denotes the conjugate complex term). The parameter $\Delta\varphi_{\text{HP}}(x,y,z_0) = \phi_\text{R}(x,y,z_0) - \phi_\text{O}(x,y,z_0)$ is the phase difference between O and R at $z = z_0$. In the presence of a sample in the optical path of O, the phase distribution represents the sum $\phi_\text{O}(x,y,z_0) = \phi_{\text{O}_0}(x,y,z_0) + \Delta\varphi_\text{S}(x,y,z_0)$, where $\phi_{\text{O}_0}(x,y,z_0)$ denotes the pure object wave phase and $\Delta\varphi_\text{S}(x,y,z_0)$ represents the optical path length change that is effected by the sample. For areas without a sample, $\Delta\varphi_{\text{HP}}(x,y,z_0)$ is estimated by a mathematical model [31, 35]:

$$\Delta\varphi_{\text{HP}}(x,y,z_0) = \phi_\text{R}(x,y,z_0) - \phi_{\text{O}_0}(x,y,z_0)$$
$$= 2\pi\left(K_x x^2 + K_y y^2 + L_x x + L_y y\right). \quad (9.8)$$

The parameters K_x, K_y in (9.8) describe the divergence of the object wave and the properties of the applied microscopy lens. The constants L_x, L_y denote the linear phase difference between O and R due to the off-axis geometry of the experimental setup. For quantitative phase measurement from $I_{\text{HP}}(x,y,z_0)$ in a first step, the complex object wave $\text{O}(x,y,z=z_0)$ in the hologram plane is determined pixel wise by solving a set of equations that is obtained from insertion of (9.8) in (9.7). For that purpose, neighboring intensity values within a square area of 5×5 pixels around a given hologram pixel are considered by application of a spatial phase shifting algorithm (for details see [9] and [23]). The utilized algorithm is based on the assumption that only $\Delta\varphi_{\text{HP}}(x,y,z_0) = \phi_\text{R}(x,y,z_0) - \phi_{\text{O}_0}(x,y,z_0)$ between the object wave $\text{O}(x,y,z_0)$ and the reference wave $\text{R}(x,y,z_0)$ varies rapidly spatially in the hologram plane. In addition, because of the spatial phase shifting algorithm, the object wave's intensity has to be assumed constant within an area of about 5×5 pixels around a given point of interest of the hologram. These requirements can be fulfilled by an adequate relation between the magnification of the microscope lens and the image recording device. Therefore, the magnification of the microscope lens is chosen in such a way that the smallest imaged structures of the sample that are restricted by the resolution of the optical imaging system due to the Abbe criterion are over sampled by the CCD sensor. In this way the lateral resolution of the reconstructed holographic phase contrast images is not decreased by the spatial phase shifting algorithm [31].

The parameters K_x, K_y, L_x, L_y in (9.8) cannot be obtained directly from the geometry of the experimental setup with an adequate accuracy and for this reason are adapted once before the measurements by an iterative fitting process in an area of the hologram without sample [31, 34].

The evaluation of digital holographic phase contrast images requires, in correspondence to microscopy with white light illumination, a sharply focused image of the sample. For the case that the object is not imaged sharply in the hologram plane HP during the hologram recording process, e.g., due to mechanical instability of the experimental setup or thermal effects, in a second evaluation step a further propagation of the object wave to the image plane can be carried out for subsequent focus correction. The propagation of $O(x, y, z_0)$ to the image plane z_{IP} that is located at $z_{\text{IP}} = z_0 + \Delta z$ in the distance Δz to HP can be carried out by a Fresnel transformation [26, 31, 35] or by a convolution algorithm [34, 36, 37]:

$$O\left(x, y, z_{\text{IP}} = z_0 + \Delta z\right) = F^{-1}\left\{F\left\{O\left(x, y, z_0\right)\right\} \exp\left(i\pi\lambda\Delta z\left(\nu^2 + \mu^2\right)\right)\right\}. \tag{9.9}$$

In (9.9) λ is the applied laser light wave length, ν, μ are the coordinates in frequency domain and F denotes a Fourier transformation. During the propagation process, the parameter Δz is chosen in such a way that the holographic amplitude image appears sharply, in correspondence to a microscopic image under white light illumination. A further criterion for a sharp image of the sample is that diffraction effects due to the coherent illumination appear minimized in the reconstructed data. As a consequence of the applied algorithms and the parameter model for the phase difference model $\Delta\varphi_{\text{HP}}$ in (9.8), the resulting reconstructed holographic images do not contain the disturbing terms "twin image" and "zero order." In addition, the method allows in comparison to propagation by Fresnel transformation, as e.g., in [31, 35], a sharply focused image of the sample in the hologram plane. The propagation of O by (9.9) enables in this way the evaluation of image plane holograms containing a sharply focused image of the sample and effects no change of the image scale during subsequent refocusing. In the special case that the image of the sample is sharply focused in the hologram plane with $\Delta z = 0$ and thus $z_{\text{IP}} = z_0$, the reconstruction process can be accelerated because no propagation of O by (9.9) is required.

From $O(x, y, z_{\text{IP}})$, in addition to the absolute amplitude $|O(x, y, z_{\text{IP}})|$ that represents the image of the sample, the phase information $\Delta\varphi_{\text{S}}(x, y, z_0)$ of the sample is reconstructed simultaneously:

$$\Delta\varphi_{\text{S}}(x, y, z_0) = \phi_{\text{O}}\left(x, y, z_0\right) - \phi_{\text{O}_0}(x, y, z_0)$$
$$= \arctan\frac{\Im\left\{O\left(x, y, z_{\text{IP}}\right)\right\}}{\Re\left\{O\left(x, y, z_{\text{IP}}\right)\right\}} \quad (\text{mod}2\pi). \tag{9.10}$$

After removal of the 2π ambiguity by a phase unwrapping process [1], the data obtained by (9.10) can be applied for quantitative phase contrast microscopy, which is the main topic of interest for the described experiments. For an incident light geometry as depicted in the left panel of Fig. 9.13, the topography z_{s} can be calculated on the phase distribution $\Delta\varphi_{\text{S}}(x, y, z_0)$:

$$z_{\text{s}}(x, y, z_0) = \frac{\lambda\Delta\varphi_{\text{S}}(x, y, z_0)}{2.2\pi} = \frac{\lambda}{4\pi}\Delta\varphi_{\text{s}}(x, y, z_0). \tag{9.11}$$

9 Coherent Laser Measurement Techniques for Medical Diagnostics

For illumination in transmission the thickness of cells d_{cell} in cell culture medium with a homogenous refractive index n_{medium} can be determined by measuring the optical path length change $\Delta\varphi_{\text{cell}}$ of the cells to the surrounding medium:

$$d_{\text{cell}}(x,y,z_0) = \frac{\lambda \Delta\varphi_{\text{cell}}(x,y,z_0)}{2\pi} \frac{1}{n_{\text{cell}} - n_{\text{medium}}}, \qquad (9.12)$$

with the integral refractive index n_{cell} and the wave length λ of the applied laser light. For fully adherently grown cells, the parameter d_{cell} is estimated in first order to describe the shape of single cells [34, 38].

Figure 9.15 illustrates the evaluation process of digital recorded holograms. Figure 9.15a,b shows a digital hologram obtained from a living human pancreas carcinoma cell (Patu8988T) with an inverse microscope arrangement in transmission mode (×40 microscope lens, NA = 0.65) and the reconstructed holographic amplitude image that corresponds to a microscopic bright field image at coherent laser light illumination. Figure 9.15c depicts the simultaneously reconstructed quantitative phase contrast image modulo 2π. The unwrapped data without 2π ambiguity, representing the optical path length changes that are affected by the sample in comparison to the surrounding medium due to the thickness and the integral refractive index, are shown in Fig. 9.15d. Figure 9.15e depicts a pseudo 3D plot of the data in Fig. 9.15d. Figure 9.15f shows the cell thickness along the marked dashed line in Fig. 9.15d, which is determined by application of (9.12) with $n_{\text{cell}} = 1.38$ and $n_{\text{medium}} = 1.337$.

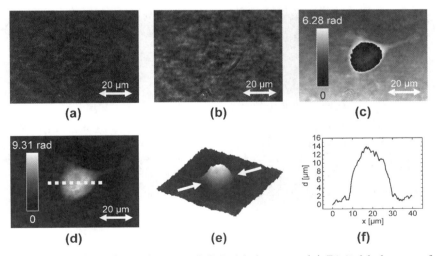

Fig. 9.15. Example for evaluation of digital holograms. (**a**) Digital hologram of a human pancreas carcinoma cell (Patu8988T); (**b**) Reconstructed holographic amplitude image; (**c**) Quantitative phase contrast image (mod 2π); (**d**) Unwrapped phase distribution; (**e**) Pseudo 3D plot of the unwrapped phase image in gray level representation; (**f**) Calculated cell thickness along the dashed white line in (**d**)

9.5.3 Resolution and Numerical Focus

The resolution of digital holographic microscopy can be characterized by investigating calibration test charts. The left panel of Fig. 9.16 shows the reconstructed amplitude (holographic image) of a negative USAF 1951 resolution test chart (illumination in transmission), recorded using a ×40 microscope lens (NA = 0.6). The magnified image section of group 9.5 (resolution limit of the test chart) represents a line width of 620 nm and is resolved clearly. The comparison with the Abbe criterion shows that the lateral resolution is diffraction limited (in correspondence to bright field microscopy) and can be increased by using microscope optics with higher numerical aperture [31]. The right panel of Fig. 9.16 demonstrates the axial resolution for incident light illumination of a reflective metal surface (nano-structured chromium on chromium surface) recorded with a ×5 microscope lens (NA = 0.1). The depicted elements represent a height of 30 nm and are resolved clearly in the reconstructed phase distribution. Because of the phase noise the axial resolution is determined to be ≈5 nm. The digital reconstruction of different object planes from a single hologram enables a variable (subsequent) numerical focus of digital holographic images ("digital holographic multi focus") without additional mechanical or optical components. Figure 9.17 demonstrates the digital holographic refocus for the example of a semitransparent USAF1951 test chart (upper panel) and for the holographic image of a living human pancreas tumor cell of the type Patu8988S (lower panel).

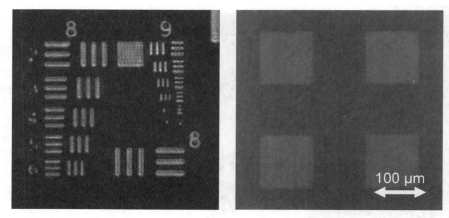

Fig. 9.16. (*Left*) Reconstructed holographic amplitude of a negative USAF 1951 test chart (digital hologram recorded in transmission mode), the elements of group 9.6 correspond to a lateral resolution of 550 nm; (**b**) Reconstructed phase contrast image of a reflective phase object (chromium on chromium sample) with steps of 30 nm in axial direction (digital hologram recorded in reflection mode) [28]

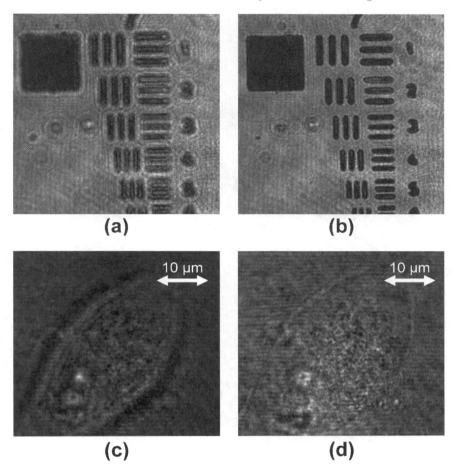

Fig. 9.17. Amplitudes reconstructed from single recorded holograms in different focus planes (*left*: unfocussed reconstruction in the hologram plane; *right*: numerically focused). (**a**),(**b**) Semitransparent USAF 1951 test chart (×40 microscope optic, NA = 0.6) and (**c**),(**d**) Living human pancreas tumor cell Patu8988S (×100 oil immersion optic, NA = 1.3) [28]

9.5.4 Digital Holographic Phase Contrast Microscopy of Living Cells

Investigations on living pancreas tumor cells [39, 40] were carried out to demonstrate the potential of digital holographic microscopy for the visualization of toxin-induced morphology changes. Therefore, pancreatic tumor cells were exposed at 37°C to taxol. Digital holograms of selected cells were recorded continuously over 4 h. Figure 9.18 shows the results. In the upper row of Fig. 9.18, the obtained unwrapped phase distributions at $t = 0$, $t = 78$,

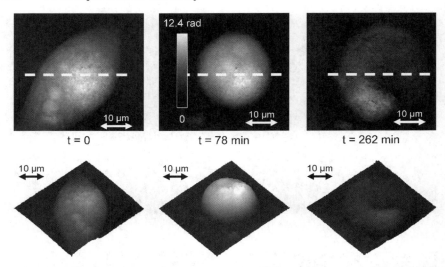

Fig. 9.18. Monitoring of a living PaTu8988S cell after adding a toxin (taxol) to the cell culture medium. Morphological changes such as cell rounding and finally the cell collapse are induced. *Upper row*: gray level coded unwrapped phase distribution at $t = 0$, $t = 78$, and $t = 262$ min after taxol addition. *Lower row*: Corresponding pseudo 3D representations of the phase data (Cooperation: Dr. Jürgen Schnekenburger, Department of Medicine B, University of Münster, Germany) [28]

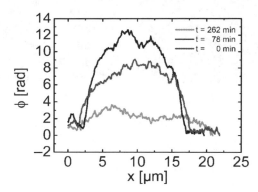

Fig. 9.19. Cross sections through the measured optical path length changes corresponding to the dashed white lines in the phase distributions of Fig. 9.18 (*upper row*) [28]

and $t = 262$ min after taxol addition are depicted. The lower row of Fig. 9.18 shows the pseudo 3D plot of the phase images. Figure 9.19 depicts cross sections through the measured optical path length changes corresponding to the dashed lines in the upper row of Fig. 9.18. From Figs. 9.18 and 9.19 it is clearly visible that the toxin induces morphological changes as cell rounding and finally the cell collapses.

9.6 Discussion and Conclusions

This contribution gives an overview of proximal and distal endoscopic ESPI systems as well as microscope (speckle) interferometric and digital holographic microscopy systems with regard to biomedical applications. The exemplarily shown results obtained from the application of the proximal and distal endoscopic ESPI systems on biological specimens demonstrate the ability of ESPI to detect displacements and differences in elasticity of biological tissues even underneath the visible surface. This could enhance conventional endoscopic investigations by the substitution of the missing tactile sense by a visual information ("endoscopic taction"). In addition, for distal endoscopic ESPI it has been shown that it is possible to develop compact speckle interferometric arrangements with a lateral resolution in microscopic range. The presented (speckle) interferometric microscopy setup provides a simple interferometer adjustment and a fast detection of microchanges even on living organisms. In connection with microscopy, digital holography allows a noncontact, fast, quantitative, minimal invasive, full field detection with high resolution (<5 nm in axial direction) by quantitative holographic phase contrast imaging of reflective and transparent samples, with the possibility to transform the obtained phase data into an illustrative pseudo 3D representation. In this way, digital holographic microscopy enables topography measurements on nanostructured surfaces, quantitative on-line detection of drug effected cellular thickness/shape variations.

In conclusion, holographic interferometric metrology provides versatile tools for medical minimally invasive diagnostics and technical intracavity inspection as well as for Life Cell Imaging in microscopy, Life Sciences, and Biophotonics.

Acknowledgments

Financial support by the German Ministry for Education and Research (BMBF) is gratefully acknowledged.

References

1. T. Kreis, *Holographic Interferometry: Principles and Methods* (Akademie Publishing, Berlin, 1996)
2. V.P. Shchepinov, V.S. Pisarev, *Strain and Stress Analysis by Holographic and Speckle Interferometry* (Wiley, Chichester, 1996)
3. M.A. Beek, W. Hentschel, Opt. Lasers Eng. **34**, 101 (2000)
4. B. Kemper, W. Avenhaus, D. Dirksen, A. Merker, G. von Bally, Appl. Opt. **39**, 3899 (2000)
5. W. Avenhaus, B. Kemper, G. von Bally, Gastrointest. Endosc. **54**, 496 (2001)
6. S. Schedin, G. Pedrini, H.J. Tiziani, Appl. Opt. **39**, 2853 (2000)

7. W. Avenhaus, B. Kemper, S. Knoche, D. Domagk, C. Poremba, G. von Bally, W. Domschke, Lasers Med. Sci. **19**, 223 (2005)
8. B. Kemper, D. Dirksen, J. Kandulla, G. von Bally, Opt. Commun. **194**, 75 (2001)
9. C. Depeursinge, in *Digital Holography and Three dimensional Display: Principles and Applications,* ed. by T.C. Poon (Springer, New York, 2006), p. 95
10. G. von Bally, B. Kemper et al., in *Biophotonics: Visions for Better Health Care,* ed. by J. Popp, M. Strehle (Wiley, Berlin, 2006), p. 301
11. J.W. Goodman, in *Laser Speckle and Related Phenomena,* ed. by J.C. Dainty (Springer, Berlin, 1975), p. 9
12. T. Bothe, J. Burke, H. Helmers, Appl. Opt. **36**, 5310 (1997)
13. B. Kemper, J. Kandulla, D. Dirksen, G. von Bally, Opt. Commun. **217**, 151 (2003)
14. K. Creath, in *Holographic Interferometry,* ed. by P.K. Rastogi (Springer, Berlin, 1994), p. 109
15. G. Pedrini, Y. Zou, H.J. Tiziani, Appl. Opt. **36**, 786 (1997)
16. T. Fricke-Begemann, J. Burke, Appl. Opt. **40**, 5011 (2001)
17. M. Takeda, H. Ina, S. Kobayashi, J. Opt. Soc. Am. **72**, 156 (1982)
18. G. von Bally, in *International Trends in Applied Optics,* ed. by A.H. Guenther (SPIE Press, 2002), p. 571
19. G. von Bally, Int. J. Optoelectronics **6**, 491 (1998)
20. S. Schedin, G. Pedrini, H.J. Tiziani, A.K. Aggarwal, Appl. Opt. **40**, 2692 (2001)
21. B. Kemper, S. Knoche, J. Kandulla, G. von Bally, VDI Rep. **1694**, 55 (2002)
22. S. Knoche, B. Kemper, J. Kandulla, G. Wernicke, G. von Bally, Opt. Commun. **270**, 68 (2007)
23. B. Kemper, S. Lai, A. Merker, G. von Bally, in *Series on Optics Within Life Sciences (OWLS V),* ed. by C. Fotakis, T. Papazoglou, C. Kalpouzos (Springer, Heidelberg, 2000), p. 29
24. O.J. Løkberg, Opt. Lasers Eng. **26**, 313 (1997)
25. G. Gülker, K.D. Hinsch, A. Kraft, Opt. Lasers Eng. **36**, 501 (2002)
26. U. Schnars, J. Opt. Soc. Am. A **11**, 2011 (1994)
27. U. Schnars, W. Jüptner, Meas. Sci. Technol. **13**, R85 (2002)
28. B. Kemper, D. Carl, A. Höink, G. von Bally, I. Bredebusch, J. Schnekenburger, Proc. SPIE **6191**, 61910 (2006)
29. T. Kreis, *Handbook of Holographic Interferometry: Optical and Digital Methods* (Wiley, New york, 2005)
30. P. Marquet, B. Rappaz, P.J. Magistretti, E. Cuche, Y. Emery, T. Colomb, C. Depeursinge, Opt. Lett. **30**, 468 (2005)
31. D. Carl, B. Kemper, G. Wernicke, G. von Bally, Appl. Opt. **43**, 6536 (2004)
32. M. Liebling, T. Blu, M. Unser, J. Opt. Soc. Am. A **21**, 367 (2004)
33. C.J. Mann, L. Yu, C.M. Lo, M.K. Kim, Opt. Exp. **13**, 8693 (2005)
34. B. Kemper, D. Carl, J. Schnekenburger, I. Bredebusch, M. Schfer, W. Domschke, G. von Bally, J. Biomed. Opt. **11**, 034005 (2006)
35. E. Cuche, P. Marquet, C. Depeursinge, Appl. Opt. **38**, 6694 (1999)
36. T.M. Kreis, M. Adams, W.P.O. Jüptner, Proc. SPIE **3098**, 224 (1997)
37. T. Demotrakooulos, R. Mittra, Appl. Opt. **13**, 665 (1974)
38. B. Rappaz, P. Marquet, E. Cuche, Y. Emery, C. Depeursinge, P.J. Magistretti, Opt. Exp. **13**, 9361 (2005)

39. J. Schnekenburger, J. Mayerle, B. Krger, I. Buchwalow, F.U. Weiss, E. Albrecht, V.E. Samoilova, W. Domschke, M.M. Lerch, Gut **54**, 1445 (2005)
40. H.P. Elsässer, U. Lehr, B. Agricola, H.F. Kern, Virchows Arch. B. Cell Pathol. Incl. Mol. Pathol. **61**, 295 (1992)

10

Biomarkers and Luminescent Probes in Quantitative Biology

M. Zamai, G. Malengo, and V.R. Caiolfa

The genome sequence of many organisms is now complete. "However, genome sequences alone lack spatial and temporal information and are therefore as dynamic and informative as census lists or telephone directories" [1]. We know the words of the genetic code, yet we need to associate each of them to a specific function. The challenge of this century is to figure out how proteins work together to make living cells and organisms. Proteins are essential for most biological processes, but they are not simply objects with chemically reactive surfaces, and it is not easy to understand their function. Proteins localize to specific environments in the cells: membranes, cytosol, organelles, or nucleoplasm, undergo diffusive or directed movement, and often are coupled to chemical events. The exceptional capability of the proteins to regulate virtually all dynamic processes in living cells depends on the fine regulation of their topology, movement, and chemistry. In fact, cells respond to stimuli through signaling pathways that orchestrate the recruitment and assembly of proteins into biomolecular *machines*. Roadmap[1] focuses on ways to solve the structures of protein machines. To achieve these goals, it is critical to understand how the cell architecture influences the formation of these protein complexes. Research is, therefore, turning to the study of protein function in their most natural context [2].

10.1 Fluorophores and Genetic Dyes

10.1.1 Small Organic Dyes and Quantum Dots

Small organic fluorophores (<1 kDa) for covalent labeling of macromolecules have been optimized for wavelength range, brightness (extinction coefficient for absorbance, fluorescence quantum yield) photostability, and reduction in self-quenching. Hundreds of such dyes are commercially available [3]. However,

---
[1] (http://www.nihroadmap.nih.gov)

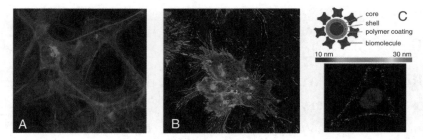

Fig. 10.1. Biomarkers gallery. (**a**) Confocal maximal projection of fixed HEK293 cells immunostained for actin (*red*) and stained for the nuclei with Hoechst 3342 (*blue*). Only one cell expresses EGFP-labeled uPAR (*green*). (**b**) Confocal maximal projection of fixed HEK293 stably transfected with EGFP-labeled uPAR (*green*), and stained with Cy5-uPA ligand (*red*). (**c**) (*Top*) Scheme of a quantum dot and (*bottom*) an example of application in cell imaging representing the distribution of GM1 gangliosides found in the plasma membrane in live HeLa cells; nucleus was stained with Hoechst 3342 (reproduced from [4])

these dyes lack specificity for any particular protein; most applications use antibodies in fixed and permeabilized cells or specifically labeled ligands (Fig. 10.1a,b).

Quantum dots (QDs) are inorganic nanocrystals that fluoresce at sharp and discrete wavelengths depending on their size [4–6]. Their extinction coefficients is 10–100 times higher than small fluorophores and fluorescent proteins (FPs). They also have good quantum yields and exceptional photostability that allow repeated imaging of single molecules [4]. Their absorbance extends from short wavelengths up to just below the emission wavelength, so that a single excitation wavelength readily excites QDs of multiple emission maxima. QDs typically contain a core of CdSe or CdTe and ZnS shell (Fig. 10.1c). For biological applications, a coating that makes QDs water soluble is necessary to prevent quenching by water, and allow conjugation to protein targeting molecules such as antibodies and streptavidin. The large size of QDs conjugated to biomolecules (10–30 nm) prevents efficient traversal of intact membranes, which restricts their use to permeabilized cells or extracellular or endocytosed proteins.

10.1.2 Fluorescent Proteins

Rapid advances in live-cell imaging technologies, combined with the use of genetically encoded fluorescent proteins (FPs), has resulted in a revolution in cell biology, as it is now possible to track the assembly of protein complexes within the organized microenvironment of the living cell. In the next sections, we discuss some of these advances, focusing on fluorescence imaging and spectroscopic techniques that are front-edge for the analysis of protein movement and interactions in living cells.

Since the discovery of the GFP in 1961 by Osamu Shimomura and colleagues [7], 30 years had to pass before the gene for GFP was cloned and the 238 amino acids sequence determined by Prasher [8]. Douglas Prasher was the first person to realize the potential of GFP as a biomarker for proteins in cells [9]. The manufacture of proteins using the instructions from the gene is called protein expression. Prasher envisioned that it would be possible to use biomolecular techniques to insert the GFP gene at the end of the gene of any protein, right before the stop codon (Fig. 10.2, left). Encoded in the DNA is some type of index that directs the molecular machinery to the start of the gene of each necessary protein. When new protein is required, the gene is read, and the protein is manufactured. At the end of the gene is a message called a stop codon, which ends protein production. When the cell needs to make that protein, it would go to the specific gene, use the information encoded in the gene to make it, but instead of stopping when the protein was made, this cell would carry on making GFP until it reached the stop codon at the end of the GFP gene. As a result, the cell would produce a "chimeric" (e.g., modified) protein with a GFP attached to it.

A variety of techniques have been developed to construct FP fusion products and enhance their expression in mammalian and other systems. The primary vehicles for introducing FP chimeric gene sequences into cells are genetically-engineered bacterial plasmids and virus that act as vectors and transfer the genetic information into the cell either transiently or stably (Fig. 10.2, right). In transient, or temporary, gene transfer experiments (often referred to as transient transfection), plasmid or viral DNA introduced into the host organism does not necessarily integrate into the chromosomes of the host cell, but can be expressed in the cytoplasm for a short period of

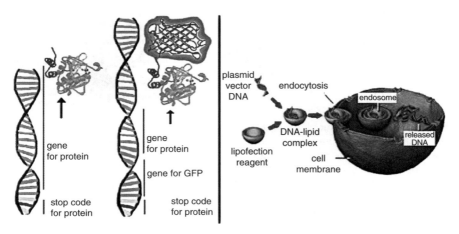

Fig. 10.2. *Left*: Fusion Construct. Adapted from [10]. *Right*: Lipid-mediated transfection in mammalian cells

time. Expression of gene fusion products usually takes place over a period of several hours after transfection and continues for 72–96 h after introduction of plasmid DNA into mammalian cells. In many cases, the plasmid DNA can be incorporated into the genome in a permanent state to form stably transformed cell lines. The choice of transient or stable transfection depends upon the target objectives of the investigation.

Structure of GFP

There are several reasons why GFP is a powerful biomarker. GFP is easy to detect by its fluorescence in cells; therefore, it is more versatile than most other bioluminescent molecules that require the addition of other substances before they glow. For example, firefly luciferase requires ATP, magnesium, and luciferin before it luminesces. It is a fairly small and compact protein (molecular weight, 27 kDa). The small size does not hinder the proper function of the protein to which it is attached, and particularly its intracellular trafficking and translocation. Among the most important aspects of the GFP to appreciate is that the entire 27 kDa native peptide structure is essential to the development and maintenance of its fluorescence. It is remarkable that the fluorophore derives from a triplet of adjacent amino acids: the serine, tyrosine, and glycine residues at locations 65, 66, and 67 (referred to as Ser65, Tyr66, and Gly67; in Fig. 10.3).

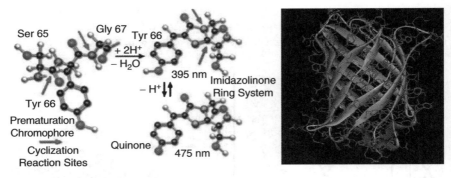

Fig. 10.3. GFP-chromophore and the β-barrel structure. Within the hydrophobic environment in the center of the GFP, a reaction occurs between the carboxyl carbon of Ser65 and the amino nitrogen of Gly67 that results in the formation of an imidazolin-5-one heterocyclic nitrogen ring system. Further oxidation results in conjugation of the imidazoline ring with Tyr66 and maturation of a fluorescent species. It is important to note that the native GFP fluorophore exists in two states. A protonated form that has an excitation maximum at 395 nm, and a unprotonated one that absorbs at approximately 475 nm. However, regardless of the excitation wavelength, fluorescence emission has a maximum peak wavelength at 507 nm, although the peak is broad and not well defined

Although this simple amino acid motif is commonly found throughout in nature, it does not generally result in fluorescence. What is unique to the fluorescent protein is that the location of this peptide triplet resides in the center of a remarkably stable barrel structure consisting of 11 β-sheets folded into a tube (Fig. 10.3). Since 1992, the FPs have become widely used as non-invasive markers in living cells, and their successful integration into variety of living systems illustrates that the expression of these proteins in cells is well tolerated. Virtually, any protein can be tagged with GFP, the resulting chimera often retains parent-protein targeting and function when expressed in cells, and therefore can be used as a fluorescent reporter to study protein dynamics.

The Future of Fluorescent Proteins

The GFP has been engineered to produce a vast number of variously colored mutants, fusion proteins, and biosensors that are broadly referred to as FPs. The GFP sequence has been modified for optimizing the expression in different cell types as well as the generation of GFP variants with more favorable spectral properties, including increased brightness, relative resistance to the effects of pH variation on fluorescence, and photostability. References [1, 3] illustrate in detail the biochemical and fluorescent properties of GFP-variants. The scientist who mostly contributed to understand how GFP works and developed new techniques and mutants is Roger Tsien. His group has obtained mutants that start fluorescing faster than wild type GFP, are brighter and have different colors (Fig. 10.4) [3].

The latest generation of jellyfish variants has solved most of the deficiencies of the first generation FPs. The search for a monomeric, bright, and fast-maturing red FP has resulted in several new and interesting classes of FPs, particularly those derived from coral species. Development of existing FPs, together with new technologies, such as insertion of unnatural amino acids, will further expand the color palette. The current trend in fluorescent probe technology is to expand the role of dyes that fluoresce into the far red and near infrared. In mammalian cells, both autofluorescence and the absorption of light are greatly reduced at the red end of the spectrum. Thus, the development of far red fluorescent probes would be extremely useful for the examination of thick specimens and entire animals. Given the success of FPs as reporters in transgenic systems (Fig. 10.5), the use of far red FPs in whole organisms will become increasingly important in the coming years. Finally, the tremendous potential in FP applications for the engineering of biosensors is just now being realized. The success of these endeavors certainly suggests that almost any biological parameter will be measurable using the appropriate FP-based biosensor.

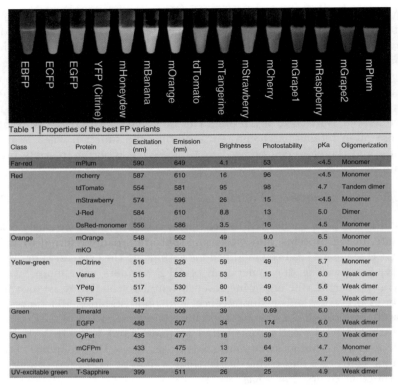

Fig. 10.4. Fluorescent protein palette. Engineered FPs cover the full visible spectrum. *Top*: Protein samples were purified from *E. coli* expression systems, excited at wavelengths up to 560 nm and photographed by their fluorescence. *Bottom*: Properties of the GFP variants (Adapted from [3])

Table 1 | Properties of the best FP variants

Class	Protein	Excitation (nm)	Emission (nm)	Brightness	Photostability	pKa	Oligomerization
Far-red	mPlum	590	649	4.1	53	<4.5	Monomer
Red	mcherry	587	610	16	96	<4.5	Monomer
	tdTomato	554	581	95	98	4.7	Tandem dimer
	mStrawberry	574	596	26	15	<4.5	Monomer
	J-Red	584	610	8.8	13	5.0	Dimer
	DsRed-monomer	556	586	3.5	16	4.5	Monomer
Orange	mOrange	548	562	49	9.0	6.5	Monomer
	mKO	548	559	31	122	5.0	Monomer
Yellow-green	mCitrine	516	529	59	49	5.7	Monomer
	Venus	515	528	53	15	6.0	Weak dimer
	YPetg	517	530	80	49	5.6	Weak dimer
	EYFP	514	527	51	60	6.9	Weak dimer
Green	Emerald	487	509	39	0.69	6.0	Weak dimer
	EGFP	488	507	34	174	6.0	Weak dimer
Cyan	CyPet	435	477	18	59	5.0	Weak dimer
	mCFPm	433	475	13	64	4.7	Monomer
	Cerulean	433	475	27	36	4.7	Weak dimer
UV-excitable green	T-Sapphire	399	511	26	25	4.9	Weak dimer

Fig. 10.5. Past and present of green fluorescent protein (GFP). 1961: The jellyfish *Aequorea* and its light-emitting organs. 1992: The amino acids sequence of GFP. 2007: Mouse with brain tumor expressing the fluorescent protein variant DsRed; the *zoomed area* shows the tumor mass (*grey*) and the blood vessels expressing GFP (*white*) that are growing in the tumor mass

10.2 Microspectroscopy in Quantitative Biology: Where and How

Static localization of the macromolecules and proteins in cells and tissues is not sufficient by itself to elucidate the key mechanisms that regulate fundamental processes of the cell. The activity of a biological molecule is not just a function of its structure but also of its *dynamic* behavior in the cell i.e., localization and interactions and their modifications in time. Thus, great importance has to be given to the spatio-temporal dynamics of the molecular interactions within cells in situ and, in particular, in vivo. In the following sections, we describe two spectro-microscopy approaches that, combined with the FPs chimera technology, are of growing importance in modern biology.

10.2.1 Fluorescence Correlation Spectroscopy

Fluorescence correlation spectroscopy (FCS) was introduced by Elson, Magde, and Webb in 1972 [11]. In 1990, Denk and Webb demonstrated a new type of microscope on the basis of two-photon excitation of molecules [12]. In 1995, Berland, So, and Gratton put together the two technologies, two-photon excitation microscopy and FCS, and demonstrated the potential of this methodology in intracellular measurements [13]. FCS is now a widely used technique. The reader is invited to refer to the many good articles published on the physics of the two-photon excitation process [14–16] and on the FCS technique [17–20]. Here, we discuss in specific the use of FCS in cell biology as it provides information about mobility (diffusion coefficients), concentration (number of particles), association (molecular brightness), and localization (image) of the target molecules. All this information helps understanding the complex molecular interaction networks, which are at the basis of cellular processes.

Chemical reactions and diffusion of molecules in solutions can be followed by a variety of methods but when we need to study them in living cells, most methods fail abruptly. The appeal of FCS is that, especially using two-photon excitation, living cells can be explored anywhere by a subfemtoliter volume without disrupting the cell integrity. The most stringent requirement for FCS to work is the possibility to observe the fluorescence signal in a small volume and at very high sensitivity and dynamic range (Fig. 10.6).

Only if the volume is so small at any instant of time, it might contain just one or few molecules. Because of these requirements, FCS is also known as a *single molecule* spectroscopy. If the number of fluorescent molecules in the volume does not change with time and if the quantum yield of the fluorophore is constant, then the average number of the emitted photon is constant. However, the instantaneous number of detected photon is not constant, because of the nature of the emission/detection process, which follows the Poisson statistics (^{2}). This added shot-noise is independent of time.

---
[2] The Poisson distribution describes events that occur rarely in a short period

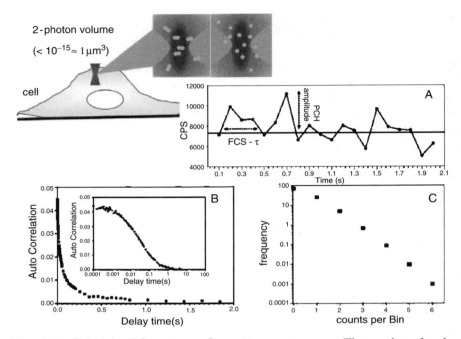

Fig. 10.6. Principle of fluorescence fluctuation spectroscopy. The number of molecules can change because of diffusion in and out of the volume, then fluorescence intensity fluctuates. (**a**) The time of the diffusion process causes characteristic frequencies to appear in the fluorescence intensity trace. (**b**) The time structure is analyzed by the autocorrelation function (ACF); *Inset*: original data are plotted in log scale. (**c**) The amplitude of the fluctuation depends on brightness (i.e., the aggregation state of a protein) and it can be analyzed by PCH

Instead, if the number of molecules in the excitation volume changes, even if the quantum yield is constant, the fluorescence intensity will change with time (Fig. 10.6a). The number of molecules can change because they can diffuse out of the volume. The time of the diffusion process causes characteristic frequencies to appear in the fluorescence intensity trace. Assume that we have two molecules in the volume and one leaves, the relative change in intensity will be one-half. However, if there are 10 molecules in the excitation volume

$$P(x,\mu) = \frac{e^{-\mu}\mu^x}{X!}, \qquad (10.1)$$

where $x = 0,1,2,3$, (number of emitted photons) and μ = mean number of successes in the given time interval or region of space (number of counts). If μ is the average number of successes occurring in a given time interval or region in the Poisson distribution, then the mean and the variance of the Poisson distribution are both equal to μ: $E(x) = m$, and $V(x) = \sigma\,e^2 = \mu$. In a Poisson distribution, only one parameter, μ, is needed.

and one of them leaves, the relative change will be only 1/10. This means that the smaller the ratio fluctuation/average-signal, the larger the number of molecules in the volume. It can be shown that this ratio is exactly proportional to the inverse of the number of molecules in the volume of excitation [21]. This relationship allows the measurements of the number of molecules in a given volume in the interior of cells [13, 22]. Molecular heterogeneity in the cell or molecular reactions can also induce the simultaneous presence of molecules of different kind in the same volume [23]. One important case is when proteins bind each other forming a molecular aggregate. Let us consider two identical proteins with one fluorescent probe each forming a molecular dimer. The dimer is different than the monomer because it carries twice the number of fluorescent moieties (Fig. 10.6). When one of these aggregates enters the volume of excitation, it will cause a larger fluctuation of the intensity than a single monomer. Clearly, the amplitude of the fluctuation carries information on the brightness of the diffusing particle (i.e., monomer vs. dimer). Thus, both time and amplitude structures are affected by underlying molecular species, and the dynamic processes that cause the change of the fluorescence intensity. The statistical analysis of the fluctuations of the fluorescence signal has to recover both information (Fig. 10.6a).

Analysis of the Time Structure of the Fluctuating Signal

The analysis of the time structure of the fluctuating signal is typically done by autocorrelation analysis. The autocorrelation function, ACF = $G(\tau)$, characterizes the time-dependent decay of the fluorescence fluctuations to their equilibrium value (Fig. 10.6b). In simple terms, ACF calculates the similarity between a signal $I(t)$, and a copy of the same signal shifted by a time lag τ, $I(t + \tau)$:

$$G(\tau) = \frac{\langle I(t)I(t+\tau)\rangle - \langle I(t)\rangle^2}{\langle I(t)\rangle^2}. \tag{10.2}$$

The autocorrelation function yields two parameters: the diffusion coefficient (D) and the average number of particles in the observation volume $<N>$ given by the inverse of $G(0)$, multiplied by a constant that depends on the illumination profile. In the case of identical molecules undergoing random diffusion in a Gaussian illuminated volume, the characteristic autocorrelation function is given by the following expression [22]:

$$G(\tau) = \frac{\gamma}{N}\left(1 + \frac{8D\tau}{\omega_r^2}\right)^{-1}\left(1 + \frac{8D\tau}{\omega_a^2}\right)^{-1/2}, \tag{10.3}$$

where D is the diffusion coefficient; ω_r and ω_a are the beam waist in the radial and in the axial directions; N is the number of molecules; γ is the numerical factor that accounts for the nonuniform illumination of the volume; and τ is the delay time. Other formulas have been derived for the Gaussian-Lorentzian illumination profile [13] and for molecules diffusing on a membrane [21].

Analysis of the Amplitude Structure of the Fluctuating Signal

The expression for the statistics of the amplitude fluctuations is generally given under the form of the histogram of the photon counts for a given sampling time Δt. This is known as the photon counting histogram (PCH) distribution. The PCH analysis calculates the probability of detecting photons per sampling time [24, 25]. The probability is experimentally determined by the histogram of the detected photons. In principle, the Poisson distribution describes the occupation number of particles that can freely go in and out a small excitation volume. However, mainly due to the diffusion of molecules in the inhomogeneous excitation volume, the distribution of photon counts deviates from the Poisson distribution. The analysis of the distribution of photons is based on the deviation of the measured PCH from the expected Poisson distribution because of the molecule occupation number (Fig. 10.7). Two parameters characterize the photon distribution: the number of molecules in the observation volume (N) and the molecular brightness (ϵ), which is defined as the average number of detected photons per molecule per second. The analytical expression for the PCH distribution for a single molecular species of a given brightness has been derived for the 3D-Gaussian illumination profile [24] and is reported below:

$$P_{3DG}(k; V_0; \varepsilon) = \frac{1}{V_0} \frac{\pi \omega_0^2 z_0}{2k!} \int_0^\infty \gamma\left(k, \varepsilon e^{-4x^2}\right) dx, \tag{10.4}$$

where V_0 is the volume of illumination; ϵ is the brightness of the molecules (counts per seconds per molecule); and k is the number of photons in a give time interval. The integral that contains the incomplete gamma function (γ) can be numerically evaluated. Similar expressions have been derived for other shapes of the illumination volume [24].

PCH is extremely useful for analysis fluctuations inside cells, where the target protein can associate in clusters or bind to some other preexisting aggregates. Diffusion coefficients for proteins in solution depend on their molecular weight ($\tau_{\text{diff}} \propto D^{-1} \propto M^{1/3}$). However, in the case of monomer–dimer species, the difference between the diffusion coefficients of the two species is

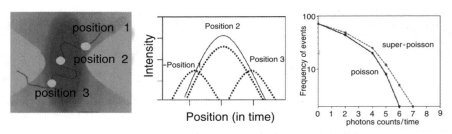

Fig. 10.7. Super-Poissonian distribution of detected photons from diffusing fluorophores due to (1) fluctuation of the particles number (Poissonian), (2) no uniform volume of the PSF, and (3) photon detection statistics (Poissonian)

just 1.2. It can be shown that a mass ratio of at least 5 is necessary to detect significant differences in the diffusion coefficient. Instead the brightness can change by a factor of 2 when the equilibrium moves from all monomers to all dimers (Fig. 10.8).

The all idea of FCS-PCH is the detection of fluorescence intensity fluctuations that can be achieved reducing the observation volume to subfemtoliters but also having few molecules in it. In practice this means that we need concentrations of the fluorophore in the nanomolar range. The consequences are twofold: (1) To study binding events, the dissociation constant of the com-

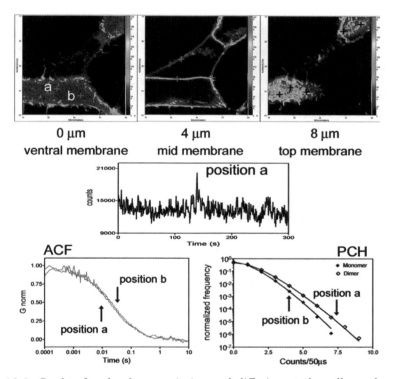

Fig. 10.8. Study of molecular association and diffusion at the cell membrane of the $\Delta D1$ form of the GPI-anchored uPAR, fused to EGFP and transfected in human kidney cells, HEK293. *Top panels*: Two-photon images of live cells showing the distribution of the receptor at the ventral, mid, and topside of the cell membrane; fluorescence in pseudo-color scale from low (blue) to high (red) intensity; (+): regions at which fluorescence fluctuation was recorded. *Mid panel*: A representative fluorescence fluctuation trace acquired in position (**a**). *Bottom panels*: Representative ACF and PCH for two fluorescence fluctuation traces acquired in regions (**a**) and (**b**). PCH indicates that the mutant receptor associates in oligomers (brightness $b = 3,900$ cpsm, brightness $a = 9,100$ cpsm). The ACFs of the two forms, however, do not reveal detectable differences in diffusion coefficient. Control experiments in solution have shown that the EGFP variant, lacking the receptor uPAR domains, does not dimerise at the concentrations found in the cells

plexes must be in the nanomolar range or below; (2) The expression in vivo of the FP-proteins systems must be low. This is not a trivial issue as many proteins are segregated in intracellular and organelle compartments often at local concentration higher than nanomolar. FCS-PCH criteria are opposite to those of classical confocal or multiphoton fluorescence microscopy and a systematic optimization of the expression levels of the FP-tagged proteins is necessary.

10.2.2 Fluorescence Lifetime Imaging (FLIM)

The development of confocal and multiphoton fluorescence microscopy has introduced enormous prospectives in biomedical sciences. These microscopies permit a variety of high-contrast and multidimensional imaging. However, the fluorescence of organic molecules is not only characterized by the emission spectrum, but also possesses a distinctive lifetime. The fluorescence lifetime is defined as the mean amount of time a fluorophore spends in the excited state after the absorption of an excitation photon. Lifetime can be measured by the time domain or by the frequency domain techniques. In the time domain (Fig. 10.9a–b), a short pulse of light excites the sample, and the subsequent fluorescence emission is recorded as a function of time. This usually occurs on the nanosecond timescale. In the frequency domain (Fig. 10.9c–d), the sample is excited by a modulated source of light. The fluorescence emitted by the sample has a similar waveform, but is modulated and phase-shifted from the excitation curve. Both modulation (M) and phase-shift ϕ are determined by the lifetime of the sample emission; that lifetime can be calculated from the observed demodulation ratio and phase-shift. Both of these domains yield equivalent data, and take advantage of the fluorescence decay law, which is based on the first-order kinetics. The decay law postulates that if a population of molecules is instantaneously excited when photons are absorbed, then the excited population, and hence the fluorescence intensity as a function of time, $I(t)$, gradually decays to the ground state. Decay kinetics can be described by: $I(t) = \alpha e^{\frac{-t}{\tau_f}}$; where α is the intensity at time $t = 0$, t is the time after the absorption, and τ_f is the lifetime, that is, when the fraction of the population of molecules in the excited state (and the fluorescence intensity) has decreased by a factor of $1/e$. Note that before absorption, $I(t) = 0$.

Time Domain

Time-domain measurements are based on the assumption that, when photons are absorbed, the molecules can be excited in an infinitely brief moment. This idea is commonly known as the delta or delta-pulse. The delta-pulse idea is used to interpret data obtained with real-pulsed light sources with measurable pulse-widths. In practice, the time-dependent profile of the light-pulse is reconvolved with the decay-law function. Reconvolution assumes that the delta-pulses are continuous functions, so that the observed decay is the convolution integral of the decays from all delta-pulses initiated during the finite

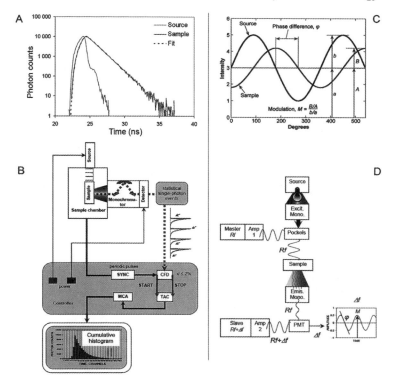

Fig. 10.9. Lifetime measurements in time and frequency domains. (a) Actual pulsed light-source (*gray*) and sample response (*black*), showing the gradual decay of fluorescence intensity with time. In this example, the decay could be fitted with a mono-exponential curve (*dotted*) giving a lifetime of 1.309 0.003 ns. (b) TCSPC-fluorometer: A pulsed light source excites the sample repetitively. The sample emission is observed by a detector, while the excitation flashes are detected by a synchronization module (SYNC). A constant fraction discriminator (CFD) responds to only the first photon detected (*small arrows*), independent of its amplitude, from the detector. This first photon from sample emission is the stop signal for the time-to-amplitude converter (TAC). The excitation pulses trigger the start signals. The multichannel analyzer (MCA) records repetitive start–stop signals of the single-photon events from the TAC, to generate a histogram of photon counts as a function of time channel units. The lifetime is calculated from this histogram. (c) Excitation (*black*) and sample response (*gray*), illustrating the phase-angle shift (ϕ) and demodulation ratio (M). (d) Multifrequency cross-correlation fluorometer. An unmodulated light source emits a spectrum of continuous-wave light. The excitation monochromator (Excit. Mono.) selects an excitation wavelength. An amplified (Amp 1) master synthesizer (Master) drives the Pockels cell (Pockels) at a base frequency, Rf, which modulates the excitation beam. The modulated beam excites the Sample, causing the sample to emit modulated fluorescence also at the base Rf. An emission monochromator (Emis. Mono.) selects one wavelength of modulated fluorescence. The photomultiplier tube (PMT) is modulated by an amplified (Amp 2) slave synthesizer (Slave) at the base Rf plus a low-frequency cross-correlation note (f). The sample emission at Rf cancels the slave $Rf + f$ frequencies to yield the f signal containing the same phase-angle shift (ϕ) and demodulation ratio (M) as the Rf fluorescence

pulse-width. The method of time-correlated single-photon counting (TCSPC) is, by far, a superior method for measuring time-domain decays (Fig. 10.9b). TCSPC uses a pulsed light-source and a circuit to detect single-photon events at a detector. In a repetitive series of many start–stop signals from the circuitry, a binned histogram in time channels of single-photon counts is gradually generated. TCSPC relies on a principle of Poissonian statistics that only one photon can be counted at a time and in any one channel, to avoid skewing the time-dependent statistics in photon-pile-up. Pile-up thus limits the data-acquisition rate of TCSPC to a few (typically 12) percent of the repetition rate. In practice, the single-photon limit is not a major hindrance because the pile-up limit can be monitored during the experiment, and decay times with sufficient photon counts in can be obtained in seconds to minutes with repetition rates in the megahertz range. In addition, the Poissonian nature of the statistics allows the data to be rigorously analyzed.

Frequency Domain

The fluorescence decay parameters in the decay law impulse function may be obtained on the basis of the relation of a sinusoidally modulated excitation beam to the fluorescence emission response (Fig. 10.9c). The emission occurs at the same frequency as the excitation. Because of the loss of electron energy (Stokes shift) between excitation and emission, the emission waveform is demodulated and phase-shifted in comparison to the excitation. Thus the demodulation ratio (M) and phase-angle shift (ϕ) constitute two separate observable parameters that are both directly related, via a Fourier transformation, to the initial fluorescence intensity, α, and lifetime, τ for a population of fluorophores. Frequency-domain measurements are best performed using multi-frequency cross-correlation phase-and-modulation (MFCC), shown in Fig. 10.9d. A modulated beam excites the sample. The fluorescence emission is detected by a photomultiplier tube (PMT) modulated at the same base radio-frequency as the master plus a low cross-correlation frequency (a few hertz). The base-frequency signals are filtered to reveal the cross-correlation frequency signal, which contains all the same demodulation (M) and phase angle shift (ϕ) information as the fluorescence emission.

Fluorescence Lifetime Imaging Microscopy (FLIM) and Förster Resonance Energy Transfer (FRET)

Fluorescence lifetime imaging microscopy (FLIM) is a technique to map the spatial distribution of nanosecond excited state lifetimes within microscopic images. FLIM has proven to be a robust and established technique in modern cell biology for lifetime contrast in ion-imaging, quantitative imaging, tissue characterization, and medical applications [26–28], and it is especially suitable for in situ protein–protein interaction studies using Förster Resonance Energy Transfer (FRET) [29–31]. FRET is a photophysical phenomenon in which energy is transferred from the first excited electronic state (S1) of a fluorophore (called donor D) to another nearby absorbing (but not necessarily emitting) molecule (called acceptor A). Thus, there is a concerted quenching of D and activation of A fluorescence (Fig. 10.10). For this reason, the acronym FRET is often used to designate *fluorescence* resonance energy transfer. The process involves the resonant coupling of emission and absorption dipoles and is thus nonradiative. That is, it competes with other radiative (fluorescence) and nonradiative pathways for deactivation, resulting in a decrease of the donor lifetime. The energy transfer rate from the donor to the acceptor decreases with the sixth power of the distance and thus is apparent only at distances shorter than 10 nm [32]. At the critical distance where 50% of the donor energy is transferred to an acceptor, the Förster radius, the donor emission and fluorescent lifetime are each reduced by 50%, and sensitized emission (acceptor emission specifically under donor excitation) is increased.

Because of its utility in reporting nanometer-scale interactions, FRET has become an important tool in cell biology [33–38]. FRET in cell biology is used commonly to verify whether labeled proteins physically interact: by measuring the FRET *efficiency*, distances on the nanometer scale (a scale within the globular radii of proteins) can thus be estimated using a light microscope. By measuring these effects, FRET microscopic imaging can verify close molecular associations between colocalized donor and acceptor-labeled fusion proteins that are far beyond the resolution of fluorescent microscopy. Obvious difficulties in intensity-based FRET measurements in cells is that the concentrations of the donor and acceptor are variable and unknown, the emission band on the donor extends into the absorption and emission band of the acceptor, and the absorption band of the acceptor commonly extends into the absorption band of the donor. A further complication is that only a fraction of the

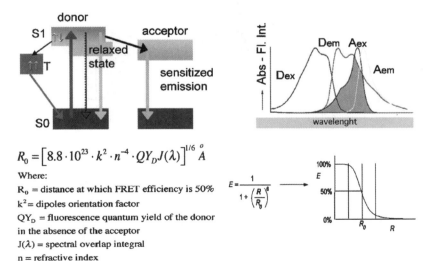

Fig. 10.10. The principle of FRET and dependence of FRET efficiency on donor–acceptor distance

donor molecules interacts with an acceptor molecule. These effects are hard to distinguish in intensity-based FRET measurements [39, 40]. In contrast, FLIM-based FRET techniques have the benefit that the results are obtained from a single-lifetime measurement of the donor. These approaches do not need calibration in different cells, and are nondestructive. The important difference between TCSPC and frequency domain FLIM resides in the source of excitation. In frequency domain FLIM (Fig. 10.9c) as the sample is excited by a sinusoidally modulated light, only single photon excitation can be applied, which in living cells induced major photobleaching effects during lifetime scans. Alternatively TCSPC-FLIM takes advantage of the two-photon laser subnanosecond pulses (Fig. 10.9a).

However, the principal strength of TCSPC–FLIM is the statistical accuracy, and this is reflected in the ability to fit two or sometimes three exponential functions to a particular fluorescence decay. The advantage is that it is often possible (depending on the quality of the data acquired) to determine the ratio of FRET vs. non-FRET lifetimes contained within a single pixel of an image, and therefore to generate a map of interacting vs. noninteracting proteins in the cell (Fig. 10.11). These results may reflect the relative bound and unbound fractions within a particular protein–protein interaction reaction. It is difficult or impossible to obtain these results directly using any other FRET or FLIM approach. Importantly for biochemists, these parameters are directly available using TCSPC–FLIM, and if the concentrations of the interacting proteins can be estimated (this is no mean feat inside cells, but it can be done), then estimates of association and dissociation constants can be

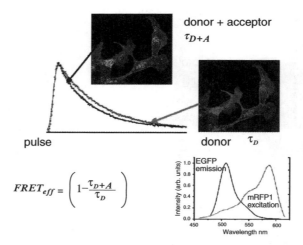

Fig. 10.11. FRET measured by TCSFC–FLIM. Representative fluorescence decays are shown for cells transfected with the donor (GPI–uPAR–EGFP) and cell cotransfected with donor and acceptor receptors (GPI–uPAR–EGFP, GPI–uPAR–mRGFP1). FRET efficiency is derived from the ratio between quenched and unqueched lifetimes

made. The resolution of a biexponential FRET system relies in part on the correct selection of fluorophore, particularly for the donor.

Many FPs have been shown to have complex fluorescence decays, often biexponential or even more complex, presumably due to protonation or some internal relaxation process, even in a non-FRET system. An example of this is ECFP (enhanced cyan fluorescent protein). This is commonly used as a FRET donor, because of its high quantum yield, long emission tail, and relatively blue emission. However, several fluorescence lifetime studies have revealed ECFP to have a biexponential lifetime [41, 42], making it difficult to resolve the interacting fraction from a noninteracting fraction in a FRET system (using FLIM); to do this, four lifetimes would need to be fitted to the data (i.e., two for the non-FRET system, and two for the FRET system), and it is beyond most FLIM measurements to acquire accurate enough data in a practical time period. However, there are FPs useful as FRET donors in FLIM measurements. EGFP (enhanced GFP) has a high quantum yield and a monoexponential decay in a non-FRET system, and has been shown to be a good donor for red FPs, such as dsRed and mRFP [monomeric RFP (red fluorescent protein)] [43] (Fig. 10.12). One advantage of EGFP for many cell

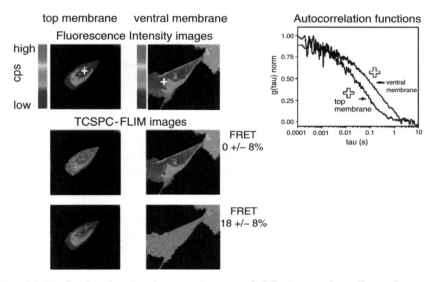

Fig. 10.12. Study of molecular association and diffusion at the cell membrane of the active GPI-anchored uPAR, fused to EGFP and transfected in human kidney cells, HEK293. *Top panels*: two-photon images of live cells showing the distribution of the receptor at the ventral and topside of the cell membrane; Representative ACF acquired in two regions at the top and ventral membrane (+) showing that the receptor has longer diffusion time at the cell adhesion site (about 0.9 and 0.1 µm²/s respectively). *Lower panels*: FLIM analysis of cells cotransfected with the GFP-mRFP1 receptor pair showing that the regions where the receptor is slow diffusing are also regions of high FRET efficiency

biologists is that it has been in use for many years, so cDNA constructs are often already made. More recently, the drive for a mono-exponentially decaying fluorescent protein led to the development of Cerulean, a variant of ECFP with all the advantages of a cyan donor, but a monoexponential fluorescent decay [44]. FLIM is still an emerging technology, of great interest to the cell biologist. One reason for lack of use of FLIM in biology laboratories is the technical difficulty associated with making reliable measurements. Although there are a number of different approaches to FLIM, some more *turn-key* than others, TCSPC can resolve additional parameters that make the technical difficulties worthwhile. In particular, the ability to make statistically accurate estimates of the bound vs. unbound fractions of fusion proteins inside living cells ought to be a very attractive prospect for cell biologists and biochemists.

10.2.3 Glossary of Molecular Biology

Vector: The DNA "vehicle" used to carry experimental DNA and to clone it. The vector provides all sequences essential for replicating the test DNA. Typical vectors include plasmids, cosmids, phages, and YACs.

Plasmid: A circular piece of DNA present in bacteria or isolated from bacteria. *Escherichia coli*, the usual bacteria in molecular genetics experiments, has a large circular genome, but it will also replicate smaller circular DNAs as long as they have an "origin of replication". Plasmids may also have other DNA inserted by the investigator. A bacterium carrying a plasmid and replicating a million-fold will produce a million identical copies of that plasmid. Common plasmids are pBR322, pGEM, pUC18.

Transfection: A method by which experimental DNA may be put into a cultured mammalian cell. Such experiments are usually performed using cloned DNA containing coding sequences and control regions (promoters, etc.) to test whether the DNA will be expressed. As the cloned DNA may have been extensively modified (for example, protein-binding sites on the promoter may have been altered or removed), this procedure is often used to test whether a particular modification affects the function of a gene.

Transient Transfection: When DNA is transfected into cultured cells, it is able to stay in those cells for about 2–3 days, but then will be lost (unless steps are taken to ensure that it is retained – see stable transfection). During those 2–3 days, the DNA is functional, and any functional genes it contains will be expressed. Investigators take advantage of this transient expression period to test gene function.

Stable Transfection: A form of transfection experiment designed to produce permanent lines of cultured cells with a new gene inserted into their genome. Usually, this is done by linking the desired gene with a "selectable" gene, i.e., a gene that confers resistance to a toxin (like G418, aka Geneticin). Upon putting the toxin into the culture medium, only those cells that incorporate the resistance gene will survive, and essentially all of those will also have incorporated the experimenter's gene.

Mutant Type Protein: Mutants are proteins lacking one or more biological activities of the wild type parent protein.

References

1. R.Y. Tsien, Building and breeding molecules to spy on cells and tumors. FEBS Lett. **579**(4), 927–932 (2005)
2. R.N. Day, F. Schaufele, Imaging molecular interactions in living cells. Mol. Endocrinol. **19**(7), 1675–1686 (2005)
3. N.C. Shaner, P.A. Steinbach, R.Y. Tsien, A guide to choosing fluorescent proteins. Nat. Meth. **2**(12), 905–909 (2005)
4. X. Michalet, F.F. Pinaud, L.A. Bentolila, J.M. Tsay, S. Doose, J.J. Li, G. Sundaresan, A.M. Wu, S.S. Gambhir, S. Weiss, Quantum dots for live cells, in vivo imaging, and diagnostics. Science **307**(5709), 538–544 (2005)
5. F. Pinaud, X. Michalet, L.A. Bentolila, J.M. Tsay, S. Doose, J.J. Li, G. Iyer, S. Weiss, Advances in fluorescence imaging with quantum dot bio-probes. Biomaterials **27**(9), 1679–1687 (2006)
6. J.K. Jaiswal, S.M. Simon, Potentials and pitfalls of fluorescent quantum dots for biological imaging. Trends Cell Biol. **14**(9), 497–504 (2004)
7. O. Shimomura, F.H. Johnson, Y. Saiga, Extraction, purification and properties of aequorin, a bioluminescent protein from the luminous hydromedusan, aequorea. J. Cell Comp. Physiol. **59**, 223–239 (1962)
8. D.C. Prasher, V.K. Eckenrode, W.W. Ward, F.G. Prendergast, M.J. Cormier, Primary structure of the aequorea victoria green-fluorescent protein. Gene **111**(2), 229–233 (1992)
9. M. Chalfie, Y. Tu, G. Euskirchen, W.W. Ward, D.C. Prasher, Green fluorescent protein as a marker for gene expression. Science **263**(5148), 802–805 (1994)
10. Marc Zimmer, *Glowing Genes: A Revolution in Biotechnology* (Prometheus Books, New York, 2005)
11. D. Magde, E. Elson, W.W. Webb, Thermodynamic fluctuations in a reacting system: measurement by fluorescence correlation spectroscopy. Phys. Rev. Lett. **29**, 705–708 (1972)
12. W. Denk, J.H. Strickler, W.W. Webb, Two-photon laser scanning fluorescence microscopy. Science **248**(4951), 73–76 (1990)
13. K.M. Berland, P.T. So, E. Gratton Two-photon fluorescence correlation spectroscopy: Method and application to the intracellular environment. Biophys. J. **68**(2), 694–701 (1995)
14. G.J. Brakenhoff, M. Muller, R.I. Ghauharali, Analysis of efficiency of two-photon versus single-photon absorption of fluorescence generation in biological objects. J. Microsc. **183**(Pt 2), 140–144 (1996)
15. W. Denk, K. Svoboda, Photon upmanship: Why multiphoton imaging is more than a gimmick. Neuron **18**(3), 351–357 (1997)
16. B.R. Masters, P.T. So, K.H. Kim, C. Buehler, E. Gratton, Multiphoton excitation microscopy, confocal microscopy, and spectroscopy of living cells and tissues; functional metabolic imaging of human skin in vivo. Meth. Enzymol. **307**, 513–536 (1999)

17. S. Breusegem, N. Barry, Q. Ruan, E. Ruan Gratton, J. Eid, in *Fluctuation correlation spectroscopy in cells: Determination of molecular aggregation*, ed. by Kluwer and Plenum Publishers. Biophotonics-Optical Science and Engineering for the 21st Century (Kluwer, Dordecht, 2004), pp. 1–16
18. S.T. Hess, W.W. Webb, Focal volume optics and experimental artifacts in confocal fluorescence correlation spectroscopy. Biophys. J. **83**(4), 2300–2317 (2002)
19. Y. Chen, J.D. Muller, K.M. Berland, E. Gratton, Fluorescence fluctuation spectroscopy. Methods **19**(2), 234–252 (1999)
20. N.L. Thompson, A.M. Lieto, N.W. Allen, Recent advances in fluorescence correlation spectroscopy. Curr. Opin. Struct. Biol. **12**(5), 634–641 (2002)
21. N.L. Thompson, in *Fluorescence Correlation Spectroscopy*, ed. by J.R. Lakowicz. Topics in Fluorescence Spectroscopy (Plenum, New York, 1991), pp. 337–378
22. H. Qian, E.L. Elson, Distribution of molecular aggregation by analysis of fluctuation moments. Proc. Natl. Acad. Sci. USA **87**(14), 5479–5483 (1990)
23. A.G. Palmer III, N.L. Thompson, Molecular aggregation characterized by high order autocorrelation in fluorescence correlation spectroscopy. Biophys. J. **52**(2), 257–270 (1987)
24. Y. Chen, J.D. Muller, P.T. So, E. Gratton, The photon counting histogram in fluorescence fluctuation spectroscopy. Biophys. J. **77**(1), 553–567 (1999)
25. J.D. Muller, Y. Chen, E. Gratton, Resolving heterogeneity on the single molecular level with the photon-counting histogram. Biophys. J. **78**(1), 474–486 (2000)
26. E.B. van Munster, T.W. Gadella, Fluorescence lifetime imaging microscopy (FLIM). Adv. Biochem. Eng. Biotechnol. **95**, 143–175 (2005)
27. D. Elson, J. Requejo-Isidro, I. Munro, F. Reavell, J. Siegel, K. Suhling, P. Tadrous, R. Benninger, P. Lanigan, J. McGinty, C. Talbot, B. Treanor, S. Webb, A. Sandison, A. Wallace, D. Davis, J. Lever, M. Neil, D. Phillips, G. Stamp, P. French, Time-domain fluorescence lifetime imaging applied to biological tissue. Photochem. Photobiol. Sci. **3**(8), 795–801 (2004)
28. I. Munro, J. McGinty, N. Galletly, J. Requejo-Isidro, P.M. Lanigan, D.S. Elson, C. Dunsby, M.A. Neil, M.J. Lever, G.W. Stamp, P.M. French, Toward the clinical application of time-domain fluorescence lifetime imaging. J. Biomed. Opt. **10**(5), 051403 (2005)
29. Y. Chen, J.D. Mills, A. Periasamy, Protein localization in living cells and tissues using FRET and FLIM. Differentiation **71**(9–10), 528–541 (2003)
30. Y. Chen, A. Periasamy, Characterization of two-photon excitation fluorescence lifetime imaging microscopy for protein localization. Microsc. Res. Tech. **63**(1), 72–80 (2004)
31. H. Wallrabe, A. Periasamy, Imaging protein molecules using FRET and FLIM microscopy. Curr. Opin. Biotechnol. **16**(1), 19–27 (2005)
32. L. Stryer, Fluorescence energy transfer as a spectroscopic ruler. Annu. Rev. Biochem. **47**, 819–846 (1978)
33. T.C. Voss, I.A. Demarco, R.N. Day, Quantitative imaging of protein interactions in the cell nucleus. Biotechniques **38**(3), 413–424 (2005)
34. R.N. Day, D.W. Piston, Spying on the hidden lives of proteins. Nat. Biotechnol. **17**(5), 425–426 (1999)

35. P.I. Bastiaens, A. Squire, Fluorescence lifetime imaging microscopy: spatial resolution of biochemical processes in the cell. Trends Cell Biol. **9**(2), 48–52 (1999)
36. R. Pepperkok, A. Squire, S. Geley, P.I. Bastiaens, Simultaneous detection of multiple green fluorescent proteins in live cells by fluorescence lifetime imaging microscopy. Curr. Biol. **9**(5), 269–272 (1999)
37. A. Squire, P.I. Bastiaens, Three dimensional image restoration in fluorescence lifetime imaging microscopy. J. Microsc. **193**(Pt 1), 36–49 (1999)
38. A. Periasamy, R.N. Day, Visualizing protein interactions in living cells using digitized GFP imaging and FRET microscopy. Meth. Cell Biol. **58**, 293–314 (1999)
39. Y. Gu, W.L. Di, D.P. Kelsell, D. Zicha, Quantitative fluorescence resonance energy transfer (FRET) measurement with acceptor photobleaching and spectral unmixing. J. Microsc. **215**(Pt 2), 162–173 (2004)
40. R.R. Duncan, Fluorescence lifetime imaging microscopy (FLIM) to quantify protein-protein interactions inside cells. Biochem. Soc. Trans. **34**(Pt 5), 679–682 (2006)
41. R.R. Duncan, A. Bergmann, M.A. Cousin, D.K. Apps, M.J. Shipston, Multi-dimensional time-correlated single photon counting (TCSPC) fluorescence lifetime imaging microscopy (FLIM) to detect fret in cells. J. Microsc. **215**(Pt 1), 1–12 (2004)
42. L.M. Felber, S.M. Cloutier, C. Kundig, T. Kishi, V. Brossard, P. Jichlinski, H.J. Leisinger, D. Deperthes, Evaluation of the CFP-substrate-YFP system for protease studies: advantages and limitations. Biotechniques **36**(5), 878–885 (2004)
43. M. Peter, S.M. Ameer-Beg, M.K. Hughes, M.D. Keppler, S. Prag, M. Marsh, B. Vojnovic, T. Ng, Multiphoton-film quantification of the EGFP-mRFP1 FRET pair for localization of membrane receptor-kinase interactions. Biophys. J. **88**(2), 1224–1237 (2005)
44. M.A. Rizzo, G.H. Springer, B. Granada, D.W. Piston, An improved cyan fluorescent protein variant useful for FRET. Nat. Biotechnol. **22**(4), 445–449 (2004)

11

Fluorescence-Based Optical Biosensors

F.S. Ligler

11.1 Introduction

Biosensors integrate biological molecules with a signal transduction device to produce a signal when a molecular recognition event occurs. This tutorial will focus entirely on biosensors that employ optical devices and incorporate fluorescent mechanisms for signal transduction. Compared with the electrochemical glucose sensors found in pharmacies world wide, optical biosensors are often more complex and more costly. However, optical biosensors are better suited for repetitive analysis or continuous monitoring, for interrogation of complex fluids, and for measuring binding events in real time. Optical imaging methods have also been widely adapted to measuring microarrays of recognition events; this experience provides a base for the development of highly multiplexed optical biosensors.

Optical biosensors not included in the following discussion are, nonetheless, worth mentioning. This group of biosensors is primarily directed at detection of a target without the requirement for a label. Thus performing the assay is simplified by requiring only the exposure of the sample (containing target) to the biological recognition molecule. Noteworthy among these methods are interferometry, surface plasmon resonance, resonant and antiresonance reflectometry, and cantilever-based systems [1, 2]. All of these sensors measure a change in optical properties, usually refractive index, at the sensing surface where the recognition molecules are immobilized. Convenient to use, they are excellent tools for measuring reactions in well-defined fluids. They tend to be less sensitive to very small targets or to very large targets that have most of their mass outside the sensing region; improvements in waveguide technology are minimizing these problems. However, nonspecific adsorption of components from complex samples can reduce the sensitivity by creating a significant background signal that must be accurately subtracted.

The following tutorial on fluorescence-based optical biosensors is organized into five parts: (1) biological recognition molecules and assay formats, (2) displacement immunosensors, (3) fiber optic biosensors, (4) bead-based biosensors, and (5) planar biosensors. These five groups are certainly not an exhaustive coverage of fluorescence-based biosensors, but should provide the reader with at least a general appreciation of the breadth of options available. All of these sensors include the same basic components: biological recognition molecules, a source of excitation light, and an optical readout device capable of discriminating excitation light from emitted fluorescence. Most of the biosensors also include a fluid transfer system to move samples and reagents over the sensing surface. The components and configurations for these systems depend on the geometry of the sensing surface and the degree of automation of the biosensor.

11.2 Biological Recognition Molecules and Assay Formats

Optical biosensors have utilized a wide variety of biological recognition molecules. With the exception of a few instances where enzyme catalysis is measured, they employ affinity or binding functions. The most often employed molecules are antibodies and oligonucleotides because they are readily available and the binding reactions have undergone extensive analysis at the molecular level. However, oligosaccharides, antibiotic peptides, siderophores, and combinatorial molecules have also been utilized.

No matter what the binding molecule is, the method by which it is immobilized on the sensing surface is very important if that binding function is to be preserved. In general, the surface is first modified with a silane or thiol to provide reactive groups for subsequent attachment of the biomolecules; if possible, this layer also helps prevent subsequent nonspecific binding of the biorecognition molecule that would result in denaturation. Next a crosslinker is attached to the free end of the silane or thiol and binds the biorecognition molecule [3]. For antibody immobilization, many studies have focused on orienting the antibody to avoid interference with the binding site. Although attachment through thiol groups in antibody fragments or the carbohydrate on the region of the molecule distal to the active sites achieves this goal, preventing secondary nonspecific adsorption is often even more critical.

In a popular variation of the silane-based immobilization approach, avidin is immobilized instead of the biorecognition molecule, and the biorecognition molecule is biotinylated and bound to the avidin through a high affinity, noncovalent bridge. When plastic is used instead of glass or silicon, nonspecific adsorption of the avidin provides a fully functional surface [4,5]. The avidin-based approach provides several advantages: (1) the avidin helps to prevent binding of the recognition molecule to the surface and subsequent denaturation, (2) the number of linkages of the recognition molecule to the surface can be limited by controlling the number (and possibly the position) of the biotins attached, and (3) the same surface can be used for immobilizing a wide variety of recognition molecules. This latter advantage is particularly useful for making arrays of different recognition molecules to capture multiple targets [6,7]. However, the biotin-avidin approach may have disadvantages if the recognition molecules used for capture are very small; in this case, a spacer may be required between the biotin and the capture molecule to make sure that it extends out of the binding pocket on the avidin and into the solution. Another situation where immobilization via an avidin bridge may not be the optimum approach is where the density of the capture molecule is critical for efficient binding – as has been documented with sugars [8] and antimicrobial peptides [9].

Assays with biological recognition molecules for fluorescence biosensors are configured in four basic formats: direct binding, sandwich, competition, and displacement assays. The direct binding format is by far the simplest to implement as it involves only the capture molecule and the target. However, it only works if either the target is inherently fluorescent or if it has somehow been prelabeled, as is the case in many DNA hybridization assays. Direct-binding assays are often used with prelabeled target standards for several reasons: to select the best capture molecules for a particular application by characterizing affinity and specificity of the binding function; to optimize the assay conditions for sensitivity, speed, and reproducibility; and to provide positive controls to monitor assay performance. A variation of the direct-binding assay utilized molecular beacons (reviewed by Yao et al. in [10]). In these assays, a fluorescence energy transfer donor is immobilized on one area of a capture molecule, while an acceptor is immobilized on an area that is adjacent only when no target is bound. When the target is bound, the two fluorophores separate, and a positive signal is generated.

The sandwich assay is used for detection of targets with at least two binding sites, including large molecules such as proteins, oligonucleotides, bacteria, and viruses. In the sandwich assay, the target binds the capture molecule (usually immobilized) and a fluorescent tracer molecule, usually in that order. The formation of the resultant fluorescent complex is measured, while free fluorescent tracer molecules are either removed or optically excluded from the sensing region. Sandwich assays are described in the majority of publications on optical biosensors.

For small molecules with only a single-binding site, the formation of such a sandwich is not possible and either a competitive or displacement assay must be utilized (Fig. 11.1). Two versions of the competitive assay are widely used. In one version, the capture molecules are immobilized, labeled target is added to the sample solution, and the labeled and unlabeled target molecules compete for the binding sites on the immobilized capture molecule. In the second version, a target molecule is immobilized, labeled antibody is added to the sample solution, and any target free in solution prevents the binding of the labeled antibody to the immobilized target. In both these versions, the fluorescent signal at the surface decreases with the increase in target in the sample solution. Although the fluorescence in solution can also be measured, the fluorescent component in solution is generally in sufficiently high concentration

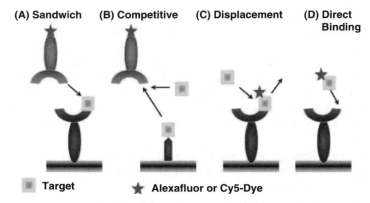

Fig. 11.1. Assay configurations. Four different types of assays are possible using fluorescence biosensors, with variations such as the binding of antibodies to the surface and labeling of antigen in competitive assays in a variation of the format shown in (**b**) or the use of molecular beacons to provide a signal upon direct binding of antigen in a variation of the format shown in (**d**). The figure was prepared by Kim Sapsford, George Mason University

that the very small decreases caused by low concentrations of target are difficult to measure. One way to avoid this problem is to use a displacement assay. In this configuration, the active sites of an immobilized monoclonal antibody are saturated with a fluorescent analog of the target molecule. Pulses of sample are introduced over the surface in a continually flowing stream. When the stream contains the target, the fluorescent analog is displaced from the immobilized capture molecule and measured downstream. Thus a signal is generated that increases in proportion to the amount of target present.

Of the most commonly described recognition molecules, oligonucleotides provide the greatest sensitivity. The principle reason is that the target genes can be amplified prior to the detection reaction using a polymerase and selective primers. This amplification step not only produces more molecules for the detection reaction, but also increases the ratio of target to background molecules in complex samples. Fluorescence biosensors have been developed that simply measure the process of amplification using dyes that bind to increasing concentrations of double stranded DNA, assuming that only target oligonucleotides are replicated. For more selective analysis, the products of the amplification have been measured using capillary electrophoresis [11] or hybridization to selective probes in solution or on a surface (reviews in [1,2,10]).

The biorecognition molecules most frequently used by the optical biosensor community have been antibodies. Antibodies have excellent selectivity and are relatively stable – unless they are exposed to too much heat or harsh chemicals. However, to find the most appropriate antibody for each application can be a time-consuming process. Genetic engineering techniques are being used to create a bigger range of specific-binding molecules. Advances in this area include the development of single-chain antibodies, in which the heavy and light chain regions of the active site are cloned and produced as a single peptide, [8] and the formation of libraries of shark and camelid active site peptides, which are naturally composed of a single chain. The latter, in particular, are very stable and usually recover from chemical or thermal denaturation to recover binding function [12].

The disadvantage of using DNA or antibodies for biorecognition is that the user must know exactly what the target is prior to assay development. To expand the repertoire of detectable targets, particularly for toxins and pathogens, molecules that recognize families of targets have been exploited as capture molecules (reviewed in [8]). Sugars and gangliosides have been used with the idea of mimicking the surface of the mammalian cells targeted for infection by pathogen. Antimicrobial peptides, used by amphibians and other organisms for protection against pathogens, have been immobilized and proven to bind families of bacteria and toxins. Other candidates for broader spectrum target recognition include siderophores, phages, and combinatorial peptides.

11.3 Displacement Immunosensors

Displacement immunosensors are some of the simplest optical biosensors in both form and function. However, why they work is not straight forward. The basic idea is as follows: First, a monoclonal antibody is immobilized on a solid surface (beads, porous membranes, capillary walls). Second, the active sites on the antibody are filled with a fluorescent version of the target of interest. Third, a continuous stream containing injected samples is passed over the immobilized antibodies. (This can continue for days.) When the target is present, it displaces a small proportion of the bound, fluorescent analog. The fluorescent molecules pass downstream and are measured using a simple fluorimeter. Simple antibody kinetics do not explain the behavior: if the fluorescent analog were bound with the dissociation constants typically measured in solution, it would come off in the flow within 20 or 30 min. If it is bound very tightly, it would not come off instantaneously in the presence of target – but it does. The general understanding is that the fluorescent analog is continually in the process of association and dissociation with the antibody. However, it tends to remain in the boundary layer at the solid surface, only diffusing very slowly into the bulk flow. When a target is present, it prevents the rebinding of the labeled analog that is dissociated at that point in time. The result is that, depending on the actual antibody–target pair selected, an operator may be able to add sample after sample and get a positive response for 50 or 60 target-containing samples in series (Fig. 11.2). The displacement immunosensor is particularly useful for the detection of low molecular weight targets such as explosives and drugs of abuse. Biosensors for these applications were commercially available for a time. Sensitivities of part per million even to part per billion have been obtained in assays that only require a minute or two

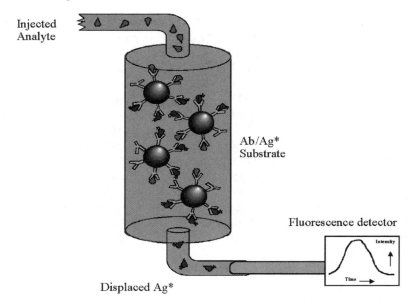

Fig. 11.2. Flow immunosensor using beads as the solid support. Target displaces labeled target analog from immobilized antibodies to generate positive fluorescence signal downstream. Reprinted from Ligler and Taitt (2002) with permission from Elsevier

to perform. For a review, see Kusterbeck in [1, 2]. Kusterbeck and colleagues have recently developed a flow immunosensor to detect unexploded ordinance in the ocean that has been tested mounted on an unmanned, underwater vehicle (personal communication).

11.4 Fiber Optic Biosensors

Fiber optic biosensors utilize two distinct assay configurations for signal generation and measurement: the optrode configuration and the evanescent wave configuration. Both configurations rely on the same principle of total internal reflection (TIR) for light propagation and guiding. However, optrodes use the light shining out the end of the fiber to generate a signal either at the distal face of the fiber or in the medium near the fiber's end, while evanescent wave sensors rely on the electromagnetic component of the reflected light at the surface of the fiber core to excite only the signal events localized at that surface. The penetration depth of the light into the surrounding medium is much more restricted than for optrodes, while the surface area interrogated is much larger in comparison to optrodes of equal diameter. The result is that evanescent wave biosensors require immobilization of the biological recognition molecules onto the longitudinal surface of the optical fiber core, primarily measure binding events, and are relatively immune from interferents in the bulk solution. Optrode-type sensors are adaptable to a much wider variety of measurements and produce a higher power excitation light in the sensing area, but the assays must be formatted to accommodate excitation light in the bulk solution and signal collection from a very small surface, i.e. the fiber tip. A history of the earliest applications of fiber optics for biosensor applications can be found in [13].

11.4.1 Fiber Optics for Biosensor Applications

Total internal reflectance (TIR) is observed at the interface between two dielectric media with different indices of refraction as described by Snell's law:

$$\frac{n_1}{n_2}\sin\theta_1 = \sin\theta_2, \tag{11.1}$$

where n_1 and n_2 are the refractive indices of the fiber optic core and the surrounding medium, respectively. The angle of light incident through the core of the optical fiber is represented by θ_1 and the angle of either the light refracting into the surrounding medium or the internal reflection back into the core is represented by θ_2. Total internal reflection requires that $n_1 > n_2$ and occurs when the angle of incidence is greater than the critical angle, θ_c, defined as:

$$\theta_c = \sin^{-1}\left(\frac{n_2}{n_1}\right). \tag{11.2}$$

This parameter must be considered when designing any biosensor on the basis of optical fibers. However, although Snell's law describes the macroscopic optical properties of waveguides, it does not account for the electromagnetic component of the reflected light, known as the evanescent wave. The evanescent wave is an electric field that extends from the fiber surface into the lower index medium and decays exponentially with distance from the surface, generally over a distance of hundred to several hundred nanometers. For multimode waveguides, the penetration depth, d_p, the distance at which the strength of the evanescent wave is $1/e$ of its value at the surface, is approximated by:

$$d_p = \frac{\lambda}{(4\pi[n_1^2\sin^2\theta - n_2^2]^{1/2})} \tag{11.3}$$

where n_1 and n_2 are refractive indices of the optical fiber and surrounding medium, respectively, and θ is the angle of incidence [14].

The importance of the evanescent wave is its ability to couple light out of the fiber into the surrounding medium, thereby providing excitation for fluorophores bound to or in proximity to the fiber core surface. This confined range of excitation is one of the major factors responsible for the relative immunity of evanescent wave-based systems to the effects of matrix components or interferents beyond the reaction surface (Fig. 11.3). Love and Button [15]

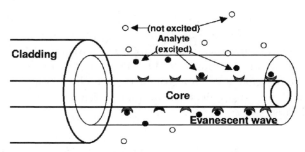

Fig. 11.3. Evanescent illumination of fluorescent complexes at the surface of the core of a partially clad waveguide. Reprinted from Ligler and Taitt (2002) with permission from Elsevier

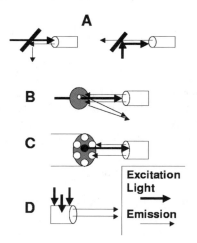

Fig. 11.4. Strategies for separation of excitation and emission light paths (Ligler and Taitt, 2002). Figure reprinted with permission from Elsevier

originally suggested that dipoles close to the surface could emit approximately 2% of their radiated power (fluorescence) into modes coupled back up the fiber; however, Polerecký et al. (2000) have more recently analyzed films of dipoles and concluded that a much higher proportion of the radiation can couple into guided modes.

A variety of configurations have been utilized with optical fibers for exciting fluorescence and collecting the emitted signal. In Fig. 11.4a, a traditional dichroic mirror is used to separate excitation and emission on the basis of the difference in wavelengths. The excitation light may pass through the mirror, while the emission light is reflected onto a detector (left) or, in a scheme that has generally proven more effective, the stronger excitation light is reflected onto the end of the fiber, while the emitted light passes straight through. In Fig. 11.4b, the excitation light passes through a hole in an off-axis parabolic mirror, while the emission light is reflected by the mirror onto a detector [16]. In Fig. 11.4c, a fiber bundle is used between a high numerical aperture sensing probe and the optics. The 635 nm excitation light is channeled down the central silica fiber while the emitted higher wavelength fluorescence is coupled back up the surrounding plastic fibers [17]. In Fig. 11.4d, the fiber is illuminated using a light source normal to the fiber, and the fluorescence is detected at the distal end [18, 19].

11.4.2 Optrode Biosensors

Optrode biosensors grew out of a wealth of applications for chemical sensing on the basis of absorbance, fluorescence, and even time-resolved fluorescence. The fibers were relatively easy to work with and facilitated the separation of excitation and emission light; frequently, the fibers themselves were used to transmit only excitation to the sensing region or only emission away from the sensing area as shown in the diagram below. Furthermore, a wide variety of light sources and detectors were used. These included lamps, LEDs, lasers, and diode lasers for excitation and photomultiplier tubes, photodiodes, and CCDs for detection. The optrode configuration is one of the most flexible in terms of the types of assays that can be performed. Optrodes have been used not only with immunoassays and DNA hybridization assays, but also with enzyme assays and for analysis of whole cell function. These assays and configurations have been reviewed by Walt [1, 2] (Fig. 11.5).

Probably the most exciting application of optrode technology today is the fiber optic bundles developed by Walt [20] and currently marked by Illumina (www.Illumina.com). In this approach, the ends of the optical fibers in the bundle are etched out, leaving the surrounding cladding to form microwells. Coded beads are placed in the microwells so that hundreds of thousands of assays can be conducted simultaneously. The assays are being used for genomics, proteomics, and biodefense.

11.4.3 Evanescent Fiber Optic Biosensors

The first use of an optical fiber as a biosensor, and one of the first optical biosensors was described by Hirschfeld and Block [21, 22] in the mid 1980s. This type of sensor was also the first biosensor to be fully automated and used remotely. In 1996–1998, Ligler, Anderson, and colleagues developed a biosensor payload for a small, unmanned plane and tested it at Dugway Proving Ground, Utah. The payload included a biosensor with four fiber probes, a RAM-air driven cyclone for collecting aerosolized bacteria, an automated fluidics system, batteries, and a radiotransmitter. They demonstrated that the airborne biosensor could collect a biothreat stimulant, identify it, and radio the data to an operator on the ground in 6–10 min [23, 24].

The best known of the commercially available fiber optic biosensors is the RAPTOR-Plus, made by Research International (http://www.resrchintl.com)

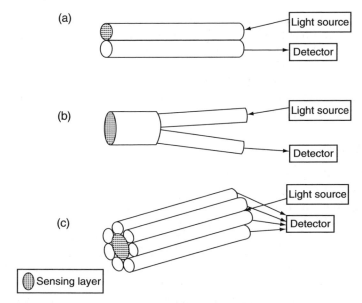

Fig. 11.5. Design principles for optrode biosensors. (**a**) Two fibers: one carries light to the sensing layer and one carries the signal to the detector. (**b**) Bifurcated fiber: the biosensing layer is placed on the fused end of the fiber. (**c**) The biosensing layer is placed on the central fiber, and the surrounding fibers are used to collect the light signals. Reprinted from [1] with permission from Elsevier

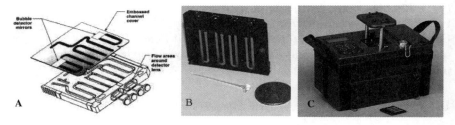

Fig. 11.6. Schematic (**a**) and photograph (**b**) of injection-molded optical fiber probes and fluidics coupon for use with the RAPTOR-Plus. (a and b from [1], reprinted with permission from Elsevier). (**c**) Photograph of the Raptor Plus with keypad control and result window on the *left* and the sample port and reagent chamber on the *right*

(Fig. 11.6). This system is proving to be very reliable in terms of long term operation (¿3 years to date, George Anderson, personal communication). In both the 4-probe Raptor and the more recently developed 8-fiber BioHawk, a disposable coupon contains the fiber probes, providing protection during long-term storage and fluidic channels for automated sample processing.

The polystyrene probes are inserted after being coated with the recognition biomolecule and dried. The probes are designed to integrate a combination tapered sensing region with a lens for signal [25]. The coupon automatically aligns the probe so that it is in the middle of the fluid channel [26]. In the Raptor, the light emitted from the end of the fiber is focused onto a collection lens in the permanent portion of the biosensor, while in the BioHawk, the light is collected normal to the fiber. The Raptor has been used to assay for a wide variety of targets, including explosives, toxins, clinical markers, and pathogens. It has proven to have sensitivities comparable to traditional enzyme-linked immunoassays (ELISA) using the same recognition molecules, but with assay times in the 10–15 min range instead of 2–4 h. Furthermore, the assays can be performed in the presence of highly complex sample matrices, including blood, urine, food homogenates, beverages, ground water, and effluents from air samplers. Results of such assays have been reviewed in [1, 2].

11.5 Bead-Based Biosensors

The use of beads as biosensors was first made popular by Raoul Kopelman with his PEBBLES (reviewed by Brasuel et al. in [1]). PEBBLES were small beads originally including ion- or pH-sensitive dyes that could be injected into cells to measure local microenvironments. Imaging was performed using a fluorescent microscope. The Kopelman group expanded the repertoire of PEBBLES to include the use of more specific recognition elements such as calcium ionophores for real-time interrogation of cell function.

Building on this approach, other groups developed highly specialized nanoparticles to measure other intracellular functions. One of the more interesting particles is composed of quantum dots modified with binding protein and a quenching dye [27]. In this system, a dye molecule is positioned on the binding protein adjacent to the active site. The dye quenches the signal from the quantum dot unless the binding protein binds its target. At that point, the environment of the dye changes so that it can no longer quench the quantum dot, and a positive signal is generated. Medintz et al. used a broad-spectrum quenching dye coupled to genetically modified maltose-binding proteins and demonstrated the capacity of different colored dots to generate signals in response to different target analytes.

The next major breakthrough in bead-based sensing was the development of coded beads that enabled simultaneous measurement of multiple analytes using beads with different amounts of two dyes. Originally developed for combinatorial chemistry applications, the fluorescent coded beads enabled highly multiplexed assays to be performed in a single fluid compartment. The bundled optrode was perfect for making such measurements (see Yu and Walt in [2]) as was flow cytometry [28]. In both the cases, beads of each code or dye ratio are coated with a distinct biorecognition molecule. When the target binds to the bead of that code, another label is introduced using either

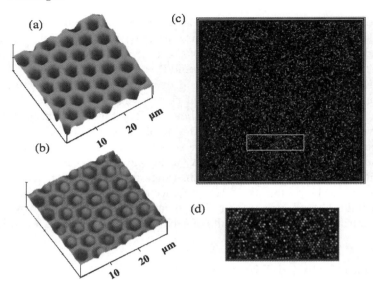

Fig. 11.7. High-density multianalyte bio-optrode composed of microsphere array on an imaging fiber. (**a**) Scanning force micrograph (SFM) of microwell array fabricated by selectively etching the cores of the individual fibers composing the imaging fiber. (**b**) The sensing microspheres are distributed in the microwell. (**c**) Fluorescence image of a DNA sensor array with ∼13,000 DNA probe microspheres. (**d**) Small region of the array showing the different fluorescence responses obtained from the different sensing microspheres [20]. Reprinted from [1] with permission from American Association for the Advancement of Science

a sandwich reaction or a third die, e.g., one that binds to double stranded DNA. Then only the beads with target have three colors and the identity can be read from the bead code (Fig. 11.7).

11.6 Planar Biosensors

Planar waveguide biosensors also usually employ evanescent illumination from a coherent source to excite fluorophores at the surface of the waveguide. The resulting fluorescence emission can be measured either by a single photodetector or by a system capable of imaging, usually a CCD or CMOS camera. A variety of focusing lenses have been used to improve detector response [7, 17, 29–32]. The introduction of bandpass and longpass filters was found to improve the rejection of scattered laser light and hence reduce the background of the system [7] (Fig. 11.8).

Unfortunately, a side effect of using bulk waveguides and collimated light is the production of sensing "hot spots" along the planar surface that occur where the light beam is reflected, illuminating only discrete regions. These

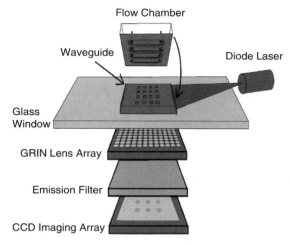

Fig. 11.8. Detection of emitted fluorescence at right angles to the waveguide interface as described by Golden et al., 1999. Reprinted with permission of the Elsevier

hot spots have been successfully utilized as sensing regions by Brecht and coworkers in the development of an immunofluorescence sensor for water analysis [33,34]. Feldstein et al. [7,45] overcame this problem by incorporating a line generator and a cylindrical lens to focus the beam into a multimode waveguide, which included a propagation and distribution region prior to the sensing surface. This resulted in uniform lateral and longitudinal excitation at the sensing region.

Another method of achieving uniform longitudinal excitation of the sensing region is to decrease the waveguide thickness [30]. When the thickness of the waveguide is much greater than the wavelength of the reflected light, the waveguide is referred to as an internal reflection element (IRE). However, if the thickness of the waveguide is decreased such that it approaches the wavelength of the incident light, and the path-length between the points of total internal reflection become increasingly shorter. At the thickness where the standing waves, created at each point of reflection, overlap and interfere with one another, a continuous streak of light appears across the waveguide, and the IRE becomes known as an integrated optical waveguide (IOW) [35].

Integrated optical waveguides, used frequently in TIRF studies, are monomode and prepared by depositing a thin film of high refractive index material onto the surface of a glass substrate. These thin films are typically 80–160 nm in thickness and consist of inorganic metal oxide compounds such as tin oxide [29], indium tin oxide [36], silicon oxynitride [37], and tantalum pentoxide [38, 39]. The light is coupled into these IOWs via a prism or grating arrangement. Studies by Brecht and coworkers compared IRE and IOW-based waveguides and concluded that the integrated optics significantly improved the sensitivity of the system by a factor of 100 [33].

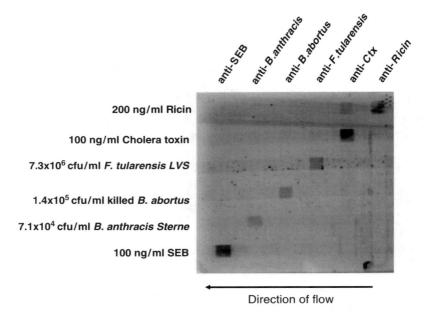

Fig. 11.9. Antibody array used to interrogate six samples for six different biothreat agents. Reprinted from [40] with permission of the Elsevier

The planar geometry provides a surface for depositing arrays of detection molecules. This makes it ideal for multianalyte detection. An example of such an array is shown below. In numerous papers reviewed by Sapsford et al. in [1,2], the limits of detection are typically about an order of magnitude better with good array biosensors, based on planar waveguides than for standard ELISAs using the same reagents. In addition to antibodies and DNA, detection molecules used with planar waveguide biosensors include siderophores, carbohydrates, antimicrobial peptides, and gangliosides [2] (Fig. 11.9).

11.7 Critical Issues and Future Opportunities

Fluorescence-based optical biosensors have inherent advantages for discriminating targets in complex samples and for increasing sensitivity compared with label-free methods. Limits of detection can be further reduced using amplification methods such as enzymes, polymerases, or multifluorophore complexes (e.g., labeled dendrimers as in [41], or virus particles carrying 60 carefully separated fluorophores as in Sapsford et al., 2006). However, labeling and amplification procedures can complicate assay processes, making automation more difficult and possibly increasing fluorescence background. Another method for increasing signal include preconcentration of target prior to analysis using filters, dielectrophoreses, or magnetic bead separations. The latter approach has been very successfully implemented in commercial systems for chemiluminescence assays (reviewed by Richter [1, 2]).

Microfluidics are becoming increasingly used for three particular facets of optical biosensors:

1. Microfluidic devices facilitate methods for automated sample processing, including the mixture of the sample with the fluorescent reagents.
2. They provide an efficient mechanism for manipulating very small volumes, which saves reagent costs.
3. They can deliver the target more efficiently to the sensing surface.

Examples of the latter are laminar fluid focusing as described by Hofmann et al. [42] and Munson et al. [43] and passive mixing to minimize target depletion at the sensing surface as described by Golden et al. [44]. However, care must be taken to make sure that fluidic channels are sufficiently large to avoid clogging by complex samples, and that they have the appropriate surface chemistry to prevent nonspecific adsorption of the target outside the sensing region. Furthermore, the issue of interrogating a sufficient volume to include targets present at low concentrations must never be neglected.

Advances in optics offer new opportunities for increased sensitivity and reductions in size and cost. Silicon technology is producing better and better integrated optical waveguides for highly multiplexed analysis. Arrays of single photon detectors are described by Eduardo Charbon in this volume that may offer the opportunity to detect single targets if the affinity of the recognition molecule is sufficiently high and if the background can be sufficiently reduced. Both scattered excitation light and stray fluorescence from molecules not bound in the detection complex can generate "noise" beyond that inherent in the optical device itself. The production of devices based on organic polymers is also very exciting because they should be relatively simple to integrate biological recognition elements and polymer-based microfluidics to form monolithic, inexpensive, or disposable sensors. Organic LEDs, transistors, and photodiodes are described in this volume in the chapter by Peter Seitz.

Finally, systems biology at the molecular level is identifying new targets for analysis that are of importance for medical, environmental, and defense applications. Furthermore, the understanding of how to make rationally designed molecules for sensing applications offers new approaches to perform the biorecognition function with appropriate specificity, increased sensitivity, and enhanced stability during sensor storage and use. New fluorophores also provide opportunities for interrogating new types of samples and systems and for generating multiple signals simultaneously.

The future for fluorescence-based optical biosensors will be rich. If there is a real limitation, it is only on the ability to synthesize all the emerging technologies into the most useful system for each customer. That ability rests on the willingness of scientists and engineers in universities, government, and industry to work together in interdisciplinary teams and the longer term vision of those that support such efforts.

Acknowledgments

The preparation of this chapter was supported by NRL 6.2 Work Unit 6006. The views expressed herein are those of the author and do not represent opinion or policy of the US Navy or Department of Defense.

References

1. Ligler, F.S. and C.R. Taitt: *Optical Biosensors: Today and Tomorrow*, (Elsevier, 2002).
2. Ligler, F.S. and C.R. Taitt: *Optical Biosensors 2008: Present and Future*, (Elsevier, 2008).
3. Cass, A.T. and F.S. Ligler: *Immobilized Biomolecules in Analysis: A Practical Approach*, (Oxford University Press, Oxford, UK 1998).
4. King, K.K., J.M. Vanniere, J.L. LeBland, K.E. Bullock, and G.P. Anderson: Environ. Sci. Technol. **34**, 2845 (2000).
5. Herron, J.N, S. zumBrunnen, J.-X. Wang, X.-L. Gao, H.-K.Wang, A.H. Terry and D.A. Christensen: Proc. SPIE **3913**, 177 (2000).
6. Silzel, J. W., B. Cercek, C. Dodson, T. Tsay and R. J. Obremski: Clin. Chem. **44**, 2036 (1998).
7. Feldstein, M. J., J. P. Golden, C. A. Rowe, B. D. MacCraith and F. S. Ligler: J. Biomed. Microdevices **1:2**, 139 (1999).
8. Ngundi, M.M., C.R. Taitt, S.A. McMurry, D. Kahne, and F.S. Ligler Biosens. Bioelectron. **21**, 1195 (2006).
9. Kulagina, N.V. K.M. Shaffer, G.P. Anderson, F.S. Ligler, and C.R. Taitt: Anal. Chim. Acta **575**, 9 (2006).
10. Thompson, R. B. In: *Fluorescence Sensors and Biosensors*, ed by Taylor & Francis, (Boca Raton, FL, 2005), p. 394
11. Lagally, E.T., J.R. Scherer, R.G. Blazej *et al*: Anal. Chem. **76**, 3162 (2004).
12. Goldman, E.R., G.P. Anderson, J.L. Liu, J.B. Delehanty, L.J. Sherwood, L.E. Osborn, L.B. Cummins, and A. Hayhurst: Anal. Chem. **79**, in press (2007).
13. Wise, D.L., and L.B. Wingard, Jr.: Biosensors with Fiberoptics, (Human Press, NJ, 1991), p. 370
14. Harrick, N.J.: *Internal Reflection Spectrosopy*, (Wiley, New York 1967), p. 327
15. Love, W.F. and L.J. Button: Proc. SPIE **990**, 175 (1988).
16. Thompson, R.B. and M. Levine: US Patent No. 5,141,312 (1992).
17. Wadkins, R.M., J.P. Golden and F.S. Ligler: J. Biomed. Optics **2**, 74 (1997).
18. Kooyman, R.P.H., H.E. de Bruijn and J. Greve: 1987, Proc. SPIE **798**, 290 (1987).
19. Ligler, F.S., M. Breimer, J.P. Golden, D.A. Nivens, J.P. Dodson, T.M. Greene, D.P. Haders and O.A. Sadik: Anal. Chem. **74**, 713 (2002).
20. Walt, D. R.: Science 287, 451 (2000).
21. Hirschfeld, T.: US Patent No. 4,447,546 (1984).
22. Hirschfeld, T.E. and M.J. Block: US Patent No. 4,558,014 (1985).
23. Ligler, F.S., G. P. Anderson, P.T. Davidson, R.J. Foch, J.T. Ives, K.D. King, G. Page, D.A. Stenger and J.P. Whelan: Environ. Sci. Technol. **32**, 2461 (1998).

24. Anderson, G.P., K.D. King, D.S. Cuttino, J.P. Whelan, F.S. Ligler, J.F. MacKrell, C.S. Bovais, D.K. Indyke and R.J. Foch: Field Anal. Chem. Technol. **3**, 307 (1999).
25. Saaski, E.W. and C.C. Jung: US Patent No. 6,136,611 (2000).
26. Saaski, E.W.: US Patent No. 6,082,185 (2000).
27. Clapp, A.R., I.L. Medintz, H.Tetsuo Uyeda, B.R.Fisher, E.R.Goldman, M.G. Bawendi, and H.Mattoussi: J. Amer. Chem. Soc. **127**, 18212 (2005).
28. Edwards, B.S., T. Oprea, E.R. Prossnitz, L.A.Sklar: Current Opinion in Chemical Biology **8**, 392 (2004).
29. Duveneck, G. L., D. Neuschafer and M. Ehrat: International Patent Go1N 21/77, 21/64 (1995).
30. Herron, J. N., D. A. Christensen, K. D. Caldwell, V. Janatova, S.-C. Huang and H.-K. Wang: US Patent 5,512,492 (1996).
31. Herron, J. N., D. A. Christensen, H.-K. Wang and K. D. Caldwell: US Patent 5,677,196 (1997).
32. Golden, J. P.: US Patent 5,827,748 (1998).
33. Brecht, A., A. Klotz, C. Barzen, G. Gauglitz, R. D. Harris, G. R. Quigley, J. S. Wilkinson, P. Sztajnbok, R. Abuknesha, J. Gascon, A. Oubina and D. Barcelo: Anal. Chim. Acta **362**, 69 (1998).
34. Koltz, A., A. Brecht, C. Barzen, G. Gauglitz, R. D. Harris, G. R. Quigley, J. S. Wilkinson and R. A. Abuknesha, 1998, Sens. Actuators B **51**, 181.
35. Plowman, T. E., S. S. Saavedra and W. H. Reichert: Biomaterials **19**, 341 (1998).
36. Asanov, A. N., W. W. Wilson and P. B. Oldham: Anal. Chem. **70**, 1156 (1998).
37. Plowman, T. E., J. D. Durstchi, H. K. Wang, D. A. Christensen, J. N. Herron and W. M. Reichert: Anal. Chem. **71**, 4344 (1999).
38. Duveneck, G. L., M. Pawlak, D. Neuschafer, E. Bar, W. Budach, U. Pieles and M. Ehrat: Sens. Actuators B **38-39**, 88 (1997).
39. Pawlak, M., E. Grell, E. Schick, D. Anselmetti and M. Ehrat: Faraday Discuss. **111**, 273 (1998).
40. Rowe-Taitt, C. A., J. P. Golden, M. J. Feldstein, J. J. Cas, K. E. Hoffman and F. S. Ligler: Biosens. Bioelectron. **14**, 785 (2000).
41. Chang, A.-C., J.B. Gillespie, and M.B. Tabacco: Anal. Chem. **73**, 467 (2001).
42. Hofmann, O., G.Voirin, P. Niedermann, and A. Manz: Anal. Chem. **74**, 5243 (2002).
43. Munson, M.S., M.S. Hasenbank, E. Fu, and P. Yager: Lab on a Chip **4**, 438 (2004).
44. J.P. Golden, T. Floyd-Smith, and F.S. Ligler: Biosens. Bioelectron., in press (2007).
45. Feldstein, M. J., B. D. MacCraith and F. S. Ligler: US Patent 6,137,117 (2000).
46. Golden, J.P., E.W. Saaski, L.C. Shriver-Lake, G.P. Anderson and F.S. Ligler: Opt. Eng. **36**, 1008 (1997).
47. Hirschfeld, T.E.: and M.J. Block: US Patent No. 4,447,546 (1984)
48. Ngundi, M.M., N.V. Kulagina, G.P. Anderson, and C.R. Taitt: Expert Rev. Proteonomics **3**, 511 (2006).
49. Polerecký, L., J. Hamrie, and B.D. MacCraith: Appl. Opt. **39**, 3968 (2000).
50. Sapsford, K.E., P.T. Charles, C.H. Patterson Jr., and F.S. Ligler: Anal. Chem. **74**, 1061 (2002).
51. Sapsford, K.E. , C.M. Soto, A.S. Blum, A. Chatterji, T. Lin, J.E. Johnson, F.S. Ligler and B.R. Ratna: Biosens. Bioelectron. **21**, 1668 (2006).

12

Optical Biochips

P. Seitz

The citizens of our modern society are exposed to an increasing number of chemical substances that are potentially harmful or that need to be monitored and kept under check. At the same time, everybody would like to know much earlier, with higher certainty and at lower cost, whether her or his health is affected in any way. For this reason there is a huge demand for very sensitive, highly specific, cost-effective, and rapid methods to detect the concentration of bio-active molecules. Because of the simplicity, efficiency, and speed with which light can be generated, manipulated, detected, and used to sense the effects of many chemical reactions, optical techniques have become the method of choice for many sensing problems involving bio-active molecules. A particularly attractive aspect is the possibility to miniaturize and to parallelize optical measurements, which has led to the very active field of integrated optical sensing [1]. Such miniaturized optical systems for the sensing of bio-active molecules are also known as biochips, although this name is used for many different levels of integration. Applications of these optical biochips include medical diagnostics (for specialized laboratories and increasingly also for home use), contaminant detection in the food industry (e.g., hormones in milk or meat), pharmaceutical research and development (drug screening), environmental and pollution monitoring (e.g., pesticides in water), security and counter-terrorism (detection of chemical and biological warfare agents), as well as process and quality control in industry.

The present contribution reviews the different principles and realizations of optical biochips, and it offers an outlook on the potential of monolithically integrating complete optical measurement systems on one single, self-contained biochip of unprecedented complexity and functionality.

12.1 Taxonomy of Optical Biochips

12.1.1 Basic Architecture of Optical Biochips

The principle architecture of an optical measurement system for bio-active molecules is illustrated in Fig. 12.1. Coordinated by an electronic control

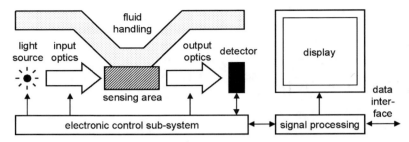

Fig. 12.1. Principle architecture of an optical measurement system for chemical and biochemical sensing, as used for the different types of optical biochips

system, which can be a single microcontroller chip in the simplest case, different optoelectronic, optomechanic, and microfluidic components interact to detect the effects of a chemical reaction, which is sensed by an optical effect in the system's transducer section [2]. A light source is emitting either modulated or unmodulated light, which is conditioned by input optics such as focusing, in-coupling, or angle-adjustment elements. In the transducer part, a chemical reaction takes place between the analyte (the target molecule) and a suitable receptor in the microfluidics part of the system. This reaction is observed by measuring a suitable change of any of the optical properties in the fluid volume where the reaction is taking place. The light exiting the interaction region is carrying the information about the analyte concentration, and it needs to be conditioned by appropriate output optics, such as imaging, out-coupling, or angle-adjustment elements, so that the light can be efficiently detected by a photosensor. The resulting electric signal is acquired and converted by an electronic analog/digital circuit, and the digital signal is the processed to extract and communicate the desired concentration information, for example, by displaying it for the user on a screen.

Although it is possible to employ many optical effects (see Sect. 12.3 below) in an extended volume of the transducer part, much more control can be exercised if this interaction volume is restricted to the immediate vicinity (typically less than 1 µm) of a sensitized surface. This principle is illustrated in Fig. 12.2. The target molecules together with other possibly bio-active molecules are in solution, and the microfluidics system has the task to transport them to and from the transducer volume. There, receptor molecules are chemically bonded to a surface where light is passing to sense the effect of the chemical bonding. These receptors should react in a highly specific way only with the analyte, so that the analyte molecules remain attached to the surface, effectively changing the mass density close to the surface, which can be subsequently measured. Since no additional measure is necessary to determine the concentration of the analyte, this approach is called label-free sensing.

A higher detection sensitivity can be obtained if the analytes are tagged with photoactive molecules. Most often, these labels are fluorescent molecules

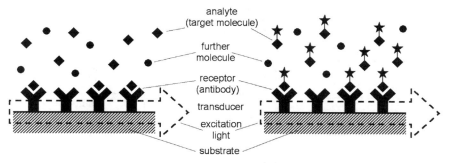

Fig. 12.2. Illustration of the transducer principles in an optical biochip. *Left*: Label-free sensing principle. *Right*: Labeled sensing principle

that emit light when excited with radiation of shorter wavelength. Alternatively, one can also make use of luminescent molecules, such as the famous luciferase, which are causing a biochemical enzymatic reaction as part of which (visible) light is emitted [3]. The term "biochip" is used very broadly in the field and does not describe the particular implementation of the principles described above precisely. Depending on the level of integration, one distinguishes usually between three different types of biochips:

Biochemical Microarray

The simplest and still the dominating type of biochip consists of a one- or two-dimensional array of transducer regions, which are sensitized with different receptors on a common substrate [4]. Neither microfluidics transportation systems nor the light sources or the detectors are part of such a biochemical microarray. This type of a biochip can consist, therefore, of up to one million individual transducer regions, each with an area as small as $5 \times 5\,\mu m^2$, on an optical biochip with an active surface of $1-2\,cm^2$ [5]. Most often, biochemical microarrays employ fluorescence tagging as the optical transducer mechanism. If they are employed for DNA analysis, optical microarrays are known as DNA-chips or Genechips^{TM}, and such DNA-chips with half a million transducer regions are commercially available for about €200 [5]. Although DNA analysis (genomics) is still the predominant application of optical microarrays, they are increasingly used for protein (proteomics) and cell analysis [6].

Lab-on-a-Chip (µ-TAS)

To simplify the practical use of biochips and to reduce the sample volumes required for an analysis, an increasing amount of functionality is placed on the same substrate. A lab-on-a-chip (LOC) combines the microfluidic sub-system, including all elements required to transport, mix, process, separate, and drain fluids, together with the optical transducer part on a single chip [7]. Such an

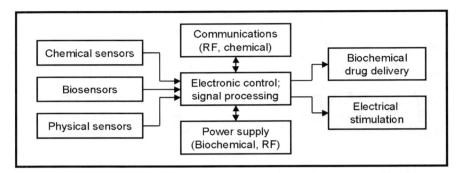

Fig. 12.3. Building blocks of an integrated diagnostic-therapeutic biochip

LOC can analyze minute sample volumes measured in femtoliters (equivalent to μm^3). In 1990, researchers at Ciba-Geigy (Basel, Switzerland) coined the alternate name Micro Total Analysis System (μ-TAS) for an LOC [8].

Since biochips are disposable items, many LOCs operate with external light sources, imaging optics, and 1D or 2D photosensor arrays (solid-state cameras). However, first LOCs have been realized making use of integrated light sources and/or photosensors, as, for example, in the OSAILS (optical sensor array and integrated light source) principle [9].

Diagnostic-Therapeutic Microsystems

The ultimate biochip consists not only of a complete LOC to perform a comprehensive analysis of biological, chemical, and physical parameters, it also includes a therapeutic BioMEMS (Micro-Electro-Mechanical System), such as a drug delivery or an electro-stimulation sub-system [10] (Fig. 12.3). Before such a vision of an integral biochip becomes reality, many practical problems have to be solved, such as the regeneration of biochemical sensing surfaces, bio-fouling resistance, robustness in the chemically aggressive environment of the body of animals and human beings, long-term power-supply using bio-fuel harvesting or inductive coupling, as well as reliable communications with other implanted systems or the outside world [11].

12.2 Analyte Classes for Optical Biochips

The analytes in our optical biochips were simply illustrated as generic molecules in Fig. 12.2. Depending on the actual biochemical function of the type of molecule, different classes of analytes are distinguished [4].

12.2.1 DNA (DNA Fragments, mRNA, cDNA)

To map or sequence the genetic information of a cell, DNA chips contain a large number (up to one million) of different single-stranded DNA fragments (so-called oligonucleotides), which are immobilized on the substrate surface in discrete spots. Each spot contains several millions of identical oligonucleotides to increase sensitivity. The sample to be tested contains single-stranded genetic chains (DNA fragments, mRNA, or cDNA), and they are usually labeled with a fluorescent dye. The genetic chains that match the immobilized oligonucleotides bind to the spots on the substrate. The biochip is then illuminated with a suitable wavelength, so that the fluorescence light pattern of the different spots allows the determination of the type and concentration of target genetic chains in the sample.

12.2.2 Proteins (Antigens)

Since the genetic information contains the instructions that proteins are produced in the body, the identification of proteins – so-called proteomics – is even more important than genomics. For this reason, the utilization of biochips for protein analysis is growing at a rate of about 29%, while the growth of DNA-chips is only about 18% [12].

A protein biochip is similar to a DNA-chip, but instead of oligonucleotides, protein probes (antibodies) for the target proteins (antigens) are immobilized on the individual spots on the biochip's surface. However, protein biochips are more difficult to handle, less specific, and less sensitive than DNA-chips for several reasons [4]: Proteins are unstable, and they can easily be denatured at solid–liquid and liquid–air interfaces. Today's production techniques for protein probes produce only low-affinity capture antibodies, making the coupling of the antigens to the antibodies not very specific. Since antibodies have usually rather large surface areas for interaction, they show important cross-reactivity between target proteins. DNA samples can readily be amplified using PCR (polymerase chain reaction), while no such amplification process is known for proteins.

The optical principles employed in protein biochips include fluorescence as well as evanescent wave sensing (see Sect. 12.4).

12.2.3 Specific Organic Molecules

The same principle used for DNA and protein biochips can be employed, in principle, to determine the concentration of various organic molecules that are of biochemical importance. Suitable probe molecules are fixed on the surface of a biochip. Corresponding target molecules then attach to the probe molecules, and their presence can be measured using the same optical techniques as used for the other types of biochips.

12.2.4 Cell Gene Products (cDNA, Proteins)

A cell microarray consists of a number of spots with different types of genetic information, such as DNA strands, plasmids, or adeno viruses. Mammalian cells are grown on these spots, where they can absorb the prepared genetic information – the cells are being transfected – and they are expressing the corresponding cDNA molecules. This genetic information, in turn, is employed by the cell to produce a specific protein. This protein can influence the functioning of the live cell, and various experiments can then be carried out to study the role of the particular protein in the cell's metabolism.

After the particular proteins are expressed in a cell, the cell can be incubated with fluorescently labeled antibodies, and the protein presence and concentration can be measured and imaged with fluorescence microscopy.

Although cell biochips are only useful for cells that are easily transfected, they overcome many of the deficiencies of conventional protein biochips described earlier, and cell biochips are adding pertinent information about the cell's metabolism to our knowledge [13].

12.2.5 Tissue

Instead of preparing a two-dimensional pattern of receptor molecules on a biochip, one can assemble a tissue microarray by combining tissue samples from a large number of specimens on the same substrate that can then be analyzed in a highly parallel fashion, for example, for specific protein expression patterns. The typical diameter of an individual tissue spot on a tissue microarray is 600 µm and its thickness is about 5–8 µm.

A particularly important application of such tissue microarrays is the high-throughput molecular profiling of tumors [14]. This technology is already offered commercially, and other applications are seen for large-scale epidemiology studies, for the development of new diagnostic and prognostics markers for tumors, for the investigation of biochemical pathways, for drug development, and for quality control of food, in particular, if genetically modified foods are involved [4].

12.3 Optical Effects for Biochemical Sensors

The many ways in which light interacts with matter can all be exploited for the detection of the concentration of an analyte by making use of suitable chemical reaction between the analyte and a receptor (or detector) molecule, either in the bulk or at the surface of an optical biochemical sensor. The taxonomy of these different optical effects that may be used for biochemical sensing is illustrated in Fig. 12.4, see also [1–4]. In the following, the most popular of these optical effects are briefly described.

12.3.1 Spectral Absorption

For a long time, colorimetric sensing principles have been used very widely because the visual color perception of the observer could be exploited, and

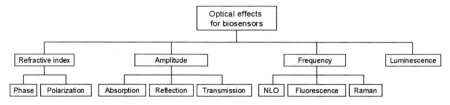

Fig. 12.4. Taxonomy of optical effects potentially useful for biochemical sensing

no additional instrument was necessary. These optical sensing methods were based on chemical reactions of the analyte with a receptor molecule that resulted in a chemical product with optical absorption properties in the visible spectral range. The resulting color change was then detected and evaluated by the observer or an instrument. Since optical measurement techniques based on spectral absorption are less sensitive than methods detecting phase changes in light, optical biochips do generally not make use of spectral absorption.

12.3.2 Phase Shift

A change in the real part of the refractive index causes a phase change of the electromagnetic wave that is used to probe the detection volume of the sensor. This phase change can either be observed in an interferometer setup, it can manifest itself as a change of the polarization properties of incident linearly polarized light, it can change the optical field distribution close to an interface, or it can change the propagation characteristics of a traveling wave. This last effect is most commonly exploited in optical biochips, in particular, through the modification of in- or out-coupling conditions, because it works very well in the confining geometries at the surface of biochips, as illustrated in Fig. 12.2, and it does not require active labeling of the analyte molecule.

12.3.3 Fluorescence

Because of the extremely high sensitivity of today's solid-state image sensors, the detection of very low light levels down to individual photons is technically not difficult to achieve [15]. As a consequence, the use of fluorescent labels to tag analyte molecules leads to a highly sensitive optical measurement method for the concentration of the analyte. Incident light at a certain wavelength is absorbed by the fluorescent labels, and the optical energy is re-emitted at a longer wavelength where it is detected by a sensitive camera. For this reason, fluorescence is one of the most widely used sensing methods in the life sciences and, in particular, in optical biochips.

12.3.4 Luminescence

Instead of attaching a fluorescent molecule to the analyte, it is also possible to bond a chemiluminescent molecule to it, such as the well-known luciferase [3].

The resulting enzymatic chemical reaction leads to the emission of (visible) light, without any need for an optical excitation, and this light can be detected by a sensitive camera. Since it is often easier to label an analyte with a fluorescent molecule than to label it with a suitable chemiluminescent molecule, luminescence is not very commonly used in optical biochips.

12.3.5 Raman Scattering

Raman scattering describes an optical process in which incident light of a certain energy is absorbed by a molecule, and it is re-emitted with a slightly different energy, where the energy difference is a characteristic function of the chemical structure of the molecule. Although Raman scattering is very specific and rich in the information it carries, its efficiency is very low: Typically, only about one in 10^5 photons is Raman scattered [4]. For this reason, Raman scattering is currently not used in optical biochips.

12.3.6 Nonlinear Optical (NLO) Effects

If the electrical field strength is not too high, the optical polarization of a molecule is linearly proportional to the amplitude of the applied electric field. Under illumination with an intense light source, however, the response of the molecule is not any more sufficiently described by a linear effect, and nonlinear optical (NLO) effects come into play.

Two NLO effects are of particular importance for biochemical sensing: higher harmonic generation and intensity-dependent multiphoton absorption. Higher harmonic generation involves the mixing of two or more light beams with frequency ν_i to produce a light beam with frequency $\nu = \Sigma \nu_i$. Multiphoton absorption describes the simultaneous absorption of two or more photons to reach a particular excited energy state [4].

Since the efficiency of both processes is a sensitive function of a molecule's environment, these NLO effects can effectively be used for biochemical sensing. However, because intense laser light is required, NLO effects are rarely used in optical biochips.

12.4 Preferred Sensing Principles for Optical Biochips

Because of the simplicity of the practical realization and the sensitivity of the methods, two basic sensing principles are preferred in optical biochips. (1) Evanescent wave sensing, where an electromagnetic wave is propagating along an optical interface, and a part of its energy is detecting changes in the refractive index in the sample volume. (2) Fluorescence sensing, where incident light is exciting the fluorescence labels on the analytes whose light emission is detected.

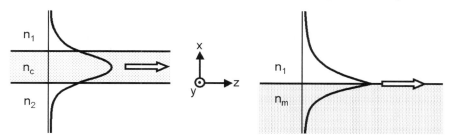

Fig. 12.5. Propagation of an electromagnetic wave in a dielectric slab waveguide (*left*) or as a surface plasmon at a metal–dielectric boundary (*right*)

12.4.1 Evanescent Wave Sensing

Electromagnetic waves can be guided by geometrically confining the space where they should propagate. Two particular configurations are commonly employed in optical biochips, as illustrated in Fig. 12.5: slab waveguides, consisting of a stack of laterally extended thin films of dielectric materials, with a core layer whose refractive index is larger than that of the neighboring layers. Alternatively, boundary layers between a dielectric and a metallic medium are used, where the dielectric constant of the metal must be negative in the used wavelength range. At such a boundary, a special guided electromagnetic mode, a so-called surface plasmon, can propagate.

In the case of a slab waveguide, the guided modes in most waveguide types can be classified in TE and in TM modes, where the electrical (TE), respectively, the magnetic (TM) field has a dominant component in the lateral direction, i.e., for the coordinate y in Fig. 12.5. Since three of the field components are zero, the TE modes consist of E_y, H_y, and H_z, and the TM modes consist of H_y, E_x, and E_z. Additionally, it is found that the field components in the propagation direction, H_z, respectively, E_z, are usually quite small [2]. Solving Maxwell's equation for the slab waveguide geometry and for a given vacuum wavelength λ of the light, one obtains for the jth TE mode an electric field component $E_j(x, y, z, t, \lambda)$ in y-direction, which is expressed as

$$E_j(x, y, z, t, \lambda) = E_j(x, y, \lambda)\, e^{i(n_{\text{eff}}(\lambda)kz - \omega t)}, \tag{12.1}$$

with the wave vector $k = 2\pi/\lambda = \omega/c$. This equation describes a harmonic wave propagating in z-direction with the velocity $v = c/n_{\text{eff}}$, where the effective refractive index n_{eff} is limited to the range

$$n_c > n_{\text{eff}} > (n_1, n_2). \tag{12.2}$$

The transverse electric field amplitude $E_j(x, y, \lambda)$ is called the field profile of the electromagnetic mode number j, and the modes are numbered according to their n_{eff} values in descending order, i.e., the mode with the highest n_{eff} corresponds to the lowest mode number j.

The number of guided modes that can propagate in a dielectric slab waveguide, their n_{eff} values and their field profiles $E_j(x, y, \lambda)$ depend on the geometry of the slab waveguide and the actual distribution of refractive indices. Configurations exist in which only one TE mode (and often one TM mode) can propagate; such structures are called monomodal or mono-mode waveguides.

If the dielectric slab waveguide is lossy, for example, through absorption or scattering, the effective refractive index n_{eff} in (12.1) has a nonzero imaginary component expressing these losses. High-quality dielectric slab waveguides with negligible losses can be fabricated, with absorption constants that are much lower than $1\,\text{cm}^{-1}$.

The surface plasmon, the special electromagnetic mode that can propagate along a metal–dielectric interface, is a single TM mode: At each wavelength only one surface plasmon can propagate, and it requires that the dielectric constant of the metal is negative in the used wavelength range. In the visible region, this is true for metals such as gold, silver, and copper. The propagation velocity in z-direction is given by $v = c/n_{\text{eff}}$, and the effective refractive index n_{eff} depends on the relative dielectric constants of the metal and of the dielectric material [4]. In contrast to the case of the dielectric slab waveguide, the propagation length of a plasmon is quite limited since the metal is strongly absorbing; the propagation length of a surface plasmon is usually smaller than $100\,\mu\text{m}$.

The importance of the dielectric slab waveguide and of the metal–dielectric interface for optical biosensors is due to the fact that in both structures a part of the wave energy is propagating in the covering dielectric material with refractive index n_1. The field strength in the dielectric materials with n_1 and n_2 is decaying exponentially as a function of the distance to the interface, with a decay length D_i ($i = 1, 2$), for which the field strength is reduced to $1/e$, given as

$$D_i = \frac{1}{k\sqrt{n_{\text{eff}}^2 - n_i^2}}. \qquad (12.3)$$

This exponentially decaying field is called evanescent field, and the covering dielectric material is actually the fluidic probe volume whose change in the refractive index is measured.

Typical values of the decay length D_1 in a dielectric slab waveguide used for optical biosensing are $0.1\,\mu\text{m}$ and slightly larger, about 0.2–$0.3\,\mu\text{m}$, for surface plasmons.

Principles of Guided Wave Coupling

Since guided waves are characterized by their property that they do not normally leave the waveguide structure, special means are necessary to couple them into or out of the waveguide. The most common coupling approaches are prism coupling, grating coupling, and end-face coupling, as illustrated in Fig. 12.6. Plasmons are almost exclusively coupled in and out with a prism in

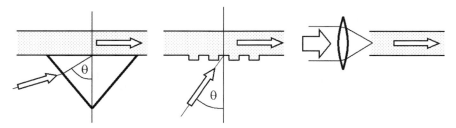

Fig. 12.6. Coupling of incident light into a waveguide using a prism (*left*), a grating coupler (*middle*), or a focusing lens at the waveguide's end face (*right*)

the so-called Kretschmann configuration [4]. The advantage is that all components are planar and easy to coat. Grating coupling is effectively used for dielectric slab waveguides, and innovative sensor configurations can be devised with grating couplers due to the freedom one has with selecting special grating parameters [1]. End-face coupling, however, is not often employed, mostly due to the alignment and stability problems encountered in practice.

Waveguides with Prism Coupler

Snell's law for the transmission and reflection of light at an optical interface states that the product of refractive index times the sine of the propagation angle with the surface normal must be identical for incident and for transmitted light [16]. Since we demand from a prism coupler that light incident under an angle Θ is guided into a waveguide parallel to the prism's surface, the sine of the output angle equals unity, and the coupling condition for the prism coupler simply requires that

$$n_{\text{eff}}(\lambda) = n_{\text{p}}(\lambda) \sin\theta, \qquad (12.4)$$

where n_{p} is the refractive index of the prism; a standard glass prism has $n_{\text{p}} \cong 1.5$. If a prism coupler is used for an optical biosensor using surface plasmons, it is important that the metal film on the prism is thin enough so that the evanescent wave on the metal's other side is formed without too many losses. In practice, the thickness of the metal film is typically 40–50 nm.

Waveguides with Grating Coupler

The same reasoning as for the prism coupling condition applies to the case of a grating coupler with period Λ. The grating coupling condition is given by

$$n_{\text{eff}}(\lambda) = n_2(\lambda) \sin\theta - m\frac{\lambda}{\Lambda}. \qquad (12.5)$$

According to (12.2), the effective refractive index n_{eff} of the guided mode is larger than the refractive index of the substrate n_2. It is immediately obvious from (12.5) that light can be coupled in from any substrate, even from air ($n_2 = 1$), since the integer diffraction order m, which can also be a negative number, gives us a large degree of freedom for our choice of coupling configuration.

In practice, coupling gratings are very shallow and the core layer of the dielectric slab waveguide is rather thin. Typical values are 100–200 nm for the core thickness, 300–500 nm for the grating period, and 5–50 nm for the grating depth [1].

Waveguides with End-Face Coupler

Since the core of the waveguide is so thin, the "obvious" end-face coupling of light with a focusing lens, as illustrated in Fig. 12.6, is very difficult to achieve. The alignment tolerances and the mechanical stability of the optical setup must be below 100 nm, which is quite difficult to achieve in practice, where optical biochips should be disposable, quick to read out, and mechanically very robust. Therefore, end-face coupling is not used in commercial optical biosensors.

Resonance Condition for Evanescent Wave Sensing

From (12.4) and (12.5) it becomes clear how measurements are carried out in optical biosensors: The presence of the analyte subtly changes the refractive index in the sample volume. This change is sensed by the evanescent wave, since it implies a change in n_{eff} of the guided wave. A change in n_{eff} alters the coupling condition, which can be adjusted either with the in- or out-coupling angle Θ or with the wavelength λ. For this reason, evanescent wave sensing in optical biochips as described above is a resonance method: The amount of light in the guided wave is maximum for the combination of Θ and λ that fulfills the coupling condition for the present value of n_{eff} which is dependent on the analyte concentration.

The most sensitive methods for measuring this resonance condition are capable of resolving changes in the refractive index of about 10^{-7} [1, 2]. This corresponds to a mass detection sensitivity of nearly $100 \, \text{fg mm}^{-2}$, and the molar concentration of the analyte can be determined with a sensitivity of about 10^{-11}, i.e., the concentration sensitivity is $10^{-11} \, \text{mol l}^{-1}$. Note that it is important to stabilize for temperature variations at these high measurement sensitivities, since $\partial n / \partial T \approx 10^{-4} \, \text{K}^{-1}$ for water at room temperature [17].

12.4.2 Fluorescence Sensing

Since fluorescence (or luminescence) sensing requires the imaging of two-dimensional distributions of excited or self-luminous light patterns, the usual sensing technique is fluorescence (or luminescence) microscopy [4]. For this purpose, either conventional optical microscopes or scanning (confocal) optical microscopes are employed, where the sample is typically illuminated through the same lens system through which the fluorescence light is collected (so-called epi-fluorescence excitation setup) [4].

The high demands on the collection and detection efficiency of the emitted light in fluorescence (and luminescence) sensing could also be met with an alternative, as yet little explored sensing approach. Instead of using an inactive substrate such as glass or plastic for the fixation of the receptor molecule, the receptors can be bonded directly to the surface of a solid-state image sensor. In this way, a large part of the light that is emitted close to the pixel surface, usually less than 1 μm from the photosensitive semiconductor material, can be collected and detected. Although this approach has been studied by several companies, no product has yet appeared on the market. This is probably due to the current cost of high-sensitivity image sensors. Since it is expected that the dropping prices and the increased sensitivity of cell phone cameras will soon have an influence on the use of these components in other areas, in particular in the life sciences, this alternate approach to fluorescence (and luminescence) sensing might become important in the future.

12.5 Readout Methods for Evanescent Wave Sensors

As mentioned earlier, the analyte concentration is measured in surface plasmon or in dielectric waveguide sensors by determining the maximum of a resonance effect. The following realizations of optical readout principles have been successfully employed in practice for this purpose:

12.5.1 Angular Scanning

The most common methods to satisfy the in-coupling condition (12.4) or (12.5) is to work with a convergent beam of light or to mechanically adjust the angle of a parallel light beam. In both cases, it is assured that a propagating mode is created at the sensor surface. While it is much simpler to employ a convergent beam of light, the efficiency is much lower compared to a mechanically adjusted system because the in-coupling condition is fulfilled only for a small fraction of the incident light.

No mechanical adjustment must be foreseen on the detection side, however, because one- or two-dimensional solid-state image sensors are employed, which can detect the resonance maximum without any loss of light.

In Fig. 12.7, a complete such system based on prism coupling is schematically illustrated, together with typical sensor signals acquired with a solid-state line-sensor. In this example, a convergent beam of light is employed, so that the complete system can be realized without any mechanically adjustable components. Such systems are commonly used with surface plasmon sensing.

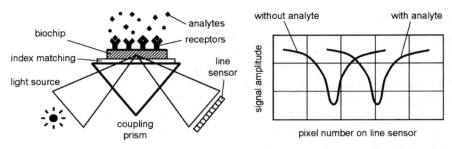

Fig. 12.7. Schematic illustration of an optical biosensor based on prism coupling (*left*) and typical line-sensor signals acquired with the system (*right*)

12.5.2 Wavelength Tuning

The light loss or the complexity of angular adjustment of the input light can be avoided if the wavelength of the incident light is modified instead. An efficient, all solid-state optical biosensor results, in which the wavelength tuning of the light source is realized by tuning the drive current of a VCSEL (vertical cavity surface emitting laser diode) [18].

12.5.3 Grating Coupler Chirping

Another approach to avoid the mechanical adjustment of the angles of the incident and the out-coupled light is to work with gratings whose period is varied laterally. The use of such chirped gratings ensures that there is always a part of the grating for which the in- or out-coupling condition (12.5) is fulfilled. This technique has the interesting property that the lateral propagation location of the guided wave is a direct measure of the resonance position. For this reason, the method has also been called the "light pointer" approach to optical biosensing [19].

A significant disadvantage of the chirped grating technique is the low efficiency with which the available light is used, similar to the convergent beam approach described in Sect. 12.5.1. Only the fraction of light for which the in-coupling condition is fulfilled can propagate in the waveguide and contributes to the sensing signal; the largest part of the optical energy is just transmitted through the grating.

12.6 Substrates for Optical Biochips

Although optical biochips are most often disposable devices, the optical sensing techniques employed put high demands on the surface quality, the mechanical stability, and the robustness of the used materials. At the same time, the materials should be of low cost, easy to process, and of low environmental concern. For these reasons, the dominant substrate materials for optical biochips are glass, fused silica (high purity synthetic amorphous silicon dioxide), or mono-crystalline silicon as used in large quantities for the semiconductor industry. Since the interface between the coupling prism and the biochip does not need to be patterned – it must only be optically flat – glass or fused silica are the preferred substrate materials for the prism coupling approach. This is particularly true for surface plasmon resonance (SPR) biosensors.

On the other hand, grating coupling requires the fabrication of coupling gratings on the substrate. Although glass is still the material of choice for this application, the use of plastic or polymeric materials such as PMMA (polymethylmethacrylate), PDMS (polydimethylsiloxane) or Ormoceres^{TM} is being investigated, since surface gratings can easily be replicated in these materials with nanometer precision. Before these materials are employed routinely in practice, however, several problems regarding their surface roughness, the mechanical stability, and the physical/chemical inertness of the surfaces need to be resolved satisfactorily. An additional problem that needs to be addressed is the thermal expansion coefficient of polymeric materials, which is typically ten times as large as the one of glass or silicon [17].

Fluorescence and luminescence sensing, in particular, using the special types of microarrays discussed in Sect. 12.2, is most often realized on glass or silicon substrates. Since the materials must neither be transparent nor of particular optical quality, the choice of cost-effective substrates is larger than that for the other optical biosensors. Because of the ease with which the microfluidic functionality can be integrated into plastic or polymeric materials (e.g., by injection molding), these materials are also of interest for fluorescence biosensors [20].

12.7 Realization Example of an Optical Biosensor/Biochip: WIOS

As a practical realization example of an optical biochip and its associated readout system, the wavelength-interrogated optical sensor ("WIOS") principle described in [18] is explained in more detail. This label-free optical biosensor is based on a dielectric waveguide with grating couplers for coupling light into and out of the waveguide.

The disposable biochip consists of a $12.5 \times 12.5\,\mathrm{mm}^2$ AF45 glass substrate of 0.7 mm thickness covered with a 150 nm thick film of Ta_2O_5 ($n = 2.13$) with a grating structure that was previously etched into the glass using dry etching. This grating has a period of $\Lambda = 360$ nm and a thickness of 13.2 nm. To minimize interference effects between the input and the output grating, the Ta_2O_5 film thickness was fabricated 150 nm thicker at the output grating, resulting in a smaller out-coupling angle. Since the complete input/output grating structure measures only $0.8 \times 1.0\,\mathrm{mm}^2$, several of them can be produced on the same biochip. The input grating of the sensing pads is selectively sensitized with suitable receptor molecules that are photo-bonded to the input

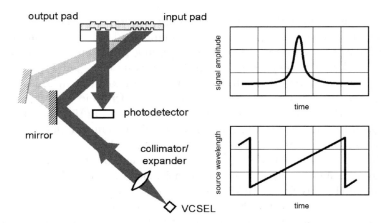

Fig. 12.8. Measurement principle of the WIOS optical biosensor [18]

grating surface using OptoDex^{TM} photolinker material. The sensor chip is then placed into a fluidic cell and sealed.

The disposable biochip is read out with a compact optical system illustrated schematically in Fig. 12.8. It consists of a VCSEL light source with center wavelength of 763 nm and an electrical wavelength-tuning range of 2 nm. The laser beam is collimated and expanded to cover the input pad of the biosensor with a suitable optical sub-system. Because of the limited wavelength tuning range, an additional deflection mirror has been placed into the input beam path, so that a deflection range of 6° is obtained. The light from the output grating is collected by an optical multi-mode fiber and transported to a silicon photosensor array for detection, amplification, and analysis. In operation, the VCSEL current is modulated with a sawtooth function, resulting in a sawtooth modulation of the VCSEL's emission wavelength. The photodetector's signal is analyzed for the precise temporal position of the resonance peak, and this time shift is a direct measure for the refractive index change and the analyte concentration in the sample volume. As described in [18], the WIOS biochip and readout instrument reach an experimental uncertainty of the refractive index measurement of 3.1×10^{-7}, corresponding to a mass coverage detection standard deviation of about 100 fg mm^{-2}. These values were obtained for the rather small Biotin molecule with a molecular weight of only 244 Da, using neutravidin as the receptor.

12.8 Outlook: Lab-on-a-Chip Using Organic Semiconductors

In the future, optical biosensors will contain more and more functionality on the same substrate for increased ease-of-use, higher reliability, and lower cost. This will lead to highly integrated lab-on-a-chip systems that will see a large number of additional uses at home, in industry, in safety/security, and for environmental applications. Current approaches employ many different materials to perform the various required functions: Optical, mechanical, fluidic, optoelectronic (light generation and detection), signal processing, information display, etc. The resulting solutions are hybrid systems, consisting of many different components, whose integration into a final product is not very cost-effective.

There is a promising class of materials, however, that might make it possible to realize a monolithic lab-on-a-chip for low-cost optical biosensors: organic semiconductors, in particular, polymer semiconductors. In the following, a few salient points of this material class and its potential for the fabrication of highly functional yet cost-effective optical biosensors are summarized.

12.8.1 Basics of Organic Semiconductors

The scientific field of organic semiconductors is barely 30 years old, and it started with the development of organic light-emitting diodes (LED) with useful conversion efficiency by the American company Kodak. Since then, this field has progressed very rapidly, due to the promise of a new class of materials offering all the optoelectronic functionality required for the realization of generic photonic microsystems: light generation, light detection, analog and digital electronic signal processing, as well as photovoltaic power generation [21]. In addition, efficient and fast production methods are available, which are often modified printing techniques such as inkjet, silk-screen, or gravure printing. Therefore, it is anticipated that the fabrication costs will be about 100 times lower than that for silicon-based semiconductor circuits in the near future (€0.1 cm^{-2} compared to €10 cm^{-2} for standard CMOS chips).

However, the performance of organic semiconductors can be significantly lower in several respects than the performance of their inorganic counterparts. The charge carrier mobility is about 1,000 times lower (about 1 cm^2 V^{-1} s^{-1} compared to 10^3 cm^2 V^{-1} s^{-1}), the diffusion length of good material is several orders of magnitude smaller (typically 10 nm compared to several 100 μm), the electrical resistivity is significantly higher (since it is much more difficult to dope organic semiconductors), and organic semiconductors deteriorate much more rapidly if exposed to ultraviolet light, to oxygen or to water vapor.

For these reasons, organic semiconductors are not a cheaper and simpler replacement of inorganic semiconductors, and it is rather necessary to evaluate the performance of each device individually for the intended application.

12.8.2 Organic LEDs

Since a large variety of organic semiconductors are available, organic LEDs (oLED) of many different colors can be fabricated [22]. Their external quantum efficiency can exceed 10%, and their power efficiency already surpasses the performance of conventional light sources. The highest luminous efficacy of a white oLED reported in mid 2006 by the Japanese company Konica Minolta is $64\,\mathrm{lm\,W^{-1}}$, which compares very favorably with the luminous efficacy of conventional light bulbs ($15\,\mathrm{lm\,W^{-1}}$), halogen lamps ($25\,\mathrm{lm\,W^{-1}}$), or compact fluorescent lamps ($60\,\mathrm{lm\,W^{-1}}$).

Because of the simplicity and the efficiency with which oLEDs – also of different color – can be produced on large surfaces, they are increasingly used for the fabrication of displays in cameras, for PCs, and for TV screens.

12.8.3 Organic Lasers

The first optically pumped organic (polymer) laser has been demonstrated already in 1996 [23]. However, because of the high electrical resistivity of organic semiconductors, the electrically pumped organic CW laser working at room temperature remains still elusive.

The fact that no intense, narrow-band or even monochromatic light sources can currently be fabricated with organic semiconductors is one of the major shortcomings for the realization of monolithic optical biochips and labs-on-a-chip using solely organic materials.

12.8.4 Organic Photodetectors and Image Sensors

Once the problem of dissociation of photogenerated, tightly bound charge pairs (Frenkel excitons) had been solved using finely dispersed polymer blend heterojunctions, the quantum efficiency of photodetectors fabricated with organic semiconductors approached that of inorganic semiconductors [24]. Total external quantum efficiencies exceeding 50% are routinely obtained today in photosensors realized with organic semiconductors. By using thin multilayer structures, it is also possible to produce high-speed photodetectors despite the low mobility of charge carriers in organic semiconductors: Maximum detection frequencies of about 500 MHz have already been demonstrated.

A particular advantage of photodetectors fabricated with organic semiconductors is the high number of sensors (pixels) that can be realized simultaneously on a very large substrate. As an example, the cost-effective fabrication of page-sized image sensors has already been demonstrated.

Since different organic semiconductors with various spectral responses can be easily integrated on the same substrate, the fabrication of color photosensors with good colorimetric performance has been achieved already early in [24].

12.8.5 Organic Photovoltaic Cells

One of the major problems of a practical, disposable optical biochip is electrical power supply. If it were possible to replace the conventional electrochemical batteries of environmental concern with another, less harmful source of electrical power, this would add considerably to the attractiveness of an all-organic optical biochip solution.

Since it is possible to fabricate efficient photodetectors using organic semiconductors, one immediately thinks of the monolithic cointegration of photovoltaic cells on an optical biochip. This is indeed feasible, and all-organic solar cells have been successfully demonstrated. Because of the relatively large fraction of power contained in the near infrared spectral range of sunlight, where organic semiconductors are notoriously insensitive, the total quantum efficiency of organic photovoltaic cells is rather limited. The maximum quantum efficiency demonstrated up to date does not exceed 5%.

12.8.6 Organic Field Effect Transistors and Circuits

The first field-effect transistors (FETs) and electronic circuits based on organic semiconductors were demonstrated in 1995 [25]. The electrical characteristics of these organic FETs are similar, in principle, to the one of inorganic FETs. However, because of the small charge carrier mobility and low electrical conductivity of organic semiconductors, the speed and the current density of organic FETs are much lower than those in their inorganic counterparts. The switching speed is currently limited to a few megahertz, and the highest frequency at which an organic circuit has worked to date is 13.56 MHz, a frequency of importance for RF-ID electronic tags and inductive data communications.

An additional problem is the difficulty of cointegrating both n-MOS and p-MOS transistors with comparable performance using compatible organic semiconductors. For this reason, power-efficient CMOS circuits cannot yet be routinely integrated with organic semiconductors, and ultra-low-power electronics still remains in the realm of inorganic semiconductors.

12.8.7 Monolithic Photonic Microsystems Using Organic Semiconductors

Organic semiconductors allow the fabrication of similar electronic and optoelectronic components as produced with inorganic semiconductors. In almost no respect are organic components superior in their performance compared to their inorganic counterparts. The huge appeal of organic semiconductors is in the ease, rapidity, and low cost of their production using modified printing techniques on large areas (several square meters) with very little material: only 1 g of organic semiconductor is sufficient to produce about $10\,m^2$ of "plastic optoelectronics." Another unique property offered by organic semiconductors is the possibility of monolithic fabrication of complete photonic microsystems on a single substrate: Analog and digital electronics, light generation, light detection, and even a photovoltaic power supply can all be integrated on the same chip [26]. Once the diffusion barrier and sealing problem is solved, it will even become possible to produce these complete photonic microsystems on flexible substrates with a lifetime of many years.

12.9 Conclusions and Summary

Optical biochips of various integration levels have become important tools in many different areas of the life sciences: Genomics, proteomics, medical diagnostics, pharmaceutical drug screening, contamination detection in the food industry, environmental and pollution monitoring, as well as counter-terrorism all make substantial use of this technology, due to its simplicity, reliability, sensitivity, and low cost. Increasing levels of integration will soon lead to complete, disposable labs-on-a-chip with even higher functionality and performance. The availability of organic semiconductors makes it possible to fabricate complete monolithic photonic microsystems on a single substrate, for example, on an injection-molded piece of polymer containing the entire microfluidics part of a lab-on-a-chip. This will open up many more application areas to optical biochips, and it will make them highly versatile and indispensable tools of our everyday lives.

Acknowledgments

This contribution could not have been prepared without the invaluable help of M. Wiki and R.E. Kunz, both at CSEM SA, whose generous support I gratefully acknowledge.

References

1. R.E. Kunz, in *Integrated Optical Circuits and Components*, ed. by E.J. Murphy (Marcel Dekker, New York, 1999), p. 335
2. P.V. Lambeck, Meas. Sci. Technol. **17**, 93 (2006)
3. J. Tschmelak et al., Talanta, **65**, 313 (2005)
4. P.N. Prasad, *Introduction to Biophotonics* (Wiley, Hoboken, NJ, 2003)
5. Affymetrix Inc., 3380 Central Expressway, Santa Clara, CA 95051, USA, http://www.affymetrix.com
6. M. Schena (ed.), *Microarray Biochip Technology* (Eaton Publishing, Natick, MA, 2000)
7. E. Oosterbroek, A. van den Berg (eds.), *Lab-on-a-Chip* (Elsevier, Amsterdam, Netherlands, 2003)
8. A. Manz, N. Graber, H.M. Widmer, Sens. Actuators B **1**, 244 (1990)
9. E.J. Cho, F.V. Bright, Anal. Chem. **73**, 3289 (2001)
10. A.C.R. Grayson et al., Proc. IEEE **92**, 6 (2004)
11. R. Bahir, Adv. Drug Deliv. Rev. **56**, 1565 (2004)

12. H.J. Fecht et al., *Nanotechnology Market and Company Report 2003* (WMtech and University of Ulm Publishing, D-Ulm, 2003)
13. J. Ziauddin, D.M. Sabatini, Nature **411**, 107 (2001)
14. J.A.B. Kononen et al., Nat. Med. **4**, 844 (1998)
15. P. Seitz, in *Computer Vision and Applications – A Guide for Students and Practicioners* (Academic Press, San Diego, CA, 2000)
16. J.D. Jackson, *Classical Electrodynamics*, 3rd edn. (Wiley, New York, 1998)
17. D.R. Lide (ed.), *CRC Handbook of Chemistry and Physics*, 87th edn. (Taylor & Francis, Bota Raton, FL, 2006)
18. K. Cottier et al., Sens. Actuators B **91**, 241 (2003)
19. R.E. Kunz et al., Sens. Actuators A **47**, 482 (1995)
20. F. Fixe et al., Nucleic Acids Res. **32**, e9 (2004)
21. S.R. Forrest, IEEE J. Sel. Top. Quantum Electron. **6**, 1072 (2000)
22. R.H. Friend et al., Nature **397**, 121 (1999)
23. M.D. McGehee, A.J. Heeger, Adv. Mater. **12**, 1655 (2000)
24. G. Yu et al., Synth. Met. **111–112**, 133 (2000)
25. E. Cantatore, in *Proceedings of ESSDERC 2001,* (Nürnberg, Germany, 11–13 Sept. 2001)
26. P. Seitz, U.S. Patent 7,038,235, 2 May 2006

13
CMOS Single-Photon Systems for Bioimaging Applications

E. Charbon

13.1 Introduction

Recent advances in neurobiology and medical imaging have put an increasing burden on conventional sensor technology. This trend is especially pronounced in time-correlated imaging and other high precision techniques, where timing accuracy and sensitivity is critical.

Techniques have been proposed to achieve high speed in conventional charge-coupled devices (CCDs) and CMOS active pixel sensor (APS) architectures [1–4]. Among some of the most successful techniques are ultra-fast low-noise electronic readout circuitries, on-pixel A/D conversion, and local analog electrical storage. However, in general a significant design effort and considerable experience is needed to achieve satisfactory results.

Alternatives to conventional CCDs and CMOS APS sensors are single photon counters (SPCs). Several types of SPCs have been known for decades. Among the most successful devices in this class are microchannel plates (MCPs) and photomultiplier tubes (PMTs) that have become the sensors of choice in many applications [5]. Even though they have been studied since the 1960s [6], silicon avalanche photodiodes (SiAPDs) have become a serious competitor to MCPs and PMTs only recently. In SiAPDs, carriers generated by the absorption of a photon in the p-n junction are multiplied by impact ionization; thus producing an avalanche. The resulting optical gain is usually in the hundreds. The main drawback of these devices, however, is a relatively complex amplification scheme and/or complex ancillary electronics. In addition, specific technologies are often required.

If biased above breakdown, a p-n junction can operate in so-called Geiger mode. Such a device is known as single photon avalanche diode (SPAD). In Geiger mode of operation, SPADs exhibit a virtually infinite optical gain; however, a mechanism must be provided to quench the avalanche. There exist several techniques to accomplish quenching, classified in active and passive quenching. The simplest approach is the use of a ballast resistance. The avalanche current causes the diode reverse bias voltage to drop below breakdown; thus pushing the junction to linear avalanching and even pure accumulation mode. After quenching, the device requires a certain recovery time to return to the initial state. The quenching and recovery times are collectively known as *dead time*.

Recently, SPADs have been integrated in CMOS achieving timing resolutions comparable to those of PMTs [7]. Current developments in more advanced CMOS technologies have demonstrated full scalability of SPAD devices, a 25 μm pitch, and dead time as low as 32 ns. The sensitivity, characterized in SPADs as photon detection probability (PDP), can exceed 25–50%. The noise, measured in SPADs as dark count rate (DCR), can be as low as a few hertz [8,9]. Thanks to these properties, SPCs based on SPADs have been proposed for imaging where speed and/or event timing accuracy are critical. Such applications range from fluorescence-based imaging, such as Förster Resonance Energy Transfer (FRET), fluorescence lifetime imaging microscopy (FLIM) [10], and fluorescence correlation spectroscopy (FCS) [11], to voltage sensitive dye (VSD) based imaging [12,13], particle image velocimetry (PIV) [14], instantaneous gas imaging [15,16], etc.

In the following sections we explore some applications of SPCs in comparison to conventional sensors, including potential fields of imaging where SPADs can be a compelling implementation aspect of SPCs. We also look at performance and architectural issues that the designer needs to take into account when approaching real problems involving the use of SPCs.

13.2 Spectroscopy

Spectroscopy-based imaging has many incarnations. One, very successful one, is known as fluorescence correlation spectroscopy (FCS); a technique used to measure transitional diffusion coefficients of macromolecules to count fluorescent transient molecules or to determine the molecular composition of a fluid being forced through a bottleneck or a gap. In FCS, a femtoliter volume is exposed to a highly focused laser beam that causes the molecules in it to emit light in a well-defined spectrum. Figure 13.1 shows an example of optical molecular response depending on the size and diffusion pattern of the molecule. The physical causes of this behavior are related to the mobility of the ligands. In the first case, free fluorescent ligands are continuously entering and leaving the detection volume. In the second case, macromolecule ligands are less mobile; thus producing slower but highly correlated intensity fluctuations. Figure 13.2 shows an example of typical autocorrelation functions simulated for different molecules [17].

A tighter correlation is observable in the case of low molecular weight ligands. A macromolecule ligand generates a much more relaxed correlation. A mixture of free and bound ligands is shown in the middle curve.

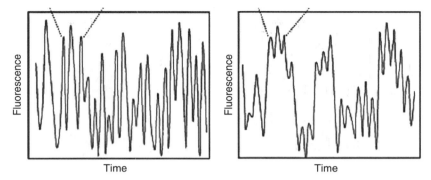

Fig. 13.1. Optical response of molecules bombarded by highly focused laser beam. Rapidly diffusing small molecule (*left*); slow, large molecule, with its large well-defined bursts of optical energy (*right*)

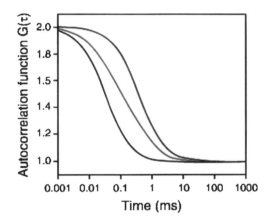

Fig. 13.2. Autocorrelation of fluorescence response. High time resolution is especially useful for sub-nanosecond response dynamics

Dual-color cross-correlation FCS measures the cross-correlation of the fluorescence intensities of two distinct dyes [18]. Thus, it becomes possible to detect different molecules without reference to their diffusion characteristics. More recently, a combination of the two methods has been reported to reduce the need for a distinct, often rather large mass ratio between the two molecule types [19]. A typical FCS setup is shown in Fig. 13.3. A highly focused laser beam is directed towards the detection volume. The reflected light is applied to the high sensitivity optical sensor through a prism. A digital correlator matches the autocorrelation function with a database of known responses. Generally, in FCS time resolutions of a few tens of picoseconds and sensitivities equivalent to a few hundred photons are needed.

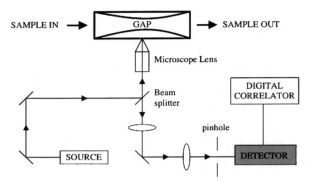

Fig. 13.3. Basic FCS setup (gap volume not to scale). Single or multipixel detectors can be used in this setup. With multipixel imaging devices information can be gained from the emissions from the surroundings of the molecule

13.3 Lifetime Imaging

Among time-correlated imaging methods, time-correlated single photon counting (TCSPC) is perhaps one of the most used in bioimaging. Multiple exposures are employed to reconstruct the statistical response of matter to sharp and powerful light pulses. The statistics are generally represented in form of a histogram, while light source repetition frequencies may vary from kilohertz to hundreds of megahertz. From the response statistics, biologists generally extract parameters that can be used to characterize the molecule under observation and/or its environment, e.g., the calcium concentration.

The study of calcium at the cellular level has made intensive use of fluorescent Ca^{2+} indicator dyes. Examples of heavily used dyes or fluorophores are Oregon Green Bapta-1 (OGB-1), Green Fluorescent Protein (GFP), and many others. Calcium concentration can be determined precisely by measuring the lifetime of the response of the corresponding fluorophore, when excited at a given wavelength.

Using FLIM, for example, one can determine the variation of calcium concentration in neural cells as a function of a given activity. Figure 13.4 shows a conceptual setup where a neural cell soma is being exposed to light via, for example, a fiber. The fluorophore molecules, previously injected into the cell, exhibit different lifetimes depending upon the calcium concentration in the vicinity of ion channels. There exist several flavors of FLIM based on how lifetime is characterized or based on the excitation mode. In one-photon FLIM, for example, only one photon is required to force a state change in the fluorophore molecule [10]. In this case, only a small shift in wavelength is observed between the excitation and response. As a consequence, filtering the excitation pulse from the measurement response may be challenging. In addition, any scattering occurring along the observation path induces photonic noise (usually at the same or close wavelength of fluorescence emissions)

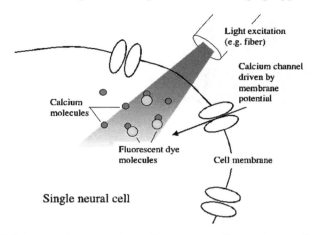

Fig. 13.4. Calcium environment in and around a cell membrane. Calcium moves through the membrane via so-called ion channels. A fluorophore can detect small variations of calcium concentration

into the measurement. Nonetheless, the electro-optical setup required by this method is generally less critical and relatively easy to build.

Multiphoton FLIM, even though conceptually known for decades, only recently has become an essential tool for neurobiology and other disciplines. Two-photon FLIM has been the method of choice, thanks to the recent advances in laser technology that can now concentrate kilowatts of light power in micrometric volumes of matter. There are several advantages to two-photon FLIM. First, fundamental spatial confinement for the excitation can be achieved; thus allowing one to isolate a single molecule or cluster of molecules. Second, because of the reduced average optical powers in play, effects such as photobleaching and photoxicity can be mitigated; thereby enhancing the suitability of the approach. Third, a better penetration in turbid media can be achieved, due to reduced scattering. Fourth, the large spectral difference between in- and out-going radiation simplifies the separation of response from excitation [10].

The main components of typical two-photon FLIM setups are a high-power femtosecond source, generally a mode-locked Ti:Sapphire laser; a SPC; time discrimination hardware; a standard laser-scanning microscope. An example of laser-scanner based FLIM image is shown in Fig. 13.5. In this experiment it was possible to detect intra-cellular chemical waves to help identify innerworkings of pathogens or the impact of certain pharmaceuticals [20]. The binding of neutrophil at the interface of two endothelial cells was monitored at high speed via calcium-triggered chemical waves propagating through the cells.

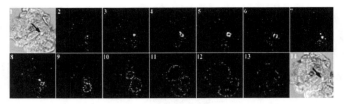

Fig. 13.5. High-speed image sequence of the binding of neutrophil. (Courtesy of H.R. Petty [20])

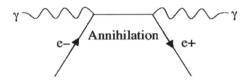

Fig. 13.6. Coplanar gamma emission where a positron annihilates with an electron. The synchronicity of the emission may be used to infer the loci of the emission using multiple synchronized detectors

13.4 Time-of-Flight in Bio- and Medical Imaging

Time-of-flight (TOF) is the time a light ray takes to propagate between two points in the three-dimensional space. There exist several applications requiring a precise measurement of TOF to image particular properties of targets and environments. Let us consider two examples of such applications: positron emission tomography (PET) and depth map imaging.

In a PET system, a positron is emitted in the tissue being imaged by a variety of substances. An example of one such substance is fluorodeoxyglucose (FDG). This compound is absorbed by human cells, in the brain for example, at different rates based on the operation – whether normal or abnormal – of the cell. After positron emission, a annihilation occurs followed by gamma ray emission.

Gamma emission is well defined spatially and temporally. In fact, annihilation results in a pair of gamma photons in opposite directions (coplanar property) at exactly the same time. Figure 13.6 shows schematically how emission occurs. To detect gamma emission one usually utilizes a crystal, e.g., lutetium oxy-orthosilicate (LSO), which converts gamma into visible radiation. Typically several thousands photons may be released in this fashion. The timing resolution of detection via LSO, however, is generally worsened due to the properties of these crystals. The lifetime of a typical LSO crystal may be as much as 39 ns; hence statistics must be used in combination with detector with high timing resolution to derive the first non-noise photon to be detected (Fig. 13.7). Such photon will give an approximation to the actual moment in time when the gamma radiation is released.

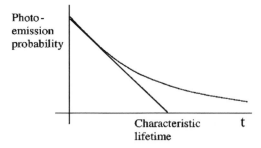

Fig. 13.7. Typical crystal photoemission probability as a function of time since gamma absorption

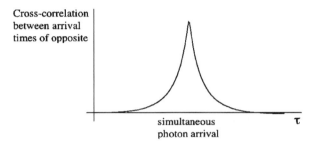

Fig. 13.8. Result of time correlation that enables the detection of the source of gamma radiation

To find the exact location of positron emission, one must monitor all gamma radiation reaching a pair of detectors on an axis at exactly the same time and then cross-correlate all estimated arrival times. The emission loci may be derived by measuring the TOF of the particle with respect to a reference point of known coordinates. The tomography of emissions may be constructed using conventional Fourier transform techniques in combination with a mechanical system where a detector pair rotates around an axis longitudinal to the cylindrical volume being probed (Fig. 13.8). The pair may be further translated after each rotation cycle to complete a cylindrical scan. The literature on TOF imaging and gated sensors (both CCD and CMOS APS) is very extensive, a few examples are found in [8, 21–23].

13.5 System Considerations

A SPC may be implemented in a number of ways. If the application requires an array of simultaneously operating single photon detectors (SPDs), then a SPAD array is a desirable alternative to PMT or SiAPD arrays in terms of cost, power consumption, and miniaturization. Several demonstrations of CMOS SPAD arrays exist in various technologies [8, 9, 24, 25]. The methods

Table 13.1. Performance parameters of a generic time discriminator implemented in a conventional IC technology

Measurement	Value	Unit
Timing resolution	$1 \sim 100$	ps
Timing accuracy	< 30	ps
Temperature stability	$10 \sim 1,000$	ppm/$°$C
Dead time	$1 \sim 1,000$	ns

to read the output of every SPD pixel to the external world range from pixel random access, similar to APS architectures, to event-driven approaches [9,25] to pipelined readout methods [26].

Time-correlated counting can be performed either via a fast clock or dedicated time discrimination circuit (Table 13.1). Time-uncorrelated counting on the contrary requires only a counter whose bandwidth is determined by the inverse of the dead time of the measurement. If counting is performed on-pixel, the operation of the sensor becomes relatively straightforward, but the silicon real estate may not be utilized efficiently. In principle, this configuration enables the best time utilization since it maximizes parallelism.

If counting is performed on-column, SPDs may be implemented in a smaller area and a smaller pitch may be obtained; however, a mechanism must be devised for pixels to share the column counter. As a consequence, the time utilization efficiency is reduced and the readout complexity increases. A good trade-off has been shown to be an event-driven readout system, which works best with low photon fluxes [9,25]. In this approach a SPD uses the column as a bus, accessing it only when it absorbs a photon. The saturation of the device is thus determined by the bandwidth of the column divided by the number of rows.

A chip-wide counting is the technique that allows the simplest architecture, while it is the least time-efficient, since only one SPD may be counting at any point in time. Hence the maximum achievable frame rate is limited by the minimum exposure time per pixel, the total number of pixels, and the speed of the switching electronics.

Time-correlated counting may be implemented using fast time discrimination circuits running at lower speed, such as time-to-digital converters (TDCs), time-to-amplitude-converters (TACs), or correlators. A TAC is a device that converts a time interval into a voltage difference. A TDC converts a time interval directly into a digital code. A correlator relates a given impulse to another time-wise, determining in effect their phase difference. There exist a wide variety of implementations for time discriminators, differentiated in terms of performance, size, cost, and power dissipation. Figure 13.9 shows some of the main parameters. In the time-correlated mode of operation, a laser trigger may be used as a reference signal, call it START. The output of a SPD may be used to terminate the time measurement, call that STOP. The main drawback of these devices is their complexity, often requiring hundreds of transistors.

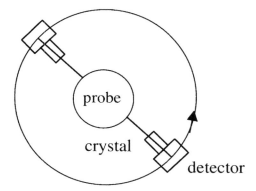

Fig. 13.9. Detector pair rotating around an axis and longitudinally translating through it

On-column and on-chip approaches are, seemingly, the sole option, along with off-chip solutions. An example of an off-chip approach was proposed in [26]. In this design, two technologies are used to independently implement SPADs and time discriminators that are subsequently connected electrically using specific techniques.

The development of architectures that support time-correlated modes with some degree of resource sharing is currently underway in many research groups. The main trade-off is at the architectural level, due to the nature of the signal generated by SPDs. In its most general implementation, an SPD generates a digital pulse when it detects a photon.

Application-specific optimal architectures are possible, provided a model of the application is built to characterize the performance of the sensor a priori. The sharing of resources may involve a number of pixels, say 4 or 16, or on-demand sharing based upon the reaction of SPDs may be used. Other trade-offs may include the complexity of the time discriminator itself.

13.6 Conclusions

With the introduction of CMOS single-photon avalanche diodes, it is possible today to achieve great levels of miniaturization. Not only large arrays of photon counters are now possible, but also very high dynamic range and timing accuracy have become feasible. Thanks to these advances, applications requiring time-resolved single photon detection have become possible. Other applications have reached unprecedented levels of accuracy. We have outlined some of these applications and we have discussed system issues related to these and novel applications in the field of bio- and medical imaging.

Acknowledgments

The author is grateful to his graduate students and to David Stoppa of FBK. This research was supported by a grant of the Swiss National Science Foundation and the Center for Integrated Systems, Lausanne.

References

1. T.G. Etoh, Proc. SPIE **3173**, 57 (1997)
2. T.G. Etoh et al., IEEE Trans. Electron Devices **50**, 144 (2003)
3. S. Kleinfelder, S. Lim, X. Liu, A. El Gamal, J. Solid-State Circuits, **36**, 2049 (2001)
4. G. Patounakis, K. Shepard, R. Levicky, in *IEEE Symposium on VLSI Circuits*, Kyoto, Japan, June 2005, p. 68
5. J. McPhate, J. Vallerga, A. Tremsin, O. Siegmund, B. Mikulec, A. Clark, Proc. SPIE **5881**, 88 (2004)
6. R.H. Haitz, Solid-State Electron. **8**, 417 (1965)
7. A. Rochas, Dissertation, EPF-Lausanne, 2003
8. C. Niclass, A. Rochas, P.A. Besse, E. Charbon, IEEE J. Solid-State Circuits, **40**, 1847 (2005)
9. C. Niclass, M. Sergio, E. Charbon, *Design and Test in Europe (DATE)*, Munich, Germany, March 2006
10. A.V. Agronskaia, L. Tertoolen, H.C. Gerritsen, J. Biomed. Opt. **9**, 1230 (2004)
11. P. Schwille, U. Haupts, S. Maiti, W.W. Webb, Biophys. J. **77**, 2251 (1999)
12. A. Grinvald et al., in *Modern Techniques in Neuroscience Research*, ed. by U. Windhorst, H. Johansson (Springer, Berlin Heidelberg New York, 2001)
13. J. Fisher, Opt. Lett. **29**, 71 (2004)
14. S. Eisenberg et al., Proc. SPIE **4948**, 671 (2002)
15. S.V. Tipinis et al., IEEE Trans. Nucl. Sci. **49**, 2415 (2002)
16. W. Reckers et al., in *Proceedings of 5th International Symposium on Internal Combustion Diagnostics*, Baden-Baden, June 2002
17. K.J. Moore, S. Turconi, S. Ashman, M. Ruediger, U. Haupts, V. Emerick, A.J. Pope, J. Biomol. Screen. **4**, 335 (1999)
18. P. Schwille, F.J. Meyer-Almes, R. Rigler, Biophys. J. **72**, 1878 (1997)
19. K.G. Heinze, A. Koltermann, P. Schwille, Proc. Natl. Acad. Sci. USA **97**, 10377 (2000)
20. H.R. Petty, Optics Photonics News **15**, 40 (2004)
21. C. Bamji, E. Charbon, U.S. Patent 6,515,740, 2003
22. R. Lange, Dissertation, University of Siegen, 2000
23. R. Jeremias, W. Brockherde, G. Doemens, B. Hosticka, L. Listl, P. Mengel, *Proceedings of IEEE ISSCC*, San Francisco, CA, USA 2001
24. D. Mosconi, D. Stoppa, L. Pacheri, L. Gonzo, A. Simoni, *Proceedings of ESSCIRC*, Montreux, Switzerland, Sept. 2006
25. C. Niclass, M. Sergio, E. Charbon, *Proceedings of ESSCIRC*, Montreux, Switzerland, Sept. 2006
26. M. Sergio, C. Niclass, E. Charbon, *Proc. ISSCC*, p. 120, 2007
27. B. Aull et al., *Proc. ISSCC*, p. 1179, 2006

14

Optical Trapping and Manipulation for Biomedical Applications

A. Chiou, M.-T. Wei, Y.-Q. Chen, T.-Y. Tseng, S.-L. Liu, A. Karmenyan, and C.-H. Lin

14.1 Introduction

The field of optical trapping and manipulation was started in 1970 by Ashkin et al. [1,2], when the group at Bell Laboratory demonstrated the following:

- A mildly focused laser beam in a sample chamber containing micron-size polystyrene beads suspended in water could attract the bead transversely towards the beam axis and drive the bead to move along the beam axis in the direction of beam propagation (Fig. 14.1a).
- A pair of mildly focused laser beams aligned coaxially and propagating in opposite direction with approximately equal optical power could stably trap a bead (suspended in water) in a three-dimensional space (Fig. 14.1b).
- A mildly focused laser beam directed vertically upward can stably levitate a bead by balancing the radiation pressure against the weight of the bead in water (Fig. 14.1c).

The tremendous growth of the field was further stimulated by another monumental paper by Ashkin et al. in 1986 [3], when the group reported that a single laser beam alone when strongly focused with high numerical aperture (NA > 0.6) could stably trap micron-sized polystyrene bead in water (Fig. 14.1d). This strongly focused single-beam configuration, known as single-beam gradient-force optical trap or optical tweezers, has since become one of the most widely used optical trapping configurations mainly due to its relative simplicity in experimental setup. Furthermore, potential biomedical applications were soon envisioned when noncontact and noninvasive optical trapping and manipulation of biological samples such as living cells, cell organelles, and bacteria by optical tweezers with a near infrared (NIR) laser beam were successfully demonstrated [4]. In addition to optical trapping and manipulation of a single particle outlined above, simultaneous optical trapping of multiple particles as well as independent manipulation of each particle has also been demonstrated by several methods, including, interferometric optical

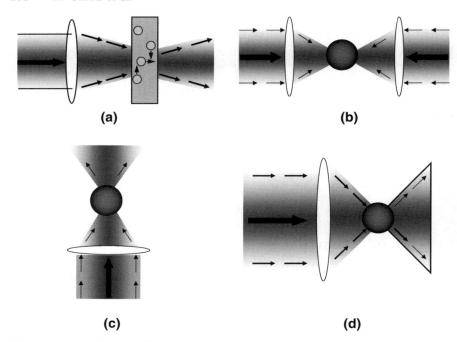

Fig. 14.1. A schematic illustration of optical trapping and manipulation of a microparticle suspended in water. (**a**) Transverse confinement and axial driving by a mildly focused laser beam; (**b**) three-dimensional trapping in a counter-propagating dual-beam trap; (**c**) optical levitation by a mildly focused laser beam pointing upward balancing against the weight of the particle; (**d**) three-dimensional trapping by a strongly focused (numerical aperture NA > 0.6) laser beam

patterning [5], interferometric optical tweezers [6], time-multiplexing with a scanning laser beam [7], and holographic optical tweezers [8–10]. In holographic optical tweezers, a set of predesigned computer generated holograms (CGH) in a spatial light modulator is often used to create dynamic multiple focal spots in a three-dimensional space to trap several particles and manipulate each independently. Likewise, a method known as generalized phase contrast (GPC) has been developed by Gluckstad et al. [11, 12] for the manipulation of multiple particles. Several very entertaining and impressive video demonstrations of multiple particle trapping and manipulation can be viewed at the website of Prof. Miles Padgett at the University of Glasgow in UK (http://www.physics.gla.ac.uk/Optics/).

Since the demonstration of optical tweezers (or single-beam gradient-force optical trap), the counter-propagating dual-beam trap has received much less attention until the demonstrations that (1) the counter-propagating dual-beam trap can be implemented fairly simply, without any focusing lens, by aligning a pair of single-mode fibers face-to-face in close proximity (approximately tens of microns to a few hundred microns) inside a sample

14 Optical Trapping and Manipulation for Biomedical Applications   251

chamber containing microparticles suspended in water [13,14], (2) the counter-propagating dual-beam trap can not only trap a red blood cell (RBC), osmotically swollen into spherical shape, but also stretch the spherical RBC to deform into an ellipsoid [15–17]. This technique has since been investigated for the measurement of the visco-elastic properties of various cells to correlate with their physiological conditions, including the feasibility of identifying a cancer cell against a normal cell [18].

Although a self-aligned counter-propagating dual-beam trap can be implemented, in principle, with a single input beam in conjunction with a nonlinear optical phase-conjugate mirror [19], it has not been further developed due to some practical limitation in the nonlinear optical properties of the materials available for optical phase-conjugation. More recently, three-dimensional stable trapping of a microparticle from a single laser beam emitting from a single-mode fiber with different structure fabricated at the tip of the fiber and without any external lens has also been successfully demonstrated [20, 21].

Another critical factor that further expands the potential biomedical applications of optical trapping is its capability to measure forces on the order of tens of femto-Newton to hundreds of pico-Newton. This topic will be discussed in greater details in Sect. 14.3 along with potential biomedical applications in Sect. 14.4, which together form the core of this chapter.

Other ramifications of optical trapping involve the integration of optical trap with one or more other optical techniques, such as near-field microscopy [22,23], Raman spectroscopy [24,25], and second and third harmonic generation microscopy [26], as well as with microfluidics for a wide range of particles or cells sorting [27–29].

In this chapter, we focus mainly on the measurement of optical forces in optical traps and potential applications of optical trapping as a force transducer for the measurement of biomolecular forces in the range of sub-pico-Newton to hundreds of pico-Newton. Partly because of the limited size of this chapter, and partly because of our lack of experiences with some ramifications of the subjects, several important topics such as those dealing with integrated photo-voltaic optical tweezers [30, 31] and optically driven rotation [32–34] of micronsized samples in optical traps are left out. Interested readers are encouraged to consult excellent accounts of these topics in the references cited above.

For those who want to learn the subject in greater depth and in more details, two excellent review papers [35, 36] each with an impressive list of references can be consulted. In addition, many leading research groups all over the world have posted incredible amount of information with very entertaining and informative video demonstrations of a wide range of features of optical trapping and manipulation. A few selected video demonstrations of various forms of optical trapping can also be viewed at our website (http://photoms.ym.edu.tw).

14.2 Theoretical Models for the Calculation of Optical Forces

Optical forces on microparticles in optical traps have been calculated theoretically mainly with the aids of two theoretical models: (1) Ray-Optics Model (or RO Model) [35–38] and (2) Electromagnetic Model (or EM Model) [35,36,39]. The former provides a good approximation for cases where the particle size is larger than the wavelength of the trapping light, whereas the latter represents a better model for cases where the particle size is smaller than the wavelength of the trapping light (often known as Rayleigh particles). For particle size comparable to the wavelength of the trapping light (often known as Mie particles), the calculation is much more involved and the results often less accurate. The key concepts of the RO Model and the EM Model are briefly outlined below further in this section.

14.2.1 The Ray-Optics (RO) Model

In the RO Model [35–38], each trapping beam is decomposed into a superposition of light rays, each with optical power proportional to the laser beam intensity distribution. For each specific light ray with optical power P in a medium with refractive index n_1, propagating along a direction represented by a unit vector $\mathbf{k}_i$, the photon momentum per second is given by $(Pn_1/c)\mathbf{k}_i$, where c is the speed of light in vacuum. When this light ray enter from a medium with refractive index n_1 into another medium with refractive index n_2, part of the optical power (RP) is reflected and the remaining part $(1-R)P$ is refracted (or transmitted) as is prescribed by the Fresnel equations, where R is the optical power reflectance at the interface due to Fresnel reflection. The photon momentum per second associated with the reflected light ray is thus $(RPn_1/c)\mathbf{k}_r$, and that associated with the refracted light ray is $[(1-R)Pn_2/c]\mathbf{k}_t$, where $\mathbf{k}_r$ and $\mathbf{k}_t$ are the unit vectors representing the propagation direction of the reflected and the transmitted rays, respectively. The net optical force imparted on the interface due to the photon momentum change associated with Fresnel reflection (and refraction) of this ray is given by

$$\mathbf{F} = (Pn_1/c)\mathbf{k}_i - \{(RPn_1/c)\mathbf{k}_r + [(1-R)Pn_2/c]\mathbf{k}_t\} \qquad (14.1)$$

or, equivalently,

$$\mathbf{F} = (Pn_1/c)\{\mathbf{k}_i - [R\mathbf{k}_r + (1-R)(n_2/n_1)\mathbf{k}_t]\}. \qquad (14.2)$$

In the RO Model, the contribution of the optical force from each constituent ray of the trapping beam, as is prescribed by (14.1) or (14.2) above, is added vectorially to obtain the net optical force on the particle by the beam at each interface. For a dielectric microsphere (such as a polystyrene bead or a biological cell) in water, the difference in refractive indices of the two media is often fairly small such that only the momentum change due to the Fresnel reflection

14 Optical Trapping and Manipulation for Biomedical Applications   253

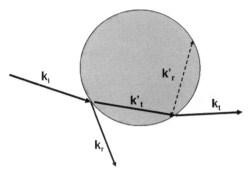

Fig. 14.2. Fresnel reflection and refraction of a light ray at the interface of a dielectric microsphere and the surrounding medium in the Ray-Optics (RO) model of optical trapping

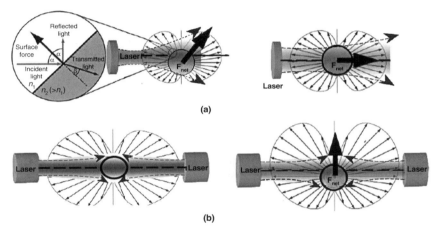

Fig. 14.3. Optical force distribution at the surface of a dielectric microsphere (adapted from Guck et al. [17]) (**a**) in a single Gaussian beam; (**b**) in a pair of counter-propagating Gaussian beams with equal optical power

at the first (or front) interface and that due to the transmission at the second (or back) interface are considered to simplify the calculation (Fig. 14.2). In general, optical forces originate from photon momentum changes due to reflection, giving rise to a net force, which mainly pushes the particle along the direction of the beam propagation; such a force is often referred to as the scattering force. In contrast, optical forces originating from photon momentum changes due to refraction often give rise to a net force pointing towards the direction of the gradient of the optical field; such a force is often referred to as the gradient force. Example of theoretical results for the optical force distribution on the surface of a dielectric microsphere due to a Gaussian beam and due to a pair of counter-propagating Gaussian beams with equal optical power are depicted in Fig. 14.3. Theoretical results from a more recent refined

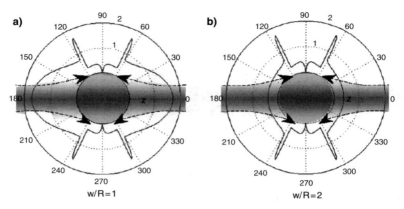

Fig. 14.4. Non-uniform force distribution with four distinguished peaks at the surface of a dielectric microsphere in a pair of counter-propagating Gaussian beams with equal optical power. Optical power $P = 100\,\text{mW}$, particle radius $R = 3.30\,\mu\text{m}$, refractive index of surrounding medium $n_1 = 1.334$, refractive index of the particle $n_2 = 1.378$, beam radius $= w$. (**a**) $w/R = 1$; (**b**) $w/R = 2$. (adapted from Bareil et al. [38])

model [38], however, indicate several distinguished spikes in the force distribution on the surface of a dielectric microsphere in a counter-propagating dual-beam trap (Fig. 14.4). In the counter-propagating dual-beam trap, the particle is stably trapped along the axial direction by the balanced scattering forces from the two beams, and in the transverse direction by the co-operative action of the transverse gradient forces from the two beams. In the single-beam gradient-force optical trap, the particle is stably trapped axially, balanced by the axial scattering force against the axial gradient force, and transversely by the transverse gradient force.

To compare the efficiency of different optical trapping configurations, the optical force as given in (14.2) above is often written as

$$F = (Pn_1/c)Q, \qquad (14.3)$$

where the parameter Q, known as the trapping efficiency, represents the fraction of photon momentum per second associated with the trapping beam, which is converted into the net trapping force. In most of the optical trapping configurations that have been demonstrated to date, the transverse trapping efficiency is typically on the order of 0.1–0.001, whereas the axial trapping efficiency is typically a factor of 3–10 smaller than the transverse trapping efficiency. The exact value of the trapping efficiency depends on many factors, including the numerical aperture (NA) of the trapping beam, the size of the particle, the refractive index of the particle and that of the surrounding fluid, etc.

14 Optical Trapping and Manipulation for Biomedical Applications

Fig. 14.5. (a) The EM model of optical trapping of a dielectric microsphere in a Gaussian beam in terms of the minimum potential energy analogous to (b) the force that pulls a dielectric block partially filling a parallel plate capacitor connected to a constant voltage source

14.2.2 Electromagnetic (EM) Model

In the EM model [35, 36, 39], stable trapping of a dielectric microparticle in one or more laser beams can be understood in terms of the potential energy minimum of the dielectric particle in the electric field associated with the optical beam. The physical mechanism that a dielectric microparticle is attracted towards the region with high optical intensity is analogous to the electrostatic force on a dielectric block partially filling a parallel plate capacitor connected to a constant voltage source (Fig. 14.5). In cases where the dielectric constant of the particle (or the block in the example of the capacitor) is smaller than that of the surrounding medium, the direction of the force is reversed, i.e., the particle is repelled away from the region of high optical intensity and attracted towards the region of lower optical intensity. For detailed mathematics of the EM model, please consult the references cited above.

14.3 Experimental Measurements of Optical Forces

Optical forces on microparticles can be measured by several methods. In general, under identical experimental condition, optical forces scale linearly with optical power. For a particle in the vicinity of the trap center of a stable three-dimensional optical trap, the net optical force along the direction of each orthogonal axis can be approximated by a Hookean optical spring; a set of optical force constants (or spring constants, k_x, k_y, and k_z) can thus be used as a convenient set of parameters that specifies the three-dimensional optical force field within a small volume surrounding each stable equilibrium position of the particle in the trap. Techniques to measure the optical forces and optical force constants are described below in this section.

14.3.1 Axial Optical Force as a Function of Position along the Optical Axis

One way to measure the axial optical force on a particle in a counter-propagating dual-beam trap as a function of position along the optical axis is to trap a particle, drive the particle to move back-and-forth along the optical axis (by alternately turning on and off one of the trapping beams), and record the position of the particle as a function of time $z(t)$ via a precalibrated CCD camera or any other position sensing device. By taking the first and the second derivatives of the experimental data $z(t)$ either numerically (or analytically after curve-fitting $z(t)$ with an appropriate polynomial), one obtains the velocity $v(t)$ and the acceleration $a(t)$ of the particle. The axial optical force $F_{ao}(t)$ can then be determined from the following equation of motion:

$$ma(t) = F_{ao}(t) - 6\pi\eta r v(t) \tag{14.4}$$

or, equivalently,

$$F_{ao}(t) = ma(t) + 6\pi\eta r v(t), \tag{14.5}$$

where r is the radius of the particle, η is the coefficient of viscosity of the fluid surrounding the particle, and the mass of the particle $m = 4\pi r^3 \rho/3$, (ρ = the density of the particle). The axial optical force as a function of position $F_{ao}(z)$ can be deduced directly from $F_{ao}(t)$ and $z(t)$ by eliminating the parameter t from these data.

If we assume a symmetric counter-propagating dual-beam trap with identical optical power from each beam, the axial optical force from each beam will be identical (except for a sign-change in both the direction of force and the direction of the relative position of the particle). The net axial optical force, when both beams simultaneously act on the particle, is simply the vector sum of the contribution from each beam. As an example, a set of experimental data of $z(t)$ is given in Fig. 14.6a and the axial force deduced by the prescription described above is given in Fig. 14.6b. In this specific example, the linear

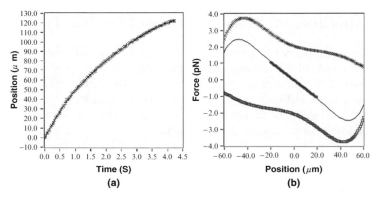

Fig. 14.6. (a) Experimental data showing the axial position of a particle as a function of time in a counter-propagating dual-beam trap as the particle was driven to move along the optical axis by switching off one of the trapping beams; (b) optical forces on the particle from each beam (represented by the upper and the lower curves) and the net force when both beams were on (represented by the middle curve) as a function of its relative position along the optical axis. The curves in (b) were deduced from (a) as is described in the text

regime where the axial optical force can be approximated by a linear optical spring is approximately 80 µm, and the axial optical force constant (given by the slope of the curve in the linear regime) is approximately 0.05 pN µm^{-1}. Since the total trapping power is approximately 25 mW, the maximum axial trapping efficiency (at the edge of the linear region) is estimated to be approximately 0.02. Although the optically driven motion method described above is relatively simple and the experiment can be easily repeated under identical experimental condition so that the measurement error can be minimized by averaging over several sets of data from repeated experiments, it suffers from several shortages: (1) it can not be used for the measurement of transverse trapping force, (2) it assumes that the particle moves linearly along the optical axis through out the course of motion, which is a good approximation only within a certain range ~100 µm; besides, the linear motion of the particle will be perturbed randomly (and unavoidably) by the thermal fluctuation (Brownian force) and systematically by the gravitation force.

14.3.2 Transverse Trapping Force Measured by Viscous Drag

An earlier method to measure the transverse optical trapping force on a particle in a single-beam gradient-force optical trap (or optical tweezers) is by trapping a particle and dragging it off transversely with increasing fluid flow speed until the particle escape from the trap [40]. The fluid flow can be implemented either by translating the sample chamber (mounted on a piezoelectric-translational stage) along with the fluid surrounding the trapped particle or with the aid of a microfluidic pump. At the threshold speed (v_{th}) of the fluid flow when the particle escapes from the optical trap, the transverse optical trapping force (F_{to}) is maximum and is equal to the viscous dragging force (F_{drag}). If the particle radius (r), the coefficient of viscosity (η) of the fluid, and the threshold flow speed (v_{th}) are known, the maximum transverse optical trapping force is given by the viscous dragging force; hence, $F_{\text{to}} = F_{\text{drag}} = 6\pi\eta r v_{\text{th}}$ (the Stokes' Law). A systematic error introduced by the presence of the side wall of the sample chamber (which is ignored in the equation above) can be corrected if the distance of the trapped particle from the wall is known [35, 36]. Even though the random error introduced by the thermal fluctuation of the particle can be reduced in principle by statistically averaging over repeated measurements, in practice, the standard deviations associated with the experimental results are relatively high (~50% or higher). A schematic diagram illustrating the underlying principle of this method is given in Fig. 14.7, and a video demonstration of this method can be viewed at our website (http://photoms.ym.edu.tw/).

14.3.3 Three-Dimensional Optical Force Field Probed by Particle Brownian Motion

In the previous sections on the measurement of optical forces, particle fluctuation due to Brownian motion was regarded as an unavoidable nuisance that

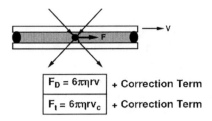

Fig. 14.7. A schematic illustration of the measurement of transverse optical force by balancing the optical force against a viscous fluid drag

leads to random errors in the experimental results. Interestingly, the three-dimensional Brownian motion of a particle in an optical trap, when tracked and measured with high precision, can be used to probe the three-dimensional optical force field on the particle with relative ease and with fairly high precision. This method, also known as photonics force microscopy [41,42], is briefly described below in this section.

Tracking and the measurement of three-dimensional Brownian motion of a particle in an optical trap are often accomplished with the aid of one or more position sensing devices such as a quadrant photo-diode (QPD). From the projection of the particle position fluctuation as a function of time [$x(t)$, $y(t)$, and $z(t)$] along each orthogonal axis, the corresponding optical force constant (k_x, k_y, and k_z) along each axis can be deduced by any one of the three methods: (1) root-mean-square (rms) fluctuation of the particle position [36], (2) the Boltzmann distribution of the particle in a parabolic potential well [41,42], (3) the temporal frequency analysis of particle position fluctuation [43,44]. To limit the length of this chapter, only the second approach based on Boltzmann distribution of the particle in a parabolic potential well is discussed below in this section.

An experimental setup used by the authors to probe the three-dimensional optical force field on a particle trapped in a fiber-optical counter-propagating dual-beam trap is shown in Fig. 14.8 [14]. A laser beam (cw, $\lambda = 532$ nm from a Nd:YVO$_4$ laser) was expanded and collimated via a 3X beam-expander (3X BE) through a half-wave plate ($\lambda/2$) and a polarizing beam splitter (PBS) cube to split into two beams with equal optical power, and each coupled into a single-mode fiber (NA = 0.11) via a single-mode fiber coupler (SMFC). The output ends of the two fibers were aligned so that the two laser beams exiting from the fibers formed a pair of counter-propagating beams along a common optical axis inside a sample chamber where microparticles or living cells were trapped in PBS solution. The distance between the two fiber end-faces was kept at 125 μm. Portions of the trapping beams scattered by the trapped particle were collected by a pair of orthogonally oriented long-working distant objectives (LOBI and LOBII, ×50, NA = 0.42) and projected onto a pair of quadrant photodiodes (QPDI and QPDII). LOBI/QPDI and LOBII/QPDII collected the scattered lights and tracked the position of the trapped particle

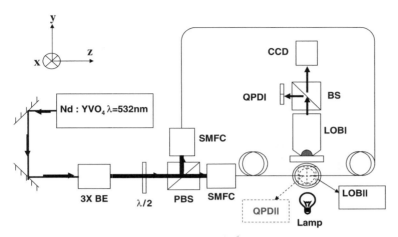

Fig. 14.8. A schematic illustration of the experimental setup for the mapping of three-dimensional optical force field on a particle in a fiber-optical dual-beam trap

projected on the xz-plane and the yz-plane, respectively (the xyz co-ordinate system is depicted in the upper left corner in Fig. 14.8). Besides, the LOBI was also used for imaging the trapped particle on a CCD camera (752×582 pixels) via incoherent illumination from a lamp. To obtain the conversion factor that converts the output voltage of the QPD into the particle displacement, we trapped the microparticle of interest on the optical axis in the middle of the end-faces of two optical fibers via equal laser power from the two fibers, and momentarily turned off one of the laser beams to drive the particle (by the remaining single beam) along the optical axis towards the opposite end-face of the optical fiber. The second beam was subsequently turned on to drive the particle back along the optical axis towards the original equilibrium position at the middle of the fiber end-faces. As the microparticle was driven back along the optical axis under the illumination of both beams, the corresponding output voltage of the quadrant photodiode and the position of the microparticle on the CCD camera were recorded simultaneously. An illustrative example of such a calibration curve (i.e., QPD output voltage vs. particle position recorded on the CCD) is depicted in Fig. 14.9, which shows a linear dependence with a slope of approximately $0.986\,\mu\text{m}\,\text{V}^{-1}$ within the region of approximately $3.4\,\mu\text{m}$. Besides the Brownian fluctuation, the accuracy of this approach is mainly limited by the determination of the particle position from the incoherent image of the illuminated particle on the pixelated digital CCD camera. The calibration method described above was carried out during the return trip of the particle (from a point offset from the center to the center equilibrium position) when both laser beams were on to ensure that the particle was evenly illuminated from both sides. Recently the authors have tracked the three-dimensional Brownian motion of polystyrene and silica

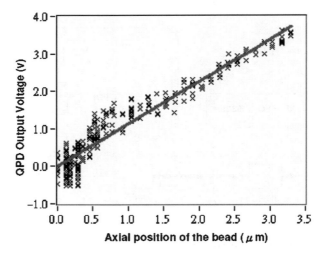

Fig. 14.9. An example of a calibration curve for the conversion of the QPD output voltage into the axial position of a particle in a fiber-optical dual-beam trap

beads of different sizes as well as Chinese hamster ovary cells trapped in a fiber-optical dual-beam trap, and analyzed their position distribution to obtain the optical force field approximated by a three-dimensional parabolic potential well [14, 15]. Under the parabolic potential approximation and the classical Boltzmann statistics, the associated optical force constant along each axis can be calculated from the following two equations [41]:

$$\rho(z) = C \exp\left[-E(z)/K_B T\right], \tag{14.6}$$

$$E(z) = -K_B T \ln \rho(z) + K_B T \ln C = k_z Z^2 / 2, \tag{14.7}$$

where $\rho(z)$ is the probability function of the trapped particle position along the z-axis, C is the normalization constant, $E(z)$ is the potential energy function along the z-axis, K_B is the Boltzmann constant, and k_z is the optical force constant along the z-axis. Identical set of equations also apply for the x-axis and the y-axis. As a specific example, a set of experimental data representing the parabolic potential $E(x)$ and $E(z)$ of the optical force field on a 2.58 μm diameter silica particle in a fiber-optical dual-beam trap (total trapping power = 22 mW, distance between the fiber end-faces = 125 μm,) is depicted in Fig. 14.10. The solid lines represent the theoretical curves based on (14.7) given above; the corresponding optical force constants, defined by (14.7), were $k_x = 0.16\,\mathrm{pN\,\mu m^{-1}}$ and $k_z = 0.04\,\mathrm{pN\,\mu m^{-1}}$. The force constant along the optical axis is weaker than those along the transverse axes, which is consistent with the theoretical results reported earlier [45]. This is also true in the case of optical tweezers [46]. For clarity sake, the experimental data for y-axis along

14 Optical Trapping and Manipulation for Biomedical Applications   261

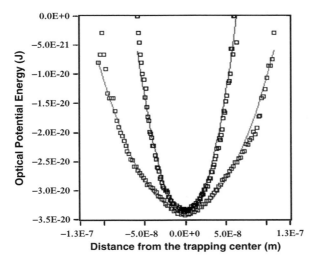

Fig. 14.10. Experimental data representing the parabolic optical potentials $E(x)$ (*the inner set of data points with a steeper slope*) and $E(z)$ (*the outer set of data points*) on a 2.58 μm silica particle. The solid lines represent the theoretical fits based on (14.7)

with the theoretical fit with $k_y = 0.15\,\mathrm{pN\,\mu m^{-1}}$ are not shown in Fig. 14.10; the data and the theoretical curve for y-axis essentially overlap with those for the x-axis.

14.3.4 Optical Forced Oscillation

Optical forced oscillation of a trapped particle is another powerful technique for the measurement of optical force and biological forces such as those associated with protein–protein interaction. Optical forced oscillation (OFO) refers to the trapping and forced oscillation of a microparticle in an optical trap [47,48]. Transverse force constants of the optical trap can be measured with fairly high precision by measuring the oscillation amplitude and the relative phase (with respective to that of the driving source) of the particle as a function of oscillation frequency. OFO can be used as a convenient tool for the measurement of protein–protein interaction [49,50] and of protein–DNA interaction with the aid of a microparticle coated with appropriate protein of interest. In this section, we introduce the basic principle and the experimental implementation of optical forced oscillation.

Optical forced oscillation has been implemented by one of the two following methods: (1) by trapping a microparticle in a conventional single-beam gradient-force optical trap (or optical tweezers) and scanning the trapping beam sinusoidally with a constant amplitude and frequency (Fig. 14.11a), (2) by trapping a microparticle in a set of twin optical tweezers and chopping one of the trapping beams on-and-off at a constant chopping frequency

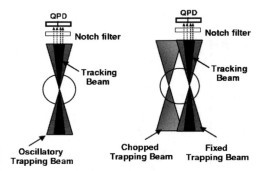

Fig. 14.11. A schematic diagram of optical forced oscillation via (**a**) oscillatory optical tweezers and (**b**) a set of twin optical tweezers

(Fig. 14.11b). In both cases, the steady-state oscillation amplitude and the relative phase (with respect to that of the driving source) of the oscillating particle can be conveniently measured with the aid of a quadrant photo-diode in conjunction with a lock-in amplifier. By changing the driving frequency, typically in the range of approximately a few hertz to a few hundred hertz, the oscillation amplitude and the relative phase of the oscillating particle can be plotted as a function of frequency. In general, the experimental data fit fairly nicely with the theoretical results deduced from a simple theoretical model of forced-oscillation with damping [48, 50]. By coating a microparticle with a protein of interest and allowing the oscillating particle to interact with protein receptors on a cellular membrane, the method described above can be used to measure the protein–protein interaction, which can be modeled as another linear spring in parallel to the optical spring (Fig. 14.12). The equation of motion of a particle, suspended in a viscous fluid, trapped, and forced to oscillate in oscillatory optical tweezers, can be written as

$$m\ddot{x} = -\beta\dot{x} - k\left[x(t) - A\left(e^{i\omega t}\right)\right], \tag{14.8}$$

where m is the mass of the particle, $\beta\dot{x} = 6\pi\eta r\dot{x}$ is the viscous drag prescribed by Stokes' law, η is the coefficient of viscosity of the surrounding fluid, r is the radius of the particle, k is the optical spring constant in the linear spring model, and A and ω are the amplitude and the frequency of the focal spot of the oscillatory optical tweezers, respectively.

In the case of a set of twin optical tweezers with the particle interacting with a cell as is illustrated schematically in Fig. 14.12, the equation of motion of the particle can be written as [48]

$$m\ddot{x} = -\beta\dot{x} - \frac{k(e^{i\omega t} + 1)}{2}(x - x_1) - (k + k_{\text{int}})(x - x_2), \tag{14.9}$$

where x_1 and x_2 represent the position of the force center of optical tweezers 1 and 2, respectively; ω is the fundamental harmonics of the chopping frequency;

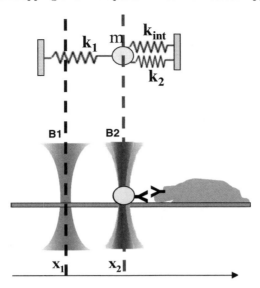

Fig. 14.12. A simplified linear spring model of a particle simultaneously acted upon by optical forces from a set of twin optical tweezers and a cellular interactive force when a cell is in contact with the particle in the equilibrium trapping position of the tweezers on the right

k_{int} is force constant of the protein–protein interaction modeled as a linear spring; and the rest of the symbols are the same as those described earlier in association with (14.8). In (14.9), we have assumed that the equilibrium position of the protein–protein interaction at the cellular membrane coincides with that of the second optical tweezers; this was accomplished experimentally by bringing the cell to touch the particle when it was stably trapped in the second tweezers as is illustrated schematically in Fig. 14.12.

The steady-state solution at the fundamental frequency of (14.8) and (14.9) above can be expressed as

$$x(t) = D\left[e^{i(\omega t - \phi)}\right] \qquad (14.10)$$

and

$$x(t) = D\left[e^{i(\omega t - \phi)}\right] - x_0, \qquad (14.11)$$

respectively, where D is the amplitude, ϕ the phase lag (with respective to that of the driving source), and x_0 the equilibrium position of the oscillating particle.

Solving (14.8) with an assumed solution in the form of (14.10) gives

$$\frac{D(\omega)}{D(0)} = \frac{k}{\sqrt{k^2 + (\beta\omega)^2 - m\omega^2}} \qquad (14.12)$$

and
$$\phi(\omega) = \tan^{-1}\left(\frac{\beta\omega}{k - m\omega^2}\right). \tag{14.13}$$

Likewise, solving (14.9) with an assumed solution in the form of (14.11) gives

$$\frac{D(\omega)}{D(0)} = \frac{\frac{3}{2}k'}{\sqrt{(k' - m\omega^2)^2 + (\beta\omega)^2}} \tag{14.14}$$

and
$$\phi(\omega) = \tan^{-1}\left\{\beta\omega/[k' - m\omega^2]\right\}, \tag{14.15}$$

where $k' = (3/2)k + k_{\text{int}}$.

Experimental data representing the amplitude and the relative phase of the oscillation of a trapped 1.5 μm polystyrene particle in water as a function of frequency along with the corresponding theoretical fits based on (14.12) and (14.13) given above are depicted in Fig. 14.13a,b, respectively. The optical spring constant k_x was deduced from the best fit to be 7.69 pN μm^{-1}, from the amplitude data, and 7.61 pN μm^{-1}, from the phase data. The optical spring constants measured by different methods described above agree to within approximately 8%. In contrast to the Brownian motion method, the results obtained by optical forced oscillation method depend only on the relative amplitude of particle oscillation and not on its absolute value. Besides, the signal-to-noise ratio is significantly enhanced by the phase locking technique. Although the transverse optical spring constant along any specific direction can be measured via oscillatory optical tweezers with reasonable accuracy by scanning the trapping beam along the selected direction, the two-dimensional

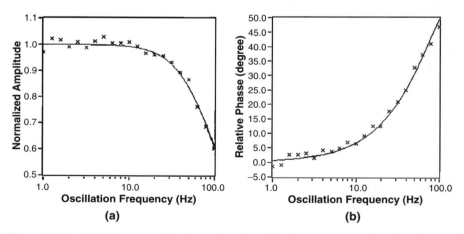

Fig. 14.13. (a) The normalized amplitude, (b) the relative phase of the optical forced oscillation of a polystyrene particle (diameter = 1.5 μm) suspended in deionized water as a function of the oscillation frequency. The solid curves are theoretical fits

14 Optical Trapping and Manipulation for Biomedical Applications   265

Table 14.1. A comparison of optical spring constants measured by different methods

Method	Analysis	$k_{OT}(x)$ (pN μm^{-1})	$k_{OT}(y)$ (pN μm^{-1})
Brownian motion	Displacement variance	7.07	6.31
	Potential well	7.28	6.84
	Power spectrum	7.21	6.88
Optical forced oscillation	Amplitude	7.69	
	Phase	7.61	

optical force field $E(x, y)$ can be simultaneously mapped, and the associated spring constants k_x and k_y conveniently measured in stationary optical tweezers via the Brownian motion method. The optical spring constants k_x and k_y on the transverse plane obtained from the Brownian motion analysis agree with each other to within approximately ±6%, which is consistent with the earlier theoretical and experimental results [46, 51, 52] and also with those reported previously for the case of a fiber-optical dual-beam trap [14, 45].

As an example, transverse optical spring constants, k_x and k_y, for a polystyrene particle (diameter = 1.5 μm) suspended in deionized water and trapped in optical tweezers (characterized by λ = 1064 nm, NA = 1.0, trapping optical power = 2 mW) obtained by different methods are compared and summarized in Table 14.1.

14.4 Potential Biomedical Applications

Potential biomedical applications of optical trapping and manipulation include (1) trapping and stretching a cell to measure its visco-elastic property and to correlate with its physiological condition, (2) trapping two cells to measure cell–cell interaction, (3) trapping two protein-coated microparticles to measure protein–protein interaction, (4) trapping one protein-coated particle to interact with membrane proteins on a cell to measure the protein–protein interaction at the cellular membrane, (5) trapping two beads with a segment of DNA molecule stretched in between to measure the interaction dynamics of the DNA with proteins injected into the sample chamber. Selected video demonstrations of some of the features listed above can be viewed at the website of the authors' lab (http://photoms.ym.edu.tw).

In general, optical trapping and manipulation promise to provide one or more of the following unique features in biomedical applications:

1. *Micropositioning*: One can guide cell–cell, cell–molecule, or molecule–molecule interactions in terms of where and when the interactions take place. The interaction can therefore be measured within a time point on the order of "second" after the initiation of the molecular interaction.

2. *Living cell capability*: The setup is readily adaptable for living cell imaging and manipulation.
3. The technique can be easily integrated with laser spectroscopy and microscopy for single cell measurement with subcellular resolution.
4. *Single molecule resolution*: Several protocols have been developed to measure the force exerted by a single biomolecule or the interaction between a molecular pair. New methods and algorithms have been developed enabling the extraction of single molecular results from the bulk measurements or other population data.
5. *The assay is sensitive to the functions of the biomolecules*: One can monitor in real-time the dynamics of molecular binding or of biological forces associated with molecular interactions to examine the functions of the biomolecules.
6. *The assay is sensitive to the conformation of the biomolecules*: One may obtain structural information of the biomolecule by integrating optical tweezers with other photonics modalities (such as time-resolved fluorescence microscopy and Raman spectro-microscopy).
7. The possibility of noncontact and noninvasive cell compliance measurements may lead to a new paradigm for assessing the physiology and the pathology of living cells with potential clinical applications.

In this section, we outline a few selected examples of the applications of optical trapping for protein–protein interactions, protein–DNA interactions, and cellular trapping and stretching.

14.4.1 Optical Forced Oscillation for the Measurement of Protein–Protein Interactions

As an example of protein–protein interaction, we present in this section the application of a set of twin optical tweezers to trap and oscillate a ConA (lectin)-coated polystyrene bead and to measure its interaction with glycoprotein receptors at the cellular plasma membrane of a Chinese hamster ovary (CHO) cell [50]. The bead was trapped between two quadratic potential wells defined by a set of twin optical tweezers and was forced to oscillate by chopping on-and-off one of the trapping beams. We tracked the oscillatory motion of the bead via a quadrant photodiode and measured with a lock-in amplifier the amplitude of the oscillation as a function of frequency at the fundamental component of the chopping frequency over a frequency range from 10 to 600 Hz. By analyzing the amplitude as a function of frequency for a free bead suspended in buffer solution without the presence of the CHO cell and compared with the corresponding data when the bead was interacting with the CHO cell, we deduced the transverse force constant associated with the optical trap and that associated with the interaction by treating both the optical trap and the interaction as linear springs. The force constants were determined to be approximately $2.15\,\text{pN}\,\mu\text{m}^{-1}$ for the trap and $2.53\,\text{pN}\,\mu\text{m}^{-1}$

14 Optical Trapping and Manipulation for Biomedical Applications

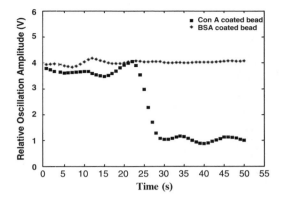

Fig. 14.14. The time-dependence of the relative amplitude of a polystyrene bead (diameter = 2.83 μm) executing optical forced oscillation at 50 Hz in the vicinity of a CHO cell. The experimental data for ConA-coated bead are denoted by solid squares and the data for a BSA-coated bead are denoted by "*"

for the lectin–glycoprotein interaction. When the CHO cell was treated with lantrunculin A, a drug that is known to destroy the cytoskeleton of the cell, the oscillation amplitude increased with time, indicating the softening of the cellular membrane, until a steady state with a smaller force constant was reached. The steady state value of the force constant depended on the drug concentration.

As an illustrative example, the time dependence of the relative amplitude of a polystyrene bead (diameter = 2.83 μm) oscillating at 50 Hz in the vicinity of a Chinese hamster ovary cell is depicted in Fig. 14.14 for the case of a BSA-coated bead and that of a ConA-coated bead. The decay of the oscillation amplitude in the case of the ConA-coated bead is a manifestation of the interaction of ConA protein with the glycol-protein on the cellular membrane of the CHO cell.

14.4.2 Protein–DNA Interaction

The interaction of proteins with DNA plays a pivotal role on a number of important biological processes, including DNA repair, replication, recombination, and segregation. Many different proteins can be used as models for such investigations. For example, one can optically trap-and-stretch a segment of dsDNA to analyze the dynamics of homologous DNA search and strand exchange reaction mediated by RecA protein.

In the study of protein–DNA interaction, micron-size polystyrene beads are often attached to the ends of each DNA sample, one at each end, to serve as handles for optical tweezers to trap and stretch the DNA sample between the beads. A schematic illustration of a dsDNA segment stretched between a fixed (large) bead (with diameter ∼20 μm) and an optically trapped small

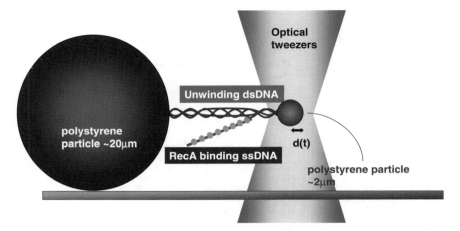

Fig. 14.15. A schematic illustration of a segment of a DNA molecule stretched between two polystyrene beads (with one bead fixed to the bottom of the sample chamber and the other bead trapped in optical tweezers) for the measurement of the dynamics of the interaction of the DNA sample with RecA proteins injected into the sample chamber

bead (with diameter ∼1 μm) for the study of the dynamic of the interaction with RecA-proteins carrying complementary segment of ssDNA is shown in Fig. 14.15. During the interaction, the stretched length and the elastic constant of the DNA can be measured in real-time by monitoring either the position of the trapped particle as a function of time (in the case of stationary optical tweezers) or the amplitude and the phase of the oscillating particle as function of frequency and time (in the case of oscillatory optical tweezers) (Fig. 14.16). One of the technical challenges of this approach is the lack of an effective method to avoid the adherence of multiple-strings of DNA molecules in parallel between the two beads. Although this can be achieved in principle by optimizing the ratio of the concentration of the DNA samples and that of the beads during the sample preparation, it is very tedious and inefficient in practice.

In the steady state, the bead is expected to wonder around an equilibrium position (slightly displaced from the optical axis) dictated by the balancing of the transverse gradient optical force and the elastic force of the stretched DNA molecule. The position fluctuation of the bead is a manifestation of the Brownian force acting on the bead and on the DNA molecule as well as the conformational change of the DNA molecule. Specifically, the stretching or the contraction as well as the winding or the unwinding of the double-strand DNA molecule is revealed by the translational and the rotational motion of the bead of which the former can be measured via a quadrant photodetector (QPD) or any other optical position sensing detector, while the later can also be measured optically by using a bead with optical birefringence in conjunction with

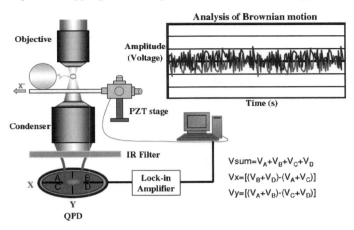

Fig. 14.16. A schematic illustration of the measurement of the dynamics of DNA–protein interaction by stretching the DNA between two beads, trapping one of the bead, and tracking its Brownian motion with a quadrant-photodiode

any polarization sensitive detection scheme. Instead of a bead with optical birefringence, any bead lacking axial symmetry illuminated by a laser beam is also expected to generate an optical scattering signal modulated (in intensity) with a frequency component at that of the rotational frequency of the bead. The basic principle of this approach is to detect (by analyzing the translational and the rotational motion of the trapped bead) the dynamic of the conformational change of a stretched double-strand DNA molecule interacting with RecA-ssDNA filaments injected into the surrounding buffer solution. The generic goal of the experiments, such as the one outlined above, is to understand the relationship between the physical properties and the biochemical (or functional) properties of DNA at a fundamental level.

14.4.3 Optical Trapping and Stretching of Red Blood Cells

As mentioned earlier, a red blood cell (RBC) can be trapped and stretched with the aid of a fiber optical dual-beam trap-and-stretch to measure its viscoelastic property. A schematic diagram to illustrate the application concept of a fiber-optical dual-beam trap-and-stretch in conjunction with a microfluidic flow chamber, fabricated with poly dimthylsiloxane (PDMS), to inject the RBC one at a time for such measurement is shown in Fig. 14.17. Photographs of an experimental set up in our laboratory are depicted in Fig. 14.18. Preliminary experimental results illustrating the morphological change of human RBC samples, osmotically swollen into spherical shape, as a function of the optical power are shown in Fig. 14.19 along with the experimental data on the fractional change in length of the major axis and the minor axis of a

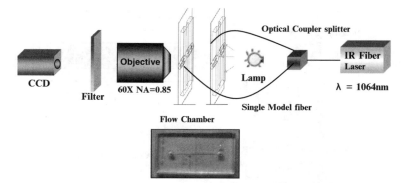

Top view of the flow chamber; material: Poly Dimthylsiloxane (PDMS)

Fig. 14.17. A schematic illustration of the experimental setup for simultaneous trapping, stretching, and morphological deformation measurement of red blood cells in a fiber-optical dual-beam trap

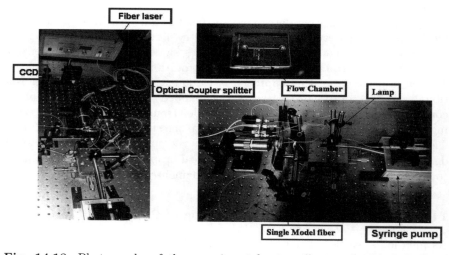

Fig. 14.18. Photographs of the experimental setup illustrated schematically in Fig. 14.17

human RBC sample as a function optical power. From these experimental data, the product (Eh) of the elasticity (E) and the thickness (h) of the RBC cell membrane was estimated to be $Eh = 2.4 \times 10^{-4}\,\mathrm{N\,m^{-1}}$, with the aid of a simple theoretical model [15–17].

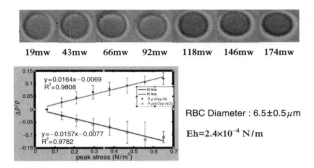

Fig. 14.19. Photographs of a human red blood cell (RBC), osmotically swollen into spherical shape, trapped and stretched at different laser power in a fiber-optical dual-beam trap, along with the experimental data showing the relative change in length along the major and the minor axes of the RBC sample as a function of laser power

14.5 Summary and Conclusion

In this chapter, we give a brief historical overview of optical trapping and manipulation followed by an outline of two theoretical models, namely the ray optics model and the electromagnetic model, along with several methods for the experimental measurement of optical forces with sub-pico-Newton resolution. Unique features of optical trapping and manipulation for biomedical applications are outlined and examples including the measurement of the dynamics of protein–protein interaction and protein–DNA interaction as well as the simultaneous trapping, stretching, and morphological deformation measurement of human red blood cells are highlighted.

Acknowledgment

Research in optical trapping and manipulation carried out by the authors is supported by the National Science Council of the Republic of China Grants NSC 95-2752-E010-001-PAE, NSC 94-2120-M-010-002, NSC 94-2627-B-010-004, NSC 94-2120-M-007-006, NSC 94-2120-M-010-002, and NSC 93-2314-B-010-003, and the Grant 95A-C-D01-PPG-01 from the Aim for the Top University Plan supported by the Ministry of Education of the Republic of China.

References

1. A. Ashkin, Phys. Rev. Lett. **24**, 156 (1970)
2. A. Ashkin, J.M. Dziedzic, Appl. Phys. Lett. **19**, 283 (1971)
3. A. Ashkin, J. Dziedzic, J. Bjorkholm, S. Chu, Opt. Lett. **11**, 288 (1986)

4. A. Ashkin, J.M. Dziedzic, Nature **330**, 769 (1987)
5. M.M. Burns, J.M. Fournier, J.A. Golovchenko, Science **249**, 749 (1990)
6. A. Chiou, W. Wang, G.J. Sonek, J. Hong, M.W. Berns, Opt. Commun. **133**, 7 (1997)
7. K. Sasaki et al., Jpn. J. Appl. Phys. **30**, L907 (1991)
8. E. R. Dufresne, D.G. Griera, Rev. Sci. Instrum. **69**, 1974 (1998)
9. E.R. Dufresne, G.C. Spalding et al., Rev. Sci. Instrum. **72**, 1810 (2001)
10. G. Sinclair, P. Jordan et al., Opt. Exp. **12**, 5475 (2004)
11. R.L. Eriksen, V.R. Daria, J. Gluckstad, Opt. Exp. **10**, 597 (2002)
12. P.J. Rodrigo, I.R. Perch-Nielsen et al., Opt. Exp. **14**, 13107 (2006)
13. A. Constable, J. Kim, J. Mervis, F. Zarinetchi, M. Prentiss, Opt. Lett. **18**, 1867 (1993)
14. M.T. Wei, K.T. Yang, A. Karmenyan, A. Chiou, Opt. Exp. **14**, 3056 (2006)
15. J. Guck, R. Ananthakrishnan, T.J. Moon, C.C. Cunningham, J. Kas, Phys. Rev. Lett. **84**, 5451 (2000)
16. J. Guck, R. Ananthakrishnan, H. Mahmood, T.J. Moon, C.C. Cunningham, J. Käs, Biophys. J. **81**, 767 (2001)
17. J. Guck, R. Ananthakrishnan, C.C. Cunningham, J. Käs, J. Phys. Cond. Mat. **14**, 4843 (2002)
18. J. Guck, S. Schinkinger, B. Lincoln et al., Biophys. J. **88**, 3689 (2005)
19. W. Wang, A. Chiou, G.J. Sonek, M.W. Berns, J. Opt. Soc. Am. B **14**, 697 (1997)
20. R.S. Taylor, C. Hnatovsky, Opt. Exp. **11**, 2775 (2003)
21. S.I. Eom, Y. Takaya, T. Miyoshi, T. Hayashi, Proc. SPIE **6326**, 70 (2006)
22. M. Gu, P. Ke, Opt. Lett. **24**, 74 (1999)
23. T. Sugiura, S. Kawata et al., J. Microsc. **194**, 291 (1999)
24. K. Ramser, K. Logg et al., J. Biomed. Opt. **9**, 593 (2004)
25. A. Fontes, K. Ajito et al., Phys. Rev. E **72**, 012903 (2005)
26. E.V. Perevedentseva, A.Y. Karmenyan, F.J. Kao, A. Chiou, Scanning **26**(Suppl. I), 178 (2004)
27. R.W. Applegate Jr., J. Squier, Opt. Exp. **12**, 4390 (2004)
28. S.C. Chapin, V. Germain et al., Opt. Exp. **14**, 13095 (2006)
29. P.T. Korda, M.B. Taylor et al., Phys. Rev. Lett. **89**, 128301 (2002)
30. P.Y. Chiou, A.T. Ohta, M.C. Wu, Nature **436**, 370 (2005)
31. A.T. Ohta, P.Y. Chiou, M.C. Wu, Proc. SPIE **6326**, 632617 (2006)
32. V. Bingelyte, J. Leach et al., Appl. Phys. Lett. **82**, 829 (2003)
33. A.I. Bishop, T.A. Nieminen et al., Phys. Rev. Lett. **92**, 198104 (2004)
34. M. Gudipati, J.S. D'Souza et al., Opt. Exp. **13**, 1555 (2005)
35. A. Ashkin, IEEE J. Sel. Top. Quantum Electron. **6**, 841 (2000)
36. K.C. Neuman, S.M. Block, Rev. Sci. Instrum. **78**, 2787 (2004)
37. A. Ashkin, Biophys. J. **61**, 569 (1992)
38. P.B. Bareil, Y. Sheng, A. Chiou, Opt. Exp. **14**, 12503 (2007)
39. Y. Harada, T. Asakura, Opt. Commun. **124**, 529 (1996)
40. W.H. Wright, G.J. Sonek, M.W. Berms, Appl. Phys. Lett. **63**, 715 (1993)
41. E.L. Florin, A. Pralle, E.H.K. Stelzer, J.K.H. Horber, Appl. Phys. A **66**, 75 (1998)
42. A. Rohrbach, H. Kress, E.H. Stelzer, Opt. Lett. **28**, 411 (2003)
43. L.P. Ghislain, W.W. Webb, Opt. Lett. **18**, 1678 (1993)
44. L.P. Ghislain, N.A. Switz, W.W. Webb, Rev. Sci. Instrum. **65**, 2762 (1994)

45. E. Sidick, S.D. Collins, A. Knoesen, Appl. Opt. **36**, 6423 (1997)
46. M.T. Wei, A. Chiou, Opt. Exp. **13**, 5798 (2005)
47. L.A. Hough, H.D. Ou-Yang, J. Nanopart. Res. **1**, 495 (1999)
48. M.T. Wei et al., Proc. SPIE **6447**, 06 (2007)
49. M.T. Wei, K.F. Hua, J. Hsu, A. Karmenyan, H.Y. Hsu, A. Chiou, Proc. SPIE **6326**, 63260 (2006)
50. S.L. Liu et al., Opt. Exp. **15**, 2713 (2007)
51. C. Xie, M.A. Dinno, Y.Q. Li, Opt. Exp. **13**, 1621 (2005)
52. A. Rohrbach, Phys. Rev. Lett. **95**, 168102 (2005)

15

Laser Tissue Welding in Minimally Invasive Surgery and Microsurgery

R. Pini, F. Rossi, P. Matteini, and F. Ratto

15.1 Introduction

Lasers have become already a widespread tool in many operative and therapeutic applications in the surgical field. The so-called "minimally invasive" laser techniques provide remarkable improvements: in these, laser surgery is performed inside the human body through small incisions by means of optical fibre probes and endoscopes, or laser tools are proposed as a replacement for conventional tools to minimize the surgical trauma, such as in the case of laser-induced suturing of biological tissues. The aim of these procedures is to improve the quality of life of patients, by decreasing healing times and the risk of postoperative complications.

Joining tissue by applying laser irradiation was first reported at the end of the 1970s, when a neodymium/yttrium–aluminium–garnet (Nd:YAG) laser was used for the microvascular anastomosis of rat carotid and femoral arteries. Ever since, laser tissue welding has been evaluated in several experimental models including blood vessels, skin, nerve, intestine, uterine tube, and so on [1,2]. Laser welding has progressively assumed increased relevance in the clinical setting, where it appears to be a valid alternative to standard surgical techniques. At present, there are many applications of tissue welding that are beginning to achieve widespread acceptance.

Many types of lasers have been proposed for laser tissue welding. Infrared and near-infrared sources include carbon dioxide (CO_2), thulium–holmium–chromium, holmium, thulium, and neodymium rare-earth-doped-garnets (THC:YAG, Ho:YAG, Tm:YAG, and Nd:YAG, respectively), and gallium aluminium arsenide diode (GaAlAs) lasers. Visible sources include potassium-titanyl phosphate (KTP) frequency-doubled Nd:YAG, and argon lasers. The laser energy is absorbed by water at the infrared wavelengths and by haemoglobin and melanin at the visible wavelengths, thereby producing heat within the target tissue. As the temperature rises, the extracellular matrix of the connective tissue undergoes thermal changes that lead to the welding of the wound (Fig. 15.1).

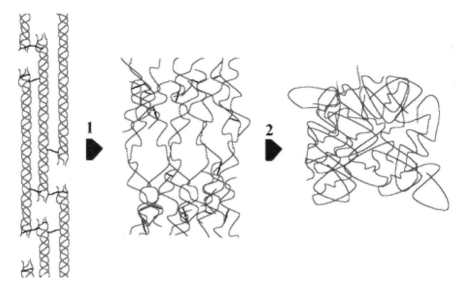

Fig. 15.1. Structural modifications induced in fibrillar collagen of connective tissue by temperature rise. Normal triple helix collagen molecules are packed in a quarter-staggered manner and connected by covalent bonds to form a collagen fibril (*left*). When heat is applied, hydrolysis of intramolecular hydrogen bonds occurs, which results in the unwinding of the triple helices (*middle*). The first step (1) leads to a shrinkage effect parallel to the axis of the fibrils. At higher temperatures, covalent cross-links connecting collagen strands break, resulting in a complete destruction of the fibrillar structure (*right*) and causing relaxation of the tissue (step 2)

Laser tissue welding has been shown to possess several advantages compared with conventional closure methods, such as reduced operation times, fewer skill requirements, decreased foreign-body reaction and therefore reduced inflammatory response, faster healing, increased ability to induce regeneration, and an improved cosmetic appearance. Laser welding also has the potential to form complete closures, thus making possible an immediate watertight anastomosis, which is particularly important in the case of vascular, genito-urinary tract, and gastrointestinal repairs. A watertight closure also discourages the exit of regenerating axons and the entry of fibroblasts. Lastly, laser welding can be used endoscopically and laparoscopically to extend the range of its applications to cases in which sutures or staples cannot be used.

However, despite the large number of experimental studies reported in the literature, very few of them have reached the clinical phase. This is mainly because of the lack of clear evidence of the advantages of laser-assisted suturing against conventional methods, and because of a low reproducibility of results. The damage induced in tissues by direct laser heating and heat diffusion and the poor strength of the resulting welding are the main problems as far as future clinical applications of the laser-assisted procedure are concerned. In fact, as water, haemoglobin, and melanin are the main absorbers of laser light within tissue, the heating effect is not selectively limited to a target area, and all irradiated tissues are heated. For instance, the CO_2 laser has been used for laser repairs of thin tissues because of its short penetration depth ($<20\,\mu m$). However, for thicker tissues, welding has been achieved only by irradiating with high laser power and longer exposure times, thus inducing high levels of heat damage [3]. The emissions of other near-infrared lasers, such as Nd:YAG and diode lasers, are more suited to the welding of thicker tissues. In any case, control of the dosimetry of laser irradiation and of corresponding temperature rise is crucial to minimize the risk of heat damage to the tissue and to generate strong welds.

Two advances have been useful in addressing the issues associated with laser tissue welding: the application of laser-wavelength-specific chromophores and the addition of endogenous and exogenous material to be used as solder.

The use of wavelength-specific chromophores enables differential absorption between the stained region and the surrounding tissue. The advantage is primarily a selective absorption of laser radiation by the target, without the need for a precise focusing of the laser beam. Moreover, lower laser irradiances can be used because of the increased absorption of stained tissues. Various chromophores have been employed as laser absorbers, including indocyanine green (ICG) [4], fluorescein [5], basic fuchsin, and fen 6 [6]. The use of a near-infrared laser – which is poorly absorbed by biological tissues – in conjunction with the topical application of a dye with an absorption peak overlapping the laser emission, is a very popular setting for the laser welding technique. Diode lasers emitting around 800 nm and ICG have been used in corneal tissue welding in cataract surgery and corneal transplant [7,8], vascular tissue welding [9–11], skin welding [4, 12], and in laryngotracheal mucosa transplant [13].

Laser welding by means of solders, namely "laser soldering," makes use of exogenous solders as topical protein preparations. This makes possible a bonding of the adjoining and underlying tissues when activated by laser light. The extrinsic agents provide a large surface area over which fusion with the tissue can occur, thus favoring the approximation of the wound edges that eventually heal together in the postoperative period. Useful welding materials include blood [14], plasma [15], fibrinogen [16] and albumin, which is the one most frequently employed [5, 17]. Several studies have demonstrated that the addition of an albumin solder to reinforce laser tissue repairs significantly improves postoperative results [5, 18]. Moreover, incorporation into the protein solder of a laser-absorbing chromophore makes it possible to confine the heat into the area of solder application, which reduces the extent of collateral heat damage to adjacent tissues. ICG-doped albumin has become an increasingly popular choice in the last decade [17]. The laser is used to denature the protein immediately after application of the protein solder to the wound site, thus yielding a bond at the solder–tissue interface.

Another improvement of traditional protein solders relies on the use of synthetic polymers mixed with serum albumin. This provides better flexibility, as well as improved repair strength over albumin-protein solders alone [19]. Poly(lactic-co-glycolic acid) (PLGA) is a class of synthetic biodegradable polymers that are easily degraded in vivo and are eliminated through normal metabolic pathways. These materials can be employed also as drug-delivery systems by adding a range of dopants including antibiotics, anaesthetics, and various growing factors that may enhance the rate and quality of wound healing [20].

A more recent development in laser tissue welding is the use of an infrared temperature feedback control of the laser device. Diagnostic feedback via surface temperature measurements has been used to control laser power delivery to achieve well-defined welding protocols [21]. Although surface temperature monitoring may be important for a primary control of the energy delivery, it does not reveal dynamic changes occurring below the tissue surface, which is where the weld actually occurs. This problem has been addressed by employing numerical models of laser-tissue interactions on the basis of directly measured surface-temperature data [21, 22]. The combination of experimental and simulated data made it possible to characterize surface and bulk tissue heating, thus allowing for a prediction of the temperature rise within the irradiated tissues.

Photochemical welding of tissues has also been investigated as an alternative method for tissue repair without direct use of heat. This technique utilizes chemical cross-linking agents applied to the cut that, when light-activated, produce covalent cross-links between collagen fibres of the native tissue structure. Agents used for photochemical welding include 1,8-naphthalimide [23], Rose Bengal, riboflavin-5-phosphate, fluorescein, and methylene blue [24]. Studies of photochemical tissue bonding have been conducted for articular cartilage bonding [23], cornea repair [24], skin graft adhesion [25] and for repairing severed tendons [26].

As far as the mechanism of tissue welding by means of laser light is concerned, a first distinction must be made between: (1) laser tissue welding with or without the addition of exogenous chromophores, (2) laser tissue soldering, and (3) photochemical tissue bonding. For all three cases, the exact underlying mechanism is not fully understood. Conversely, several hypotheses do exist that are based on a few electron and optical microscopy observations and on some in vitro studies.

It is widely accepted that laser tissue welding is primarily a thermal process [27]. Thermal modifications of biological components within the connective tissue have been mainly monitored by means of microscopy. The microscopic data reported in the literature on laser-welded tissues can be schematically split into two groups on the basis of different modifications of the collagen matrix observed upon laser welding.

In certain studies, a full homogenization of the tissue was observed, in which the loose structure of the collagen fibrils was lost following laser welding

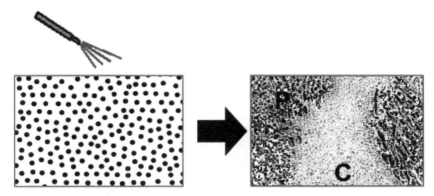

Fig. 15.2. "Hard" laser welding approach resulting in a complete homogenization of the tissue. The extracellular matrix is schematically represented by an array of *black dots* simulating cross-sectioned collagen fibrils embedded in the ground substance, as in a typical connective tissue (*left*). Laser irradiation leads to hardly recognizable collagen structures completely (C) or partially (P) coagulated (*right*). Temperatures above 75°C are usually recorded in these cases. The mechanism proposed relies on the adhesion upon cooling between proteins degradated by laser heat, which, acting as microsolders, seal the wound

[4, 10, 28, 29] (Fig. 15.2). In this context, fibrils fused together and morphologically altered revealed a complete denaturation of the collagen matrix [3, 30]. As a result of laser irradiation, cell membranes were also disrupted causing leakage of the cellular proteins. In these cases, operative temperatures of over 75°C were induced at the welded area [31]. The wound sealing mechanism has been attributed to the photocoagulation of collagen and other intracellular proteins, which act like micro-solders or endogenous (biological) glue, which form new molecular bonds upon cooling [28].

In other studies, less severe modifications of the collagen matrix have been observed (Fig. 15.3). In these cases, the collagen fibrils were still recognizable and not or only partially swollen [3, 32, 33]. Operative temperatures around 60–65°C were generated at the welded site inducing no (or only partial) denaturation of fibrillar collagen [33–35]. In water-bath experiments, opposed tendon specimens were thermally bonded, obtaining maximum tensile strength at similar temperature values [36, 37]. The most common interpretation of the working mechanism is an unravelling of the collagen triple-helix followed by "interdigitation" between fibers upon cooling, with generation of new bonds [9, 32]. Some researchers have suggested the formation of a certain type of noncovalent bonding between collagen strands on both sides of the weld [38], while others have found that new covalent cross-links were created upon laser irradiation [39]. A secondary role of fibrillar type I collagen in the laser welding mechanism has been pointed out in some recent studies [33, 40], suggesting the involvement of some other extracellular matrix components, and being in agreement with earlier studies on welded tissue extracts analyzed by gel-electrophoresis [34, 41].

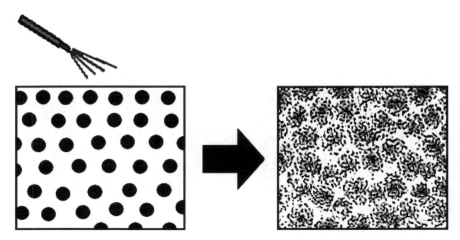

Fig. 15.3. "Mild" laser welding approach resulting in a partial modification of the tissue morphology. Native connective tissue (*left*) subjected to laser irradiation is characterized by collagen fibrils still recognizable and not or only partially swollen (*right*). Temperature values in the range 60–70°C are usually induced at the weld site. The welding mechanism hypothesized is based on "interdigitation" upon cooling between collagen fibres unraveled by laser heat

Photothermal soldering relies on the coagulation of a protein solder by means of a laser-induced temperature increase in the tissue. Upon cooling, noncovalent interactions between the solder and the collagen matrix within the tissue are supposed to be responsible for the strength of the weld. Evidence of albumin intertwining within the collagen matrix was found during scanning electron microscopy analyzes of specimens irradiated at temperatures above 70°C [17]. Such a threshold value is fully in agreement with the threshold temperature of albumin coagulation (around 65°C), as reported in several spectroscopic and calorimetric studies. Evidence of extracellular matrix infiltration of solder within the tissue was also found by using standard histological analysis [10, 42].

In photochemical tissue bonding photosensitive dyes applied to the wound edges behave as reactive species when irradiated by laser light. They react with potential electron donors and acceptors such as amino acids (e.g. tryptophan, tyrosine, cysteine) of proteins. Strong covalent bonds are produced between the approximated surfaces of the wound, forming instantaneous protein cross-links [26]. The formation of cross-links in collagen type I molecules by means of photochemical activation has been confirmed by using gel electrophoresis [43].

Here as follows, we will review some of the main surgical applications of laser tissue welding, in particular with regard to the fields of ophthalmology and microvascular surgery, in which the clinical use of this technique seems closer to being accepted.

15.2 Laser Welding in Ophthalmology

The first attempt to join biological tissues were proposed in ophthalmology, for the treatment of retinal, corneal, and scleral samples [44–46]. The first studies were unsuccessful, resulting in no tissue fusion, while with the improvement of laser technique several research groups adopted different approaches to induce welding of scleral and corneal tissue. Retina fusion was achieved by inducing photocoagulation of the tissue, while the technique used to treat other ocular tissues is based on a soft thermal treatment, properly defined as laser welding. Successful experimental studies of laser-induced suturing of ocular tissues on animal models have been reported since 1992 by different authors [35, 42, 47–51] on the basis of the use of near and far-infrared lasers, directly absorbed by the water content of the cornea. Various laser types with wavelengths exhibiting high optical absorption in water have been used, such as CO_2 (emitting at 10.6 µm) [42, 50, 51], Erbium:YAG (1455 nm) [35], and diode lasers (1,900 nm) [48, 49]. The main problem with such laser wavelengths is that, without an adequate control of the laser dosimetry, the direct absorption of laser light in a short penetration depth of the tissue outer portion caused a high temperature rise at the irradiated surface, followed by collagen shrinkage and denaturation; on the contrary, the deeper layers are hardly heated at all, resulting in a weak bonding, as the full thickness of the tissue is not involved in the welding process. Improved results in tissue welding were observed by using exogenous chromophores to absorb laser light, sometimes in association with protein solders. Addition of highly absorbing dyes allowed fusion of wounds at lower irradiation fluences, thus avoiding excessive thermal damage to surrounding tissues. In fact, the usage of a chromophore was found to induce a controllable temperature rise only in the area where it had been previously applied, resulting in a selective thermal effect.

Recently a new approach was proposed and tested by some of us for the closure of corneal wounds [8, 52–54] and for the treatment of anterior lens capsule bags [55]. It is based on the use of a near-infrared diode laser, in association with the topical application of a water solution of ICG. The welding procedure, which was optimized so that it could be used in ophthalmic surgery applications, has been proposed as a valid alternative to, or as a supporting tool for, the traditional suturing technique used for the closure of corneal wounds, such as in cataract surgery, penetrating keratoplasty (i.e., transplant of the cornea), and in the treatment of accidental corneal perforations. It can also be used for closure of the lens capsule (to repair capsular breaks caused by accidental traumas or produced intraoperatively), as well as to provide closure of the capsulorhexis in lens refilling procedures.

15.2.1 Laser Welding of the Cornea

The cornea is an avascularised connective tissue on the outer surface of the eye, and forms the outer shell of the eyeball together with the sclera. It acts as one of the main refractive components, assuring good vision with its clarity and shape. It also acts as a mechanical barrier and as a biological defence system. The cornea is composed of different layers: epithelium, stroma, and endothelium are the principal ones, proceeding from the external surface toward the inner part of the eye. More than 90% of the cornea is stroma, and consists of extracellular matrices (mainly type I collagen and glycosaminoglycans), keratocytes, and nerve fibers. The collagen is regularly arranged in fibers, thus contributing to corneal transparency; the collagen fibers are organized in lamellae, i.e., in planes running parallel to the corneal surface. These particular structures and architecture confer unique properties to corneal tissue, e.g., in the reaction process to external injuries, such as incidental traumas, surgical incisions, or ulcers. The injured tissues heal by repair and do not recover the "normal" configuration, because of an induced disorganization in the ordered array of the collagen fibers. This results in an opaque scar with less tensile strength than that of an unwounded cornea, and in a subsequent impairment of the main corneal functions. Moreover, the healing of corneal stroma is slower than that of other connective tissues because of the lack of blood vessels. Clinical changes in scar formation may, in fact, be detected years after surgery has taken place [56].

For all these reasons, the characteristics of laser welding procedures may be very useful in practical surgery, offering the possibility of avoiding many postoperative complications. The immediate watertight closure of wound edges provides protection from external inflammation, and may prevent endophthalmitis, which sometimes occurs after cataract surgery. The position of the apposed margins has been found to be stable in time, thus assuring optimal results in terms of postoperative induced astigmatism after cataract [54] and keratoplasty surgery. The absence or the reduction in the number of stitches does not induce foreign body reaction, thus improving the healing process. Histological analyzes performed on animals and morphological observations on treated patients have shown that, in a laser-welded wound, tissue regains an architecture similar to that of the intact tissue, thus supporting its main functions (clarity and good mechanical load resistance).

The technique recently proposed for welding corneal tissue [7, 8, 52–54] has been tested and optimized on animal models. Experimental analyzes were first performed ex vivo on pig eyes; the healing process was then studied in vivo in rabbits. When a satisfactory result was achieved, it was proposed clinically and is currently being used for the closure of corneal tissue after penetrating keratoplasty, supporting the application of eight stitches instead of a continuous suturing. The main advantages of this clinical practice are an increased patient comfort during the healing process and a reduction of hospitalization costs.

The technique is based on the use of near-infrared continuous-wave AlGaAs diode laser radiation at 810 nm, in association with the topical application of a sterile water solution (10% weight/weight) of ICG to the corneal wound to be repaired. This dye is characterized by high optical absorption around 800 nm

[57], while the stroma is almost transparent at this wavelength. ICG is a frequently used ophthalmic dye, with a history of safety in humans. It is being used increasingly as an intraocular tissue stain in cataract and vitreoretinal surgery, as well as in staining of the retinal internal-limiting membrane [58–60]. Furthermore, ICG is commonly used as a chromophore in laser welding or laser soldering [1], to induce differential absorption between the dyed region and the surrounding tissue. Photothermal activation of stromal collagen is thus induced by laser radiation only in the presence of ICG, resulting in a selective welding effect, which produces an immediate sealing of the wound edges and good mechanical strength. In addition, with the use of ICG, very low laser power is required (below 100 mW), and this generally means much safer operation with respect to the use of other laser types without the association of dyes.

The procedure used to weld human corneal tissue is as follows: the chromophore solution is placed inside the corneal cut, using an anterior chamber cannula, in an attempt to stain the walls of the cut in depth. A bubble of air is injected into the anterior chamber prior to the application of the staining solution, so as to avoid perfusion of the dye. A few minutes after application, the solution is washed out with abundant water. The stained walls of the cut appear greenish, indicating that the concentration of ICG absorbed by the stroma is much lower than that of the applied solution. Lastly, the whole length of the cut is subjected to laser treatment. Laser energy is transferred to the tissue in a noncontact configuration, through a 300-µm core diameter fibre optic terminating in a hand piece, which enables easy handling under a surgical microscope. A typical value of the laser energy density is about $13\,\mathrm{W\,cm^{-2}}$ in humans, which results in a good welding effect. During irradiation, the fiber tip is kept at a working distance of about 1 mm, and at a small angle with respect to the corneal surface (side irradiation technique). This particular position provides in-depth homogenous irradiation of the wound and prevents from accidental irradiation of deeper ocular structures. The fiber tip is continuously moved over the tissue to be welded, with an overall laser irradiation time of about 120 s for a 25-mm cut length (the typical perimeter of a transplanted corneal button).

Experimental studies on laser-induced heating effects on ocular tissues were performed both ex vivo on animal models [21] and in vivo during surgery. An IR thermo-camera provided information on the heating of the external surface of the irradiated tissues and on heat confinement, during the treatment. A partial differential equation modeling of the process enabled us to investigate temperature dynamics inside the tissue. From this study, it was possible to point out that the optimal welding temperature is about 60°C inside the treated wound, i.e., in correspondence with a thermally induced phase transition of stromal collagen. The heating effect was found to be selectively localized within the cut, with no heat damage to the adjacent tissue, and to induce a controlled welding of the stromal collagen. After laser welding, collagen fibrils appeared differently oriented among themselves in

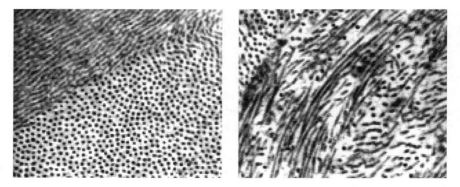

Fig. 15.4. Transmission electron microscopy images of the fibrillar arrangement observed in a control (*left*) and in a laser-welded (*right*) corneal stroma. Differently oriented and interwoven fibrils across the cut are visible at the weld site upon laser irradiation (×13,500)

comparison with those of the untreated samples, but with similar mean fibril diameters [33] (Fig. 15.4). Moreover, thanks to the characteristics of the ICG solution, it seems that laser welding is a self-terminating process, which thus avoids accidental excessive temperature rises inside the tissue.

A follow up study on animal models [7] and clinical application of the technique provided evidence that laser-welded tissues exhibits good adhesion and good mechanical resistance (Fig. 15.5). A thorough study of the healing process – on the basis of morphological observations, standard histology, and multispectral imaging auto-fluorescence microscopy (MIAM) [61] – proved experimentally that this takes place in shorter time and with lower inflammatory reaction, when compared with conventionally sutured wounds. Objective observations 2 weeks after surgery showed a good morphology of laser-treated corneas, with almost restored cuts, generally characterized by better adhesion and less edematous appearance compared with sutured ones. These features were confirmed by means of histological examinations of rabbit corneas, which revealed a well-developed repair process involving the epithelium, which almost regained its physiological continuity and thickness and a partially reorganized architecture of the stroma. Histological analyzes on longer follow-up times indicated that the healing of laser-welded wounds was completed in about 30–60 days, while in sutured wounds the healing process was still in progress. This result is particularly important as far as corneal tissue is concerned, as it typically requires much longer times to be repaired than do other types of tissue. Furthermore, the restored tissue regains a stromal architecture that is very close to the native one, which is crucial to regaining of correct vision.

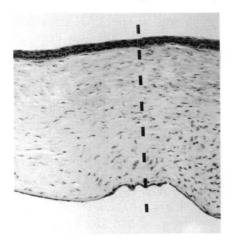 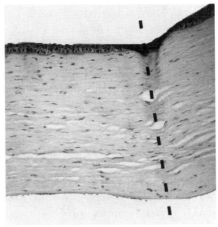

Fig. 15.5. *Left*: histological section of a rabbit cornea after laser welding on postoperative day 30 (Hematoxillin and Eosin, ×50); the architecture of the cornea regained an almost physiological appearance (the original cut is indicated by the *dashed line*). *Right*: histological section of a rabbit cornea after conventional stay suturing on postoperative day 30; the cut is still clearly detectable and large lacunae are evident in the corneal stroma

15.2.2 Combing Femtosecond Laser Microsculpturing of the Cornea with Laser Welding

In recent decades, femtosecond (FS) lasers have been developed, tested, and optimized for miniinvasive intratissue surgery and cell manipulation [62–64]. In particular, eye tissue characteristics are well suited to FS lasers applications, as they provide precise intrastromal cuts that are the basis of the new corneal and refractive surgery techniques. With the term FS lasers, we indicate lasers that emit pulses in the near-infrared spectral region with durations ranging between a few femtoseconds and hundreds of femtoseconds. With the use of these ultra-short pulses, it is possible to induce photo-disruption inside semitransparent and transparent media, such as corneal tissue. To induce this effect, the laser beam is focused in a focal region of a few micrometers in diameter. High intensities $\sim 10^{13}$ W cm^{-2} are achieved, which induce multiphoton nonlinear absorption, followed by plasma formation. The laser-induced plasma expands rapidly with high pressure (GPa), and this is followed by development and further collapse of cavitation bubbles, accompanied by formation of destructive shock waves. To obtain a highly-localized destructive effect without significant collateral damage, small bubble diameters and low photomechanical effects in the surroundings are required. These features are satisfied in FS laser-induced photo-disruption, first because initiation of nonlinear absorption of laser radiation requires tight focusing of the light, which assures effect confinement in the central portion of the focus volume (where really high intensities are achieved). Second, the multiphoton absorption coefficient has a high value when compared with the linear absorption value: the disruptive effect may be achieved with very low intensity levels (μJ cm^{-2} or nJ cm^{-2}). These nonlinear effects have been used to ablate and to modify corneal tissue, with very high spatial precision and minimal side effects.

Currently, FS laser cutting cells are characterized by high numerical apertures (NA) (>0.9 in water and glass) that make possible submicrometrical precision. Long series of pulses from FS lasers at 80 MHz repetition rate imply accumulative effects, which may cause tissue ablation at pulse energy below the optical breakdown threshold in presence of low density plasmas [64]. The advantages are that nonlinear propagation effects are reduced, highly localized energy deposition occurs, and subsequently nanosurgery on a cellular and subcellular level is possible. Moreover, the optical breakdown threshold weakly depends on the target absorption coefficient. Thus, any arbitrary cellular structure may be manipulated. For these reasons, near-infrared FS lasers have been considered the innovation for overcoming problems associated with the use of UV nanosecond lasers in refractive surgery: UV mutation effects on cells, low light penetration depth, collateral damage outside the focal volume, the risk of photon-damage to living cells due to absorption, and probable induction of oxidative stress leading to apoptosis. Moreover, a focused nanosecond-pulsed laser beam can cause thermal damage and denaturation around the laser focus.

Typically, the systems proposed for performing corneal manipulation against refractive problems are based on mode-locked diode-pumped FS Nd:Glass lasers providing pulses of 500–800 fs duration and a few μJ energy, at repetition rates of some tens of kilohertz. Each individual laser pulse is focused on a specific location inside the cornea, which is fully transparent at the laser wavelength. A micro-plasma is created, and this generates a microcavitation bubble of 5–15 μm in diameter, which separates the corneal lamellae. Thus, a resection plane can be created by delivering, in a prescribed pattern, thousands of laser pulses connected together. The cut can be performed with micrometrical precision at different depths inside the stroma, thus allowing for corneal flaps with a preset, constant thickness. Laser pulses can also be stacked on the top of each other, to create a vertical or angled cleavage plane to precisely sculpture the border of the lamellar flap. The possibility of performing the same resection procedure on the donor cornea as well as on the patient's recipient eye, allows to match the transplanted flap precisely with the recipient corneal bed.

By exploiting previous clinical experiences in diode laser welding of corneal wounds, some of us have recently designed a new laser-assisted technique for lamellar keratoplasty (i.e., a corneal transplant involving replacement of only the anterior corneal stroma). It is performed by using a FS laser to prepare donor button and recipient corneal bed, and then suturing the edges of the wound by means of diode laser-induced corneal welding, without the application of conventional suture material. This minimally invasive procedure

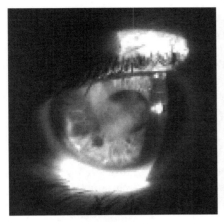 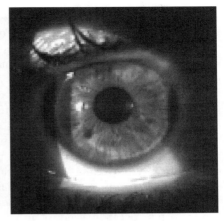

Fig. 15.6. Aspect of a human cornea affected by leucoma (*left*): before surgery, and (*right*): three days after corneal transplant, performed according to the so-called ALSL-LK, which combines the use of a femtosecond laser for corneal sculpturing and a diode laser for corneal welding

was called "all-laser" sutureless lamellar keratoplasty (ALSL-LK) (Fig. 15.6). Intraoperative observations and follow-up results for up to 6 months indicated the formation of a smooth stromal interface, the total absence of edema and/or inflammation, and a reduction in postoperative astigmatism, when compared with conventional suturing procedures [65, 66].

15.2.3 Laser Closure of Capsular Tissue

The lens, or crystalline lens, is a transparent, biconvex structure in the eye: after the cornea, it is the second refractive component in the eye. The lens is flexible and its curvature is controlled by ciliary muscles through the zonules. It is included within the capsular bag, maintained by the zonules of Zinn, which are filament structures connected to the ciliary muscles fulfilling lens accommodation (i.e., focusing of light rays into the retina in order to assure good vision). This capsule is a very thin (about 10 μm thick [67]), transparent acellular membrane that maintains the shape of the lens. This tissue is a collagenous meshwork mainly composed of type IV collagen and other noncollagenous components such as laminin and fibronectin. Type I and type III collagen are also present [68]. The function of the lens capsule is primarily mechanical: in the accommodation process it has load-transmitting function. With ageing, the lens loses its ability to accommodate, thus requiring cataract surgery, which consists of replacing the native lens with a nonaccommodating plastic prosthetic one. The ultimate goal of this surgery is ideally the restoration of the accommodative function by refilling the capsular bag with an artificial polymer [69], after the endocapsular aspiration of nuclear and cortical material. This technique may become a viable lens-replacing procedure, as soon as experimental tests are able to prove preservation of capsular mechanical functions and clarity of refilled lens. Moreover, it would be important for the feasibility of this technique, to demonstrate that a biocompatible valve on the anterior lens capsule tissue could be set up to facilitate lens-refilling operations.

To improve cataract surgery, thus providing a surgical solution to presbyopia, some of us proposed a solution for performing a flap valve with the use of a patch of capsular tissue obtained from a donor lens, to be laser-welded onto the recipient capsule. The procedure may also be used to repair accidental traumas, such as capsular breaks or perforations during intraocular lenses implantation [70]. Because of its particular fragility and elasticity, it is quite impossible to suture capsular tissue using standard techniques; however, at present there are no alternative methods. Laser welding could be used to accomplish this goal. The study is in progress, and preliminary evidence of the feasibility of this technique has recently been obtained [55, 70].

Experimental tests were carried out ex vivo, on freshly-enucleated porcine eyes (Fig. 15.7). Closure tests were performed by means of patches of donor capsulae (mean diameter: 3 mm). The inner side of the patch was stained with an ICG-saturated solution in sterile water (7% weight/weight). The staining solution was left in place for 5 min. The sample was then washed with abundant water, to remove any excess of ICG. The stained patch was then applied to the anterior lens capsule, through a previously performed corneal incision.

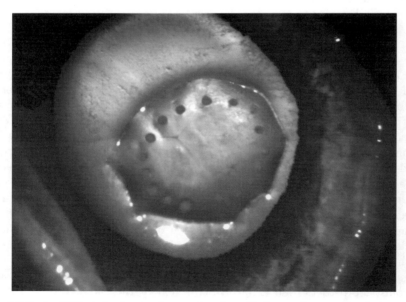

Fig. 15.7. An ICG-stained capsular patch was welded onto the anterior lens capsule of a pig eye. Laser spots are clearly evident at the periphery of the patch

The stained inner side of the patches were positioned on the exterior surface of the recipient capsule, so as to maintain the original orientation and curvature. This procedure facilitated adhesion between the tissues to be welded. The patch was then irradiated along its external perimeter by means of contiguous laser spots emitted by a 200-μm-core fiber, whose tip was gently pressed onto the patch surface (contact welding technique). Exposure times were found to be critical to avoid heat damage. Continuous wave irradiation, which is typically employed in other laser welding applications, was unsuitable, while pulses around 100 ms (with energies of 30–50 mJ) provided the best results. Once welded, the capsular patch showed good adhesion to the recipient anterior capsular surface. Preliminary biomechanical tests performed on laser-welded anterior capsule flaps showed that the load resistance of welded specimens was comparable to that of healthy tissues. Standard histology analysis indicated good adhesion between the apposed samples and thermal damage localized in the treated area.

15.3 Applications in Microvascular Surgery

Microvascular anastomosis is a surgical technique for the connection of two small-calibre blood vessels (both arteries and veins) with typical diameters of a few hundreds of micrometers. It is commonly used in various surgical fields, such as plastic and reconstructive surgery to restore traumatized or thrombotic vessels, as well as in neurosurgery in the treatment of cerebral ischemia, vascular malformations, or skull base tumors [71,72]. In this regard, conventional suturing methods are associated with various degrees of vascular wall damage, which can ultimately predispose to thrombosis and occlusion at the anastomotic site [73]. To minimize vascular wall damage and improve long-term patency, various alternative nonsuture methods have been investigated experimentally and, in some cases, in the clinical practice [74–76].

Laser welding of arteries was reported in 1979 by Jain and Gorisch [77] to perform vascular anastomosis by the use of a Nd:YAG laser. Then, other lasers (e.g., argon [78] and CO_2 [79]) were used experimentally with questionable results. Further improvements came with the introduction of low-energy diode lasers in association with exogenous chromophores and solders [80,81]. More recently, in vitro and in vivo acute studies have better defined some technical aspects of end-to-end arterial laser welding [82,83]. Despite the large number of experimental studies reported in the literature, very few have reached the clinical phase, mainly due to the lack of clear evidence of the advantages offered by laser-assisted suturing (when compared with conventional methods), and of reproducibility of results.

Experimental studies on diode laser-assisted end-to-end microvascular anastomosis (LAMA) in association with ICG topical application in femoral arteries and veins of rats were reported by some of us since 1993 [80,81]. In the design and development phases of these studies [80], the number of suture stays supporting the anastomosis was progressively reduced, to minimize foreign body reaction, and eventually to achieve simpler and faster operation. In conventional suturing procedures of microvascular anastomosis, a number of suture stays from 8 to 12 are applied to support the anastomosis. On the other hand, laser-assisted procedures to weld the vessel edges were first accomplished with the support of four conventional stays; then successful LAMAs were performed with only two sutures stays, and lastly with no permanent stays at all. These results demonstrated experimentally that diode laser welding alone could withstand both blood pressure and vessel tensile strength, providing mechanical and functional support to the anastomosis. Moreover, the most significant result of these studies emerged from histological examination on longer follow-up times (up to three months), by comparison of the healing of laser-welded and conventionally sutured vessels. Specimens from both arteries and veins yielded unquestionable evidence of faster and superior healing induced by the laser treatment, very similar to a *"restitutio ad integrum"* of vessel structures (Fig. 15.8). The repair mechanism of vessel walls in LAMAs seemed significantly favored by reduction or absence of suture material, which limited the occurrence of foreign body reaction, granulomas and inflammation, and ultimately resulted in a more effective and faster restoration of the architecture of the vessel walls.

In a more recent study, diode laser welding was experimentally tested to perform end-to-side anastomosis in carotid bypass surgery [11]. This application may be particularly important in neurosurgery in order to minimize the occlusion time of the carotid artery during the application of a venous bypass graft, thus reducing the risk of post-operative brain damages. With regards to technical aspects of the laser-assisted welding procedure, the study

Fig. 15.8. *Left*: aspect of femorary artery of a Wistar rat after end-to-end LAMA performed by staining the vessel edges with ICG and then welding them by using a low power diode laser emitting at 810 nm, without conventional stitches. *Right*: histological section of the artery wall 3 months after the surgery showing the complete restoration of the wall architecture (Wiegert elastica-Van Gieson stain, ×250)

demonstrated the feasibility of laser-assisted end-to-side anastomosis in bypass surgery, which – to our knowledge – had never been tested before in vivo and presents peculiar technical features. In fact, beside a more complex geometry of the junction, it requires the accomplishment of an effective welding between the artery and the vein graft, which exhibit different wall structures. Moreover, surgical advantages were observed in laser-assisted when compared with conventionally sutured anastomoses, such as a simplification of the procedure, a reduction or suppression of bleeding, and a shortening of the operative time, which may be potentially reduced by up to a factor of three. Again, histological, ultrastructural and immunohistochemical analyzes confirmed the occurrence of lesser inflammation and of better preservation of the endothelium and of the inner wall structures of both artery and vein in laser-treated segments. This is expected to reduce the occurrence of thrombosis and favor an optimal restoration process.

15.4 Potentials in Other Surgical Fields

15.4.1 Laser Welding of the Gastrointestinal Tract

Repair of the gastrointestinal tract by means of laser welding has been found to be a technique that is much easier, faster, and yielding better healing response and no stone formation, when compared with suturing closure techniques. The first published study regarding laser bowel welding was reported in 1986 by Sauer et al. They used a CO_2 laser to repair longitudinal transmural incisions in an otherwise-intact rabbit ileum, producing strong hermetic closures [84]. The same authors later proposed the use of a biocompatible, water-soluble, intraluminal stent in conjunction with India ink as an exogenous chromophore, to perform a suture-free end-to-end small bowel anastomosis [85]. In 1994 Rabau et al. described the healing process of CO_2 laser intestinal welding in a rat model, which evidenced a higher probability of dehiscence in the first 10 postoperative days compared with control sutured repairs [86]. Moreover, comparable healing response among argon and Ho:YAG laser-welded and control-sutured anastomoses have been found by using a temperature-controlled laser system [87]. Oz et al. investigated the feasibility of using a pulsed THC:YAG and a cw argon laser for the welding of biliary tissue. They found better histological response and higher pressure prior to breaking in the latter case [30]. Lastly, laser soldering with ICG-doped liquid albumin-solder in conjunction with a diode laser was successfully evaluated for the purpose of sealing liver injuries with minimal heat damage [88].

15.4.2 Laser Welding in Gynaecology

Laser welding was experimented in 1978 for reconstructive surgery in the fallopian tubes, using a CO_2 laser with good results [89]. Contrasting results appeared in the 1980s up until the study by Vilos et al., in which a series of intramural-isthmic fallopian-tube anastomoses were performed with a patency rate of 100% [90]. The main problem claimed in these studies appeared to be the nonuniform heating of the tissue. Improved results were obtained by introduction of a low-power diode laser in conjunction with a protein solder in a rabbit model [91]. In this study, no thermal damage was detected, while a significant reduction in the operative time was evidenced.

15.4.3 Laser Welding in Neurosurgery

The major advantage of using laser welding on nerves is the much shorter time-consumption for operation when compared with conventional microsuturing techniques. However, as high power densities are required to coapt the nerve edges, thermal damage in the epineurium is easily induced [92]. In contrast, low power does not result in sufficient tensile strength. To improve nerve welding, usage of sealing material was proposed. The most popular solders are dye-enhanced protein solders, which were employed either with CO_2 or diode laser [28,93]. The bonding rate and the functional recovery of the nerves subject to laser soldering are superior to those of manual suturing techniques, while the risk of inflammation and foreign body reaction are minimized [93].

15.4.4 Laser Welding in Orthopaedic Surgery

Other investigations of the laser tissue welding technique have been carried out in the case of orthopaedic surgery. The welding of tendinous tissue was investigated using a Nd:YAG laser alone [94] or in conjunction with an albumin solder, and an argon laser in conjunction with a fluorescein-dye-doped albumin solder [95]. These studies pointed out the inability of the laser technique to produce a weld of sufficient strength to withstand the significant tensile loads, which tendons are subjected to from immediately after the repair. Laser welding of meniscus tears using an argon laser in conjunction with a fibrinogen solder gave similar results [16].

15.4.5 Laser Welding of the Skin

The principal advantage claimed for the laser welding of skin is a superior cosmetic result. However, control of the temperature enhancement at the weld site is challenging. It is nearly impossible to obtain full-thickness welds and limit thermal denaturation laterally around the weld site in the epidermis and papillary dermis. Although thermal denaturation of tissue is necessary in order to produce a strong weld, excessive thermal damage may result in scarring and dehiscence [96]. The use of chromophores has been proposed to minimize unwanted collateral thermal injury ensuring selective laser absorption at the same time [4, 96]. The application of a variety of solders has also

15 Laser Tissue Welding in Minimally Invasive Surgery and Microsurgery   293

been investigated for improving the strength of cutaneous repairs [19, 97]. To minimize the effects of collateral thermal damage dynamic cryogen cooling was successfully employed [98]. A more recent proposal was the use of gold nanoparticles as exogenous absorbers, which allowed for the application of light sources undergoing minimal absorption from tissue components, thereby minimizing the damage to surrounding tissues [99]. Previously, the ability of laser welding to accelerate and improve the skin wound healing process was convincingly demonstrated [100]. However, it should be emphasized that – until now – clinical work on laser skin welding has never been reported.

15.4.6 Laser Welding in Urology

Urological surgery requires watertight closures, because of the continuous flow of urine. As urine lacks the clotting features of blood, producing effective closures of the urinary tract using traditional suturing techniques tends to be technically demanding and time-consuming. Moreover, suture material itself can induce the formation of stones. Conversely, laser welding can provide immediate watertight, non-lithogenic anastomoses, yielding a tensile strength that is superior to that of conventional closure techniques [5]. Laser tissue welding has been tested on several tissues of the genitourinary tract and for different applications, including urologic applications, which gave the best clinical achievements.

Laser vasal anastomosis using CO_2 and noncontact Nd:YAG laser was initially investigated in several experimental studies, followed by clinical trials with good postoperative results [101]. Afterwards, laser soldering with ICG-doped albumin in conjunction with an 808-nm diode laser was investigated for hypospadia repair on 138 children [18]. Results were compared with conventional suturing. The study emphasized the occurrence of fewer postsurgical complications in the laser group compared with the sutured one, and an easier operation in the laser set-up.

15.5 Perspectives of Nanostructured Chromophores for Laser Welding

Laser welding of biological tissues has received substantial momentum from coupling with exogenous chromophores with enhanced absorbance in the near-infrared, applied topically at the edges of the wounds prior to irradiation. As mentioned earlier, suitable exogenous chromophores absorb efficiently and selectively the near-infrared light from a laser, which immediately translates into well-localized hyperthermia and an overall decrease of the power thresholds required to achieve closure of wounds. This in turn minimizes collateral thermal damage to healthy tissues, which has ultimately backed the emergence of laser-welding as a minimally invasive and convenient alternative to traditional suturing or grafting. The ultimate exogenous chromophore should display high absorption coefficient in the near-infrared and enable high localization of power deposition, ideally down to the scale of individual biological structures. Further desirable features include good chemical and thermal stability and high photo-bleaching threshold.

Conventional chromophores of common use in laser-welding are organic molecules such as ICG. These have given outstanding experimental and clinical achievements in a number of medical fields, and that in spite of relatively poor performances with respect to the aforementioned criteria [102, 103]. The absorption efficiency and photo-stability of organic molecules are limited. Their optical properties depend strongly on biochemical environment and temperature, and generally deteriorate rapidly with time [104]. The range of chemical functionalities accessible is narrow, which is incompatible with flexible and selective targeting of distinct biological structures. Overcoming of these limitations would represent a real breakthrough in the practice of laser-welding.

Possible ways forward are disclosed by the advent of nanotechnology, as a powerful paradigm to develop new functionalities, by manipulation of self-organization processes at the nanoscale. Here, we mention the introduction of a new class of nanostructured chromophores, which is attracting much attention in view of many applications, including the laser-welding of biological tissues: colloidal gold nanoparticles (nano-gold). Whereas the optical response of organic molecules stems from electronic transitions between molecular states, light absorption, and scattering in nano-gold originates from excitation of collective oscillations of mobile electrons, i.e., surface plasmon resonances [102]. This translates into molar extinction coefficients higher by 4–5 orders of magnitude with respect to those of organic chromophores [102], enhanced thermal stability and photo-bleaching threshold, lower dependence on surrounding chemical environment (although surface plasmons shift with variations in dielectric constant and electron donating/withdrawing tendency of embedding tissues [105, 106]). Inspiring perspectives arise from the surface chemistry of nano-gold. As a traditional material for implants, gold is believed to ensure good biocompatibility [107], which is a critical prerequisite in front of clinical applications. The possibility of flexible conjugation of gold surfaces with biochemical functionalities opens a wealth of novel opportunities [108], such as selective targeting against desired and well-defined biological structures. In summary nano-gold may become the ideal substitute of organic chromophores.

The utilization of nano-gold dates back to the ancient Romans, when employed for decorative purposes in the staining of glass artifacts (e.g., the Lycurgus cup). Synthesis of stable aqueous colloidal preparations of nano-gold was first achieved by M. Faraday toward the 1850s by use of phosphorous to reduce a solution of gold chloride. Subsequent developments led to a number of variants especially based on reduction of chloroauric acid in sodium citrate (Turkevich method) [109, 110]. Conventional nano-gold is composed of spherical nanoparticles of variable and controllable radii [110]. Their optical properties are well-understood in the framework of classical Mie theory. The plasmon resonance of spherical nano-gold in aqueous environment is found within 500–600 nm, almost independent of geometrical volume [102].

Because of absorption in the visible, conventional nano-gold is not regarded as a candidate ideal chromophore for laser-welding of biological tissues. Localization of power deposition requires preferential use of near-infrared radiation. Theoretical calculations based on different approaches (including Gans theory [102], dipolar approximations [102, 105], or hybridisation of Mie resonances [111]) agree on the possibility to steer the absorption of nano-gold to well within the near-infrared by introduction of nonspherical morphologies. The experimental synthesis of gold nanoparticles with unconventional shape is a very active field of research [112]. By intrusive modification of existing procedures for spherical nano-gold, a number of nonspherical gold nanoparticles have been demonstrated, including dielectric-core/metal-shell silica or gold-sulphide/gold nano-shells [113–115], complex hollow shells as gold nano-cages [116], or high aspect ratio gold nano-rods [106, 117, 118]. Tuning of size and shape of these nanoparticles allows for tuning of plasmon resonances in good agreement with theoretical calculations. In particular, absorption within the near-infrared range of interest is becoming a mature achievement. Near-infrared-light irradiation of biological media dispersed with nonspherical nano-gold was proven to result in selective, controllable, and significant heating [119, 120].

Photo-activated nonspherical nano-gold holds the promise of manifold applications in the emerging field of nano-medicine. Proposals of extreme interest are e.g., in the treatment of tumors by selective ablation of individual malignant cells [120–122]. In our context, the potential of silica/gold nano-shells in the laser welding of tissues has recently been demonstrated in combination with an albumin solder [99]. Absorption of 820 nm diode laser radiation by a low concentration of nano-shells was shown to induce successfully the coagulation of albumin proteins and the ensuing soldering of muscles ex vivo and of skin in vivo (rat model). Preliminary results are very promising. However, much progress is still required. We estimate that future innovation will take special advantage of the adaptable functionalization of nano-gold. This may for example enable the selective targeting of individual biological structures, which may in turn result in the engineering of tissue power absorption profiles with resolution in the nano-range. This is a completely novel and powerful perspective in the laser-welding of biological tissues.

Currently, the replacement of traditional organic molecules with nanostructured chromophores is an inspiring possibility, which is yet far from the clinical application. As a very recent concept and technology, there exists at present basically no experimental evidence of the superiority of these new materials in the welding of tissues. Additional aspects of practical concern include their biocompatibility and overall sustainability (e.g., economical). In short, nano-medicine is a broad and thriving context, offering a wealth of novel opportunities. This is true also in the laser-welding of tissues, which is yet a poorly explored frontier. We foresee exciting evolution in the very next future, possibly along the guidelines sketched in this section.

References

1. K.M. McNally, in *Biomedical Photonics Handbook*, ed. by T. Vo-Dihn (CRC Press, Boca Raton, 2003), p. 1
2. L.S. Bass, M.R. Treat, Lasers Surg. Med. **17**, 315 (1995)
3. G.E. Kopchok, R.A. White, G.H. White, et al., Lasers Surg. Med. **8**, 584 (1988)
4. S.D. DeCoste, W. Farinelli, T. Flotte, et al., Lasers Surg. Med. **12**, 25 (1992)
5. D.P. Poppas, D.S. Scherr, Haemophilia **4**, 456 (1998)
6. S.G. Brooks, S. Ashley, H. Wright, et al., Lasers Med. Sci. **6**, 399 (1991)
7. F. Rossi, R. Pini, L. Menabuoni, et al., J. Biomed. Opt. **10**, article 024004 (2005)
8. R. Pini, L. Menabuoni, L. Starnotti, Proc. SPIE **4244**, 266 (2001)
9. R.A. White, G.E. Kopchok, C.E. Donayre, et al., Lasers Surg. Med. **8**, 83 (1988)
10. B. Ott, B.J. Zuger, D. Erni, et al., Lasers Med. Sci. **16**, 260 (2001)
11. A. Puca, A. Albanese, G. Esposito, et al., Neurosurgery **59**, 1286 (2006)
12. C. Chiarugi, L. Martini, L. Borgognoni, et al., Proc. SPIE **2623**, 407 (1996)
13. Z. Wang, M.M. Pankratov, L.L. Gleich, et al., Arch. Otolaryngol. Head Neck Surg. **121**, 773 (1995)
14. S. Wang, P.E. Grubbs, S. Basu, et al., Microsurgery **9**, 10 (1988)
15. D.F. Cikrit, M.C. Dalsing, T.S. Weinstein, et al., Lasers Surg. Med. **10**, 584 (1990)
16. S.K. Forman, M.C. Oz, J.F. Lontz, et al., Clin. Orthop. Relat. Res. **310**, 37 (1995)
17. K.M. McNally, B.S. Sorg, A.J. Welch, et al., Phys. Med. Biol. **44**, 983 (1999)
18. A.J. Kirsch, C.S. Cooper, J. Gatti, et al., J. Urol. **165**, 574 (2001)
19. B.S. Sorg, A.J. Welch, Lasers Surg. Med. **31**, 339 (2002)
20. D.P. Poppas, J.M. Massicotte, R.B. Stewart, et al., Lasers Surg. Med. **19**, 360 (1996)
21. F. Rossi, R. Pini, L. Menabuoni, J. Biomed. Opt. **12**, 014031 (2007)
22. W. Small, D.J. Maitland, N.J. Heredia, et al., J. Clin. Laser Med. Surg. **15**, 3 (1997)
23. M.M. Judy, R.W. Jackson, H.R. Nosir, et al., Proc. SPIE **2970**, 257 (1997)
24. L. Mulroy, J. Kim, I. Wu, et al., Invest. Ophthalmol. Vis. Sci. **41**, 3335 (2000)
25. B.P. Chan, I.E. Kochevar, R.W. Redmond, J. Surg. Res. **108**, 77 (2002)
26. B.P. Chan, C. Amann, A.N. Yaroslavsky, et al., J. Surg. Res. **124**, 274 (2005)
27. S.A. Prahl, S.D. Pearson, Proc. SPIE **2975**, 245 (1997)
28. T. Menovsky, J.F. Beek, M.J.C. van Gemert, Lasers Surg. Med. **19**, 152 (1996)
29. J. Tang, G. Godlewski, S. Rouy, et al., Lasers Surg. Med. **21**, 438 (1997)
30. M.C. Oz, L.S. Bass, H.W. Popp, et al., Lasers Surg. Med. **9**, 248 (1989)
31. V.L. Martinot, S.R. Mordon, V.A. Mitchell, et al., Lasers Surg. Med. **15**, 168 (1994)
32. R. Schober, F. Ulrich, T. Sander, et al., Science **232**, 1421 (1986)

33. P. Matteini, F. Rossi, L. Menabuoni, et al., Proc. SPIE **6426**, article no. 642614 (2007)
34. L.W. Murray, L. Su, G.E. Kopchok, et al., Lasers Surg. Med. **9**, 490 (1989)
35. H.E. Savage, R.K. Halder, U. Kartazayeu, et al., Lasers Surg. Med. **35**, 293 (2004)
36. G.M. Lemole, R.R. Anderson, S. DeCoste, Proc. SPIE **1422**, 116 (1991)
37. P.J. Drew, A. Watkins, A.D. McGregor, et al., Lasers Med. Sci. **16**, 291 (2001)
38. L.S. Bass, N. Moazami, J. Pocsidio, et al., Lasers Surg. Med. **12**, 500 (1992)
39. W. Small, P.M. Celliers, G.E. Kopchok, et al., Lasers Med. Sci. **13**, 98 (1998)
40. M. Constantinescu, A. Alfieri, G. Mihalache, et al., Lasers Med. Sci. **22**, 10 (2007)
41. C.R. Guthrie, L.W. Murray, G.E. Kopchok, et al., J. Invest. Surg. **4**, 3 (1991)
42. E. Strassmann, N. Loya, D.D. Gaton, et al., Proc. SPIE **4244**, 253 (2001)
43. M.M. Judy, L. Fuh, J.L. Matthews, et al., Proc. SPIE **2128**, 506 (1994)
44. H.C. Zweng, M. Flocks, Tran. Am. Acad. Ophthalmol. Otolaryngol. **71**, 39 (1967)
45. R.H. Keates, S.N. Levy, S. Fried, et al., J. Cataract Refract. Surg. **13**, 290 (1987)
46. R.P. Gailitis, K.P. Thompson, Q.S. Ren, et al., Refract. Corneal Surg. **6**, 430 (1990)
47. N.L. Burnstein, J.M. Williams, M.J. Nowicki, et al., Arch. Ophthalmol. **110**, 12 (1992)
48. T.J. Desmettre, S.R. Mordon, V. Mitchell, Proc. SPIE **2623**, 372 (1996)
49. G. Trabucchi, P.G. Gobbi, R. Brancato, et al., Proc. SPIE **2623**, 380 (1996)
50. A. Barak, O. Eyal, M. Rosner, et al., Surv. Ophthalmol. **42**, S77 (1997)
51. E. Strassmann, N. Loya, D.D. Gaton, et al., Proc. SPIE **4609**, 222 (2002)
52. L. Menabuoni, B. Dragoni, R. Pini, Proc. SPIE **2922**, 449 (1996)
53. L. Menabuoni, F. Mincione, B. Dragoni, et al., Proc. SPIE **3195**, 25 (1998)
54. L. Menabuoni, R. Pini, F. Rossi, et al., J. Cataract Refract. Surg. (in press)
55. R. Pini, F. Rossi, L. Menabuoni, et al., Proc. SPIE **6138**, 307 (2006)
56. G.S. Schultz, in *Cornea. Fundamentals of Cornea and External Disease*, vol. 1, ed. by J.H. Krachmer, M.J. Mannis, E.J. Holland (Mosby, St. Louis, MO, 1997) p. 183
57. M.L.J. Landsman, G. Kwant, G.A. Mook, et al., J. Appl. Physiol. **40**, 575 (1976)
58. L.A. Yannuzzi, M.D. Ober, J.S. Slakter, et al., Am. J. Ophthalmol. **137**, 511 (2004)
59. G.P. Holley, A. Alam, A. Kiri, et al., J. Cataract Refract. Surg. **28**, 1027 (2002)
60. T. John, J. Cataract Refract. Surg. **29**, 437 (2003)
61. R. Pini, V. Basile, S. Ambrosini, et al., Proc. SPIE **6138**, 404 (2006)
62. M. Han, G. Giese, L. Zickler, et al., Optics Exp. **12**, 4275 (2004)
63. K. Koenig, O. Krauss, I. Riemann, Optics Exp. **10**, 171 (2002)
64. A. Vogel, J. Noack, G. Httman, et al., Appl. Phys. B **81**, 1015 (2005)
65. L. Menabuoni, R. Pini, M. Fantozzi, et al., Invest. Ophthalmol. Vis. Sci. **47**, E (2006)
66. R. Pini, L. Menabuoni, I. Lenzetti, et al., *VII International Symposium on Ocular Trauma*, Rome (2006)
67. N.M. Ziebarth, F. Manns, S.R. Uhlhorn, et al., Invest. Ophthalmol. Vis. Sci. **46**, 1690 (2005)

68. C.C. Danielsen, Exp. Eye Res. **79**, 343 (2004)
69. E. Haefliger, J-M. Parel, Exp. Eye Res. **10**, 550 (1994)
70. R. Pini, F. Rossi, L. Menabuoni, et al., Opthalmic Surg. Laser Imaging (in press)
71. M.T. Lawton, M.G. Hamilton, J.J. Morcos, et al., Neurosurgery **38**, 83 (1996)
72. L.N. Sekhar, C. Kalavakonda, Neurosurgery **50**, 321 (2002)
73. C.J. Zeebregts, R.H. Heijmen, J.J. van den Dungen, et al., Br. J. Surg. **90**, 261 (2003)
74. I.A. Aksik, R.P. Kikut, D.L. Apshkalne, Microsurgery **7**, 2 (1986)
75. E.S. Ang, K.C. Tan, L.H. Tan, et al., J. Reconstr. Microsurg. **17**, 193 (2001)
76. D.P. Falconer, T.W. Lewis, E.G. Lamprecht, et al., J. Reconstr. Microsurg. **6**, 215 (1990)
77. K.K. Jain, W. Gorisch, Surgery **85**, 684 (1979)
78. O.M. Gomes, R. Macruz, E. Armelin, et al., Tex. Heart Inst. J. **10**, 145 (1983)
79. M.R. Quigley, J.E. Bailes, H.C. Kwaan, et al., Microsurgery **6**, 229 (1985)
80. R. Gelli, R. Pini, F. Toncelli, et al., J. Reconstr. Microsurg. **13**, 199 (1997)
81. U.M. Reali, R. Gelli, V. Giannotti, et al., J. Reconstr. Microsurg. **9**, 203 (1993)
82. A. Lauto, A.H. Hamawy, A.B. Phillips, et al., Lasers Surg. Med. **28**, 50 (2001)
83. A.B. Phillips, B.Y. Ginsburg, S.J. Shin, et al., Lasers Surg. Med. **24**, 264 (1999)
84. J.S. Sauer, D.W. Rogers, J.R. Hinshaw, Lasers Surg. Med. **6**, 106 (1986)
85. J.S. Sauer, J.R. Hinshaw, K.P. McGuire, Lasers Surg. Med. **9**, 70 (1989)
86. M.Y. Rabau, I. Wasserman, S. Shoshan, Lasers Surg. Med. **14**, 13 (1994)
87. I. Cilesiz, S. Thomsen, A.J. Welch, et al., Lasers Surg. Med. **21**, 278 (1997)
88. Y. Wadia, H. Xie, M. Kajitani, J. Trauma **51**, 51 (2001)
89. L. von Klitzing, R. Grosspietzsch, F. Klink, et al., Fortschr. Med. **96**, 357 (1978)
90. G.A. Vilos, Fertil. Steril. **56**, 571 (1991)
91. L.W. Kao, H.R. Giles, J. Reprod. Med. **40**, 585 (1995)
92. H. Maragh, R.S. Hawn, J.D. Gould, et al., J. Reconstr. Microsurg. **4**, 189 (1988)
93. A. Lauto, R. Trickett, R. Malik, et al., Lasers Surg. Med. **21**, 134 (1997)
94. J.D. Burt, M. Siddins, W.A. Morrison, Plast. Reconstr. Surg. **108**, 688 (2001)
95. F.X. Kilkelly, T.J. Choma, N. Popovic, et al., Lasers Surg. Med. **19**, 487 (1996)
96. N.M. Fried, J.T.J. Walsh, Lasers Surg. Med. **27**, 55 (2000)
97. D. Simhon, T. Brosh, M. Halpern, et al., Lasers Surg. Med. **35**, 1 (2004)
98. N.M. Fried, J.T.J. Walsh, Phys. Med. Biol. **45**, 753 (2000)
99. A.M. Gobin, D.P. O'Neal, D.M. Watkins, et al., Lasers Surg. Med. **37**, 123 (2005)
100. A. Capon, S. Mordon, Am. J. Clin. Dermatol. **4**, 1 (2003)
101. P.T. Gilbert, R. Beckert, Lasers Surg. Med. **9**, 42 (1989)
102. P.K. Jain, K.S. Lee, I.H. El-Sayed, et al., J. Phys. Chem. B **110**, 7238 (2006)
103. F. Wang, W.B. Tan, Y. Zhang, et al., Nanotechnology **17**, R1 (2006)
104. V. Saxena, M. Sadoqi, J. Shao, Int. J. Pharm. **308**, 200 (2006)
105. M.M. Miller, A.A. Lazarides, J. Opt. A: Pure Appl. Opt. **8**, S239 (2006)
106. P. Mulvaney, J. Perez-Juste, M. Giersig, et al., Plasmonics **1**, 61 (2006)
107. R. Shukla, V. Bansal, M. Chaudhary, et al., Langmuir **21**, 10644 (2005)
108. M.C. Daniel, D. Astruc, Chem. Rev. **104**, 293 (2004)
109. L.M. Liz-Marzan, Materialstoday **7**(2), 26 (2004)
110. J. Kimling, M. Maier, B. Okenve, et al., J. Phys. Chem. B **110**, 15700 (2006)
111. E. Prodan, C. Radloff, N.J. Halas, et al., Science **302**, 419 (2003)

112. M. Hu, J. Chen, Z.Y. Li, et al., Chem. Soc. Rev. **35**, 1084 (2006)
113. S. Kalele, S.W. Gosavi, J. Urban, et al., Curr. Sci. **91**, 1038 (2006)
114. H.S. Zhou, I. Honma, H. Komiyama, et al., Phys. Rev. B **50**, 12052 (1994)
115. L.R. Hirsch, A.M. Gobin, A.R. Lowery, et al., Ann. Biomed. Eng. **34**, 15 (2006)
116. Y. Sun, Y. Xia, Science **298**, 2176 (2002)
117. J. Perez-Juste, I. Pastoriza-Santos, L.M. Liz-Marzan, et al., Coord. Chem. Rev. **249**, 1870 (2005)
118. N.R. Jana, L. Gearheart, C.J. Murphy, J. Phys. Chem. B **105**, 4065 (2001)
119. C.H. Chou, C.D. Chen, C.R.C. Wang, J. Phys. Chem. B **109**, 11135 (2005)
120. L.R. Hirsch, R.J. Stafford, J.A. Bankson, et al., Proc. Natl. Acad. Sci. **100**, 13549 (2003)
121. J.Y. Chen, F. Saeki, B.J. Wiley, et al., Nano Lett. **5**, 473 (2005)
122. X. Huang, I.H. El-Sayed, W. Qian, et al., J. Am. Chem. Soc. **128**, 2115 (2006)

16

Photobiology of the Skin

A.P. Pentland

16.1 Basics of Skin Structure: Cell Types, Skin Structures, and Their Function

Everyone is familiar with the skin because it is the organ that is primarily responsible for the appearance of human beings. In addition to this important role, skin also has a myriad of important maintenance functions. The specialized cells of the skin work together to form the protective and supportive structures necessary for our bodily integrity [1]. The main structures of the skin include the epidermis, which is the most superficial layer of the skin, and forms its outer barrier [2] (Fig. 16.1). The epidermis is supported underneath by the dermis, the portion of the skin that is responsible for the physical integrity of the skin, and gives it the capacity to undergo stretching and flexing and resist trauma [3–5]. Interspersed in this tough matrix is the skin's vascular network and the nerves that permit the transmission of signals indicating pain or itch. Lastly, there are many adnexae situated in skin, such as the hair, nails, sebaceous, and sweat glands. Collectively, the key function of these structures is to keep toxic environmental influences out, and prevent loss of bodily fluids [1]. Some specific aspects of these functions are:

- Temperature regulation; sweating and vasodilation for heat loss [6]
- Physical integrity; dermal toughness and elasticity [4, 5]
- Maintenance of water balance; providing an intact evaporative barrier [2]
- Microorganism defense; skin has active innate and acquired immunity [7]
- Appearance [8]
- Use of light to synthesize Vitamin D [9]
- Protection from hazardous UV radiation [10]

The cell types present in the epidermis are keratinocytes, melanocytes, and langerhans cells. Keratinocytes constitute approximately 90–95% of the cells in epidermis, and are responsible for its structure [1]. Keratinocytes are produced in the innermost layer of the epidermis adjacent to the dermis. They

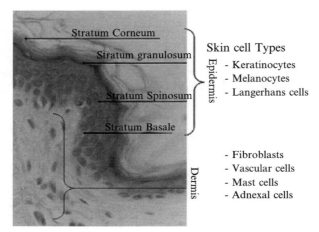

Fig. 16.1. Skin is a multilayered organ. Epidermis comprises the most superficial layer. The cells of the epidermis divide in the basal layer, then differentiate, moving upward to the skin surface to slough. The dermis underlies the epidermis, providing structural support for blood vessels and adnexae, as well as the epidermis

then differentiate, flattening and progressing toward the surface of the skin as the process of differentiation proceeds, until they are no longer viable. During this differentiation process, keratinocytes synthesize and secrete an array of lipids intercellularly in organelles termed lamellar bodies. As differentiation is completed, these organelles are secreted into the extracellular space where their lipid contents spread and disperse among the cells of the stratum corneum, creating a "brick and mortar" motif that is extraordinarily impermeable [2,11]. The barrier is resistant to moisture, invasion of microorganisms, and even oxygen. The constant sloughing of fully differentiated epidermal cells also is an adaptive strategy that prevents microorganism invasion and provides a constant new supply of skin as we wash and scrub the skin surface in the course of work and play. Keratinocytes are also capable of innate immune responses, secreting inflammatory cytokines and chemotactic mediators that help to repel invading organisms [7], and are responsible for the synthesis of the vitamin D precursor, Vitamin D_3 [9, 12].

A second important cell type located in the epidermis is the melanocyte, which represents ∼4% of the cells in the epidermis [13]. It is responsible for our skin color because of its capacity to synthesize melanin granules and secrete them to be taken up by keratinocytes [14,15]. The melanin produced by melanocytes is the chief photoprotective substance within the skin, as it is capable of absorbing light throughout the visible and ultraviolet light spectrum. One melanocyte provides melanin for about 36 keratinocytes (Fig. 16.2).

In addition to these two key cell types, the epidermis also contains ∼4% Langerhans cells, which are the resident immune cell of the epidermis. When microorganisms do successfully penetrate the epidermal stratum corneum

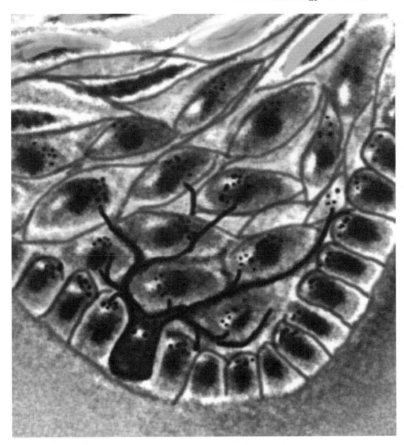

Fig. 16.2. The melanocyte is responsible for the synthesis of melanin. Melanocytes reside in the basal layer of the epidermis. They extend dendrites to many nearby keratinocytes. Melanin is delivered to keratinocytes by transfer of pigment-containing melanosomes to these nearby keratinocytes, where it is situated above the nucleus. Melanin is responsible for skin color

barrier, it is the Langerhans cell that identifies and transports microorganism proteins to regional lymph nodes so an adaptive, long-term, and specific immune response can be initiated.

The dermis underlies the epidermis, creating its physical support structure. It contains fibroblasts, vascular cells, adnexal cells, mast cells, and dermal dendrocytes [4,5]. Fibroblasts are responsible for the synthesis of dermal proteins. The most abundant of these is type 1 collagen, which is present in the dermis as a highly cross-linked protein network. In addition to collagen, fibroblasts synthesize and secrete elastic fibrils, which impart resiliency and elasticity to skin and comprise ∼1–2% of skin protein. Mature elastic fibers in the dermis are composed of an amorphous elastin core surrounded by microfibrils, also produced by fibroblasts. Interestingly, most synthesis of elastin occurs in the first trimester, so that as chronic UV exposure degrades this protein, our skin loses its stretchiness, causing wrinkles [8, 16].

In addition to fibroblasts, the dermis contains a vascular network that provides nourishment to the skin, and which can dilate and contract as needed to adjust the core temperature [6]. Mast cells are located at the intersections of the small vessels in the dermis where they can readily detect minor trauma and invading microorganisms. The mast cell is the key cell responsible for hives observed in allergic reactions [17]. The mast cell also contributes to the skin's innate immune defenses. When stimulated physically or by proteins present on invading organisms, it releases granules that contain vasodilator substances and mediators that can activate the immune system. These substances "kick start" the host immune response.

Interposed between the epidermis and the dermis is a thin layer of elegantly assembled attachment proteins, termed the basement membrane [18, 19]. There are approximately 15 different proteins present in the basement membrane zone, which work in concert to permit the rapidly proliferating epidermis to maintain its attachment to the underlying dermis.

16.2 Effects of Light Exposure on Skin

As in any other system, light interacting with the epidermis is reflected, re-mitted, scattered, or absorbed. Only absorbed solar radiation has a biological effect on the skin. The most bioactive wavelengths in skin are those in the ultraviolet range, from 280 to 400 nm [10]. Although some work has been done to demonstrate that 670 nm light may improve wound repair [20, 21], most literature on the interaction of light with skin focuses on wavelengths in the ultraviolet range, because of their capacity to promote synthesis of vitamin D [9], cause photoaging and skin cancer [22], as well as their therapeutic utility.

Radiation in the ultraviolet range is divided into two types, on the basis of the associated biology. Wavelengths from 280 to 320 nm are termed UVB light, while those from 320 to 400 nm are termed UVA light. UVB light is primarily responsible for initiation and promotion of skin cancer, and vitamin D synthesis [10, 12]. UVA wavelengths also contribute to skin cancer, but are much less effective. Their primary impact is on photoaging. Both UVA and UVB wavelengths stimulate melanin synthesis by melanocytes to protect against the unwanted effects of light [23]. The biological effects of these two types of light are dictated by the cellular structures that absorb light. DNA absorbs light readily in the UVB range, and therefore UVB light is far more mutagenic than UVA light [24].

When skin is exposed to UV light, it produces both acute and chronic effects. In the acute response, observed 4–48 h after exposure to UV light, skin can develop redness due to vasodilation, swell, become painful, or even blister if the dose of light is sufficiently large. Somewhat after these changes occur, in 2–7 days, tanning and stratum corneum thickening occur. Those individuals who have very little capacity to tan will not produce much extra melanin after UV exposure, but will still produce thickening of the stratum corneum [9]. The time course of the acute changes resulting from UV exposure is also dependent on the dose of light. Doses of light that are in large excess of the photoprotective capacity of an individual's skin will result in much more long-lasting and intense erythema [9]. Thus, the changes produced in skin are very dependent on the genetic capacity for melanin synthesis, which is termed the skin type [25]. Skin types are rated on a scale of 1 through 6 (Fig. 16.3). Individuals with type 1 skin always burn and never tan when exposed to UV light. Individuals with type 2 skin burn initially, but then can tan modestly after recurrent exposure to light. Those with type 3 skin may burn initially in the spring, but will reliably tan thereafter. Those with type 4 skin have significant coloration without sun exposure, and always tan readily. Those with type 5 skin have deep coloration without sun exposure, but tanning is evident after sun exposure. Lastly, individuals with type 6 skin have constant very deep pigmentation in which tanning is difficult to detect. Light-related skin cancer is prevalent primarily in skin types 1–3. Over the course of years, recurrent exposure of individuals to UV light produces chronic changes in susceptible, light skinned, individuals. Most evident is photoaging. Characterized by deep wrinkling, uneven pigmentation, vascular dilatation, and hyperkeratotic changes in epidermis, photoaging is what most people think

Skin Type	Response to Sun	Skin Cancer Risk
I	Always burns, Never tans Example: Irish	High
II	Burns readily, some tanning with repeated exposure Example: French	High
III	Burns sometimes in spring, tans readily Example: Light asian or Italian	Moderate - High
IV	Moderate visible constitutive color, difficult to sunburn Example: Indian, Arabic	Low
V	Brown constitutive color, does not burn Example: African	Very Low
VI	Extremely dark constitutive color Example: Australian aborigine	Very Low

Fig. 16.3. Skin types: capacity to tan, burn, and risk of skin cancer

of as aging because of the passage of time [8]. However, vigilant protection against exposure to UV light or a naturally dark complexion are highly effective in preventing these skin changes [9]. For those individuals who are light skinned, chronic exposure to light can also result in immunosuppression and skin cancer formation [22]. Despite these negative consequences of UV light exposure, UV light can be useful in the treatment of skin disease. The response of skin to light is shaped by the presence of many chromophores in skin. These include aromatic amino acids, DNA, NADH, beta-carotene, porphyrins, and most importantly melanin, which has an absorptive range from 280 nm through the whole visible spectrum of light [22, 26]. Although absorption of light by melanin is generally harmless, absorption of light by DNA, amino acids, and membrane lipids produces immediate cellular damage. Therefore, intricate defensive mechanisms are in place to assess the extent of UV damage that has been produced by a particular exposure to UV light and to initiate appropriate cellular responses to it.

Immediately after exposure to UV light, the cell division process is paused (Fig. 16.4). UVA wavelengths produce damaging free radical production, while both UVA and UVB can produce photoproducts in DNA [10,27]. The key protein initiating this halt in cell cycling is p53, sometimes called "the guardian of the genome" because of its role in response to UV injury. At about the same time, a transcriptional activation occurs. This permits induction of DNA excision/repair enzymes so that damaged DNA can be repaired. Efforts at DNA repair are focused mostly on actively transcribed genes. In addition, oxidized or denatured proteins are degraded, and the synthesis of their replacements is initiated. Similar mechanisms are carried out to repair damaged lipids. Synthesis of inflammatory cytokines is also initiated in response to light exposure, including TNFα, IL-1, INFγ and TGFα, which act to regulate repair responses and also are immunomodulators [24, 28, 29]. As new antigens and DNA fragments can be produced by exposure of the skin to light, these cytokines help to ensure that sun exposure does not result in sun allergy or autoimmune disease. About 72 h after sun exposure, melanin synthesis and increased differentiation of keratinocytes is prominent [30]. By 7 days after exposure to light, the repair of DNA and proteins is back to baseline values, and the photoprotective capacity of the exposed epidermis has been increased as needed. In some instances after UV exposure, the damage to cells is so severe, that repair is not a viable option. In this case, cellular apotosis is induced. Cells undergoing apoptosis after UV injury were initially described as sunburn cells. Apoptotic destruction of injured cells allows their orderly removal from injured tissue [31].

The process of repair that occurs after UV exposure is imperfect, making chronic UV exposure problematic. Mutations in DNA can escape repair processes, an error that is key for cancer initiation. Mutations may increase expression of genes that support proliferation or block genes that inhibit it. The capacity of UV light to both cause mutations and to stimulate a proliferative response, as described earlier, is the reason that UV light is

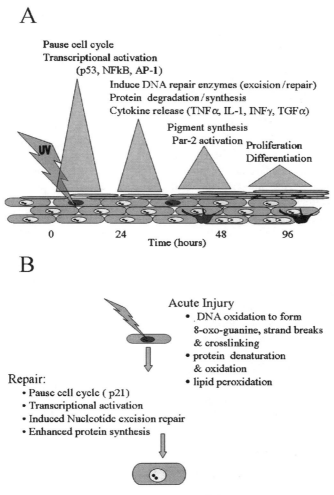

Fig. 16.4. After exposure to UVB radiation, a cellular process to assess the response of epidermis to UV injury is activated. Cell division is halted, the extent of injury is assessed, and repair processes are initiated. Concurrently, those individuals capable of tanning induce melanin synthesis and pigment transfer to keratinocytes. Those cells that have been severely injured are removed by the process of apoptosis

a complete carcinogen, both initiating and promoting cancerous growth in skin [24]. Clones of these initiated cells proliferate as recurrent UV exposure occurs over time, increasing the opportunity for a second mutation to occur in a cell, and thereby causing cancer [10]. The added twist that UV light is immunosuppressive further increases the capacity of UV light to cause cancer, as the immune surveillance capability of irradiated skin is reduced, permitting survival of more initiated cells. After UV exposure, Langerhans cells, the resident immune cell in epidermis, are depleted [24]. Immunosuppressive cytokines are released, including IL-10, TNF, αMSH and PGE_2. PGE_2 also powerfully stimulates keratinocyte cell division [32]. Although this immunosuppression is necessary to prevent sun allergy and autoimmune disease, it decreases the ability of individuals with poor native photoprotection to endure sun exposure over the course of their lifetime without developing skin tumors.

The mutations in DNA induced by UV exposure have been well characterized. Adjacent thymine residues in DNA form cyclobutane pyrimidine dimers. Also formed are pyrimidine–pyrimidone photoproducts, which can isomerize to a Dewar isomer. Should replication of DNA occur before repair of these photoproducts, base substitution and therefore mutation, can occur [10]. In addition to this type of DNA damage, UV light can also produce strand breaks. Additionally, UVA light is capable of producing an 8-oxo-guanine photoproduct in DNA. Once DNA photoproducts occur, they are removed over time. Those that are most likely to be mutagenic are removed more efficiently than those that are not. The removal of photoproducts is increased in areas of actively transcribed DNA. It is also much quicker in newborns than in older individuals (Fig. 16.5).

With so much evident harm resulting from exposure of the skin to light, it is not surprising that the skin is well endowed with defenses against UV injury. In addition to the repair pathways already described, and the broad UV absorbing capacity of melanin, the epidermis also has other substances in abundance that can absorb UV energy harmlessly. Glutathione is present in the cell cytosol in large amounts, as is urocanic acid, which isomerizes when absorbing energy in UV wavelengths. In addition, epidermis contains catalase and superoxide dismutase to detoxify UV-induced free radicals [9].

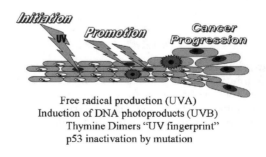

Free radical production (UVA)
Induction of DNA photoproducts (UVB)
Thymine Dimers "UV fingerprint"
p53 inactivation by mutation

Cells must repair or be removed

Fig. 16.5. UV light is a complete carcinogen, capable of initiating mutations in DNA that also decrease the capability of cells to detect these alterations. It also promotes mutated cells to divide, increasing the chance of a second mutation within a cell, and therefore increasing the chance of cancer progression

One *good* activity that occurs when skin is exposed to UV light is the synthesis of Vitamin D [9,12]. Provitamin D (7-dehydrocholesterol) is synthesized in epidermis. When this compound is exposed to light wavelengths between 280 and 315 nm, it is altered to become previtamin D3. This compound is then isomerized by the warmth of the skin to become Vitamin D3 (cholecalciferol). Vitamin D3 is sufficiently soluble that it can then enter circulation to be metabolized to 25(OH)D3 termed calcediol, or 25-hydroxy vitamin D3. This vitamin is important in calcium deposition in bones and calcium homeostasis in the kidney. UV exposure is not a requirement for adequate supplies of Vitamin D; however, it is readily obtained from dietary sources.

16.3 Sun Protection and Sunscreens

Skin cancer is the most common form of cancer, with nearly a million new cases occurring every year. The most dangerous type of skin cancer, melanoma, is rapidly rising in incidence, with lifetime risk now approaching 1 in 50. All forms of skin cancer are linked to sun exposure [22,33]. It is, therefore, quite important to be cautious about sun exposure, particularly for those with light complexions, and to adopt sun protection practices to minimize skin cancer risk and also to decrease the effects of photoaging. The simplest and most effective method for minimizing sun exposure is to avoid outdoor activities during the period of the day when UVB exposure is most intense, from 10 AM to 4 PM. In addition, when outdoors, good sun protection can be achieved by wearing protective clothing, a hat with a 4″ brim, and a sunscreen [9].

Some clothing is also now rated by its sun protection factor, so an individual can determine the degree of protection provided by a garment [34,35]. Some garments provide more protection than others. The style of the garment matters – it must cover the skin. Two layers are better than one, and newer garments and thicker garments are more UV resistant than older ones. Interestingly, even the type of fiber makes a difference. The UV blocking capacity of fibers is as follows: polyester > wool > silk > nylon > cotton and rayon. Of course wet, see-through fabric also does a poorer job filtering the light than dry fabric.

As individuals vary so widely in their sunburn responses to UV light, the common method to rate protection is called the sun protection factor, or *SPF*. The SPF of a sun protection device or lotion is determined empirically. An initial exposure of a subject's skin to graded doses of a UV light source is made to determine the minimum amount of light that can produce detectable erythema with a distinct border 24 h later. Clearly, the more melanin an individual has in their skin, the larger the dose of light necessary to produce redness. Once the minimal erythema dose (MED) is determined for an individual, the protective measure to be tested is applied to the area to be irradiated and the skin again exposed to light (Fig. 16.6). The capacity of the protective measure to block erythema production 24 h later is expressed as

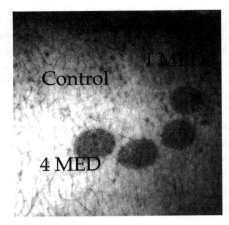

Fig. 16.6. The sensitivity of an individual to damage from sun exposure is measured by determining their minimal erythema dose. The least amount of light capable of producing a clearly outlined spot of redness, (termed erythema) is called the minimal erythema dose for that individual. The amount of light required to produce erythema varies nearly 100-fold because of the differences in constitutive skin color

the sun protection factor, or SPF. For instance, if a garment has an SPF of 15, then an individual who ordinarily develops erythema after 10 min of light exposure without protection will now be able to tolerate 150 min of exposure when wearing the garment [9].

Use of sunscreens as a method of protection from skin cancer risk has been actively debated in recent years. There is definitive evidence that their use protects animals from skin cancer in the laboratory. Similar trials have not been conducted in humans, but there is strong supportive evidence. Some debate has arisen about whether sunscreens may interfere with Vitamin D metabolism, but it is clear that the amount of Vitamin D needed can be readily obtained from dietary sources. In addition, very little light is needed to produce Vitamin D from endogenous sources. It would be the rare individual who could be sufficiently assiduous in their sun protective measures to produce Vitamin D deficiency.

Most currently available sunscreens were initially formulated to protect against UVB wavelengths, from 280 to 320 nm. There are two major classes of these compounds: UVB absorbing and physical sunscreens [9]. With recent rapid advances in material science, many new sunscreen compounds are being created and tested. Some of these new sunscreens have already been tested for safety and met standards for European approval, while some aspects of safety for new sunscreen materials are still being evaluated.

Physical sunscreens are effective across a broad range of wavelengths. They are made up of particulates that can scatter and reflect UV light. In general, they contain zinc oxide, titanium dioxide, magnesium oxide, talc, or kaolin (up to 25%). Sunscreens in this category provide the best block available in the UVA range (up to 380 nm), and are also very effective at blocking UVB. Their chief drawback is that when formulated to have a high sun protection factor, they are visible as a white substance when applied to skin. Much work has been conducted lately to alter their physical properties to decrease scattering such that visibility is decreased. Some of this alteration has been done by making nanoparticle formulations. Unfortunately, their capacity to protect against light in the UVA range is decreased when the particle size is decreased into the nanometer range. This problem is being addressed by grouping some nanoparticles in the sunscreen vehicle, and also by coating the particles with dimethicone or silica. A recent concern has been raised that the capacity of nanoparticles to penetrate the skin may be different from that of larger particles. Because of the character of the stratum corneum, it may be they cannot penetrate readily into viable tissue in significant quantities. Research is ongoing in this area [9].

UVB absorbing sunscreens are capable of harmlessly absorbing UV light, and are generally degraded over time by their absorption of light. Sunscreens of this type are wavelength selective, unlike the physical sunscreens. Therefore, it is necessary to combine several chemical sunscreens to gain optimal coverage across the UV spectrum. More consideration is also now being given to grouping sunscreen ingredients so that their absorption of UV can be transferred to heat, thereby leaving the sunscreen ingredient intact and able to function in its absorbing capacity longer. UVB blockers include paraaminobenzoic acid (up to 15%), Padimate O, a PABA ester (up to 8%), Octocrylene (up to 10%), Cinnamates (such as octyl methoxycinnamate, up to 8%), and Salicylates (up to 5%). Some chemical sunscreens are broader in their protective capacity, such as the benzophenones and oxybenzone (up to 6%), which are useful for UVB protection, but also protect against the shorter wavelengths in the UVA spectrum. Until recently, the only available absorbing sunscreen protective in most of the short UVA spectrum was Parsol 1789, which is utilized in concentrations up to 3%. New formulations of UV absorbing molecules are now becoming available, which will greatly expand the capacity of sunscreen manufacturers to make a product that protects broadly across UV wavelengths.

16.4 Phototherapy: Use of Light for Treatment for Skin Disease

Despite the problems associated with exposure to UV light, it has been recognized for centuries that it is effective in the treatment of skin disease. Diseases such as psoriasis, eczema, and vitiligo have been managed with moderate efficacy, using sun exposure for centuries [9]. With the advent of artificial light sources, it became practical to expand the use of light as a treatment option, as therapy could be administered throughout the year and in cold climates.

Three types of phototherapy are in common use today, on the basis of the light source in use for treatment. The type of lamp selected for use in treatment depends upon the patient's diagnosis, the disease location and the patient's skin type. All types of phototherapy are less effective in dark-skinned individuals because melanin blocks some of the therapeutic light transmission. The most widely available is UVB therapy, using lamps that emit light primarily in the wavelengths from 280 to 320 nm. Because of its similarity to solar UVB emission, treatment can be administered daily, taking advantage of the body's adaptive responses to UVB light. In use since the early part of the last century, it is a well-characterized treatment method. Because the wavelengths of light emitted are the same wavelengths responsible for photocarcinogenesis, it is used sparingly. Studies examining whether UVB phototherapy produces increased cancer risk suggest that the increase is modest, due to physician supervision of light treatment and the fact that its use is generally intermittent for disease therapy. Tanning also occurs when patients are treated with this light source, which decreases the efficacy of light treatment. Disease located deep within the skin is difficult to treat because of absorption of UVB wavelengths primarily in the epidermis [9].

In addition to UVB phototherapy, two other light sources have been introduced in the more recent past for phototherapy treatment. The creation of UVA light sources (320–400 nm) triggered an investigation of whether this light source could be useful for disease located deep in the skin, which is resistant to UVB treatment because of poor skin penetrating capacity. To make UVA wavelengths efficacious as a treatment method, light exposure occurs after patients have taken the photosensitizing drug psoralen. The treatment method is, therefore, termed "PUVA," meaning psoralen + UVA. When psoralen is present in the skin during UVA irradiation, psoralen becomes cross-linked to DNA. This initiates the repair pathways discussed earlier, and has a beneficial effect on a number of diseases. Because the treatment produces more inflammatory change than does UVB therapy, it is only given 2–3 times per week to allow ample time for repair. As this treatment also produces crosslinks in DNA, the risk of photocarcinogenesis is quite high; studies have demonstrated that the rate of squamous cell skin cancer occurring after PUVA increased eightfold and that melanoma rates are increased as well. Nonetheless, PUVA treatment is significantly more efficacious than UVB treatment, and therefore in cases of refractory disease, it is used [9]. The lamps used in tanning parlors are the same as those used for PUVA treatment, but the public does not take a photosensitizing drug prior to their visit!

To try and reduce the undesirable side effects of UVB and PUVA treatment, lamps that emit light most intensely at 312 nm have been brought to market, termed as narrow-band UV lamps. This light source avoids some of the cancer risk of broad-band UVB lamps, and can work well without producing intense tanning of patients. This form of treatment has been found to be nearly as effective as PUVA, and is becoming the standard of care in many treatment centers. The diseases most responsive to phototherapy using any of the light sources mentioned earlier are psoriasis, atopic dermatitis, vitiligo, refractory itching, some forms of lymphoma, and many more [9].

In summary, the skin is a complex organ with many functions. Although we are constantly reminded of its role in appearance, it has important utility in protecting us from the environment. There are many special adaptations of skin that are involved in protecting us from UV light, some of which we are able to exploit to treat disease.

References

1. D. Chu, A. Haake, K. Holbrook, L. Loomis, *The Structure and Development of Skin*, 6th edn. (McGraw-Hill, New York, 2003)
2. R.R. Wickett, M.O. Visscher, Am. J. Infect. Control **34**, S98 (2006)
3. J.M. Waller, H.I. Maibach, Skin Res. Technol. **12**, 145 (2006)
4. J. Uitto, L. Pulkkinen, M. Chu, *Collagen*, 6th edn. (McGraw-Hill, New York, 2003)
5. J. Uitto, M. Chu, *Elastic Fibers*, 6th edn. (McGraw-Hill, New York, 2003)
6. B. Wenger, *Thermoregulation*, 6th edn. (McGraw-Hill, New York, 2003)
7. G. Stingl, D. Maurer, C. Hauser, K. Wolff, *The Skin: An Immunologic Barrier*, 6th edn. (McGraw-Hill, New York, 2003)
8. M. Yaar, B. Gilchrest, *Aging of Skin*, 6th edn. (McGraw-Hill, New York, 2003)
9. H. Lim, H. Honigsman, J. Hawk, *Photodermatology* (Informa Healthcare, New York, 2007)
10. D.I. Pattison, M.J. Davies, Experientia Supplementatum **96**, 131 (2006)
11. P. Elias, K. Feingold, J. Fluhr, *Skin as An Organ of Protection*, 6th edn. (McGraw-Hill, New York, 2003)
12. R. Ebert, N. Schutze, J. Adamski, F. Jakob, Mol. Cell Endocrinol. **248**, 149 (2006)
13. R. Halaban, D. Hebert, D. Fisher, *Biology of Melanocytes*, 6th edn. (McGraw-Hill, New York, 2003)
14. G. Scott, S. Leopardi, S. Printup, N. Malhi, M. Seiberg, R. Lapoint, J. Invest. Dermatol. **122**, 1214 (2004)
15. Z. Abdel-Malek, M.C. Scott, I. Suzuki, A. Tada, S. Im, L. Lamoreux, S. Ito, G. Barsh, V.J. Hearing, Pigment Cell Res. **13**(Suppl 8), 156 (2000)
16. A. Oba, C. Edwards, Skin Res. Technol. **12**, 283 (2006)
17. F. Hseih, C. Bingham, K. Austen, *The Molecular and Cellular Biology of the Mast Cell*, 6th edn. (McGraw-Hill, New York, 2003)
18. T. Masunaga, Connect. Tissue Res. **47**, 55 (2006)
19. L. Bruckner-Tuderman, *Basement Membranes*, 6th edn. (McGraw-Hill, New York, 2003)
20. M.T. Wong-Riley, X. Bai, E. Buchmann, H.T. Whelan, NeuroReport **12**, 3033 (2001)
21. H.T. Whelan, R.L. Smits Jr., E.V. Buchman, N.T. Whelan, S.G. Turner, D.A. Margolis, V. Cevenini, H. Stinson, R. Ignatius, T. Martin, J. Cwiklinski, A.F. Philippi, W.R. Graf, B. Hodgson, L. Gould, M. Kane, G. Chen, J. Caviness, J. Clin. Laser Med. Surg. **19**, 305 (2001)
22. R.N. Saladi, A.N. Persaud, Drugs Today **41**, 37 (2005)

23. W. Westerhof, O. Estevez-Uscanga, J. Meens, A. Kammeyer, M. Durocq, I. Cario, J. Invest. Dermatol. **94**, 812 (1990)
24. M. Kripke, H. Ananthaswamy, *Carcinogenesis: Ultravioloet Radiation*, 6th edn. (McGraw-Hill, New York, 2003)
25. T.B. Fitzpatrick, Arch. Dermatol. **124**, 869 (1988)
26. F.R. de Gruijl, H.J. van Kranen, L.H. Mullenders, J. Photochem. Photobiol. B **63**, 19 (2001)
27. X. Chen, A. Gresham, A. Morrison, A.P. Pentland, Biochim. Biophys. Acta **1299**, 23 (1996)
28. V.E. Reeve, Methods **28**, 20 (2002)
29. T. Nakamura, I. Kurimoto, S. Itami, K. Yoshikawa, J.W. Streilein, J. Dermatol. Sci. **23**(Suppl 1), S13 (2000)
30. G. Scott, A. Deng, C. Rodriguez-Burford, M. Seiberg, R. Han, L. Babiarz, W. Grizzle, W. Bell, A. Pentland, J. Invest. Dermatol. **117**, 1412 (2001)
31. B.J. Nickoloff, J.Z. Qin, V. Chaturvedi, P. Bacon, J. Panella, M.F. Denning, J. Investig. Dermatol. Symp. Proc. **7**, 27 (2002)
32. R.L. Konger, R. Malaviya, A.P. Pentland, Biochim. Biophys. Acta **1401**, 221 (1998)
33. A.C. Geller, G.D. Annas, Semin. Oncol. Nurs. **19**, 2 (2003)
34. V.E. Reeve, M. Bosnic, D. Domanski, Photochem. Photobiol. **74**, 765 (2001)
35. D. Domanski, M. Bosnic, V.E. Reeve, Redox Rep. **4**, 309 (1999)

17
Advanced Photodynamic Therapy

B.C. Wilson

17.1 Introduction

Photodynamic therapy (PDT) is the use of drugs (photosensitizers) that are activated by visible or near infrared light to produce specific biological effects in cells or tissues [1]. The basic steps in a PDT treatment are application of the photosensitizer (systemically or topically), a time interval to allow for photosensitizer accumulation in the target diseased tissue or cells, and illumination of the target area or volume with light of an appropriate wavelength to activate the sensitizer. PDT is a highly multidisciplinary topic, involving optical biophysics and bioengineering, synthetic chemistry, pharmacology, photophysics and photochemistry, photobiology, and different clinical specialties. The main emphasis in this chapter is on those aspects of greatest interest to specialists in biophotonics.

The term "photodynamic" was first coined a century ago with the observation that light and an acridine dye in the presence of oxygen could kill microorganisms [2]. At around the same time, light therapy was being used in patients but without any administered photosensitizing agent. The first reported clinical use of PDT was in 1904 when eosin was applied locally to a tumor on the lip and exposed to light. There was limited follow up to this early work, probably due to the lack of potent photosensitizers and of suitable light generation/delivery technologies. The modern era of PDT started with the discovery of hematoporphyrin derivative (HpD) in the 1950s/1960s, and this was first used in patients in the 1970s to treat bladder cancer. Substantial preclinical and clinical studies were started in the late 1970s and the first government approval for PDT was in 1993, with a purified version of HpD (Photofrin®).

In addition to being further developed since as a treatment for solid tumors and precancerous conditions, PDT is currently approved in various countries for nononcological applications [3,4]. These include actinic keratosis (sun-damaged skin) and age-related macular degeneration, in which abnormal blood vessel growth in the retina causes central vision loss, particularly in the elderly. Recently, there has also been a large interest in PDT to treat localized infection [5], in part driven by the rise of bacteria and other microorganisms that are multidrug resistant. For example, PDT using the agent methylene blue was recently approved for treating gum infection (periodontitis) and is also approved for treating acne using the agent ALA (see later): in both cases the sensitizer is applied topically.

These applications exploit the different biological mechanisms that can be selectively activated, depending on the PDT treatment parameters: cell killing, blood vessel shut down, and possibly immunological effects [6, 7].

The invention of the laser and optical fibers in the 1960s was also an important driver for the development of PDT, since these enable light of adequate intensity to be delivered to almost anywhere in the body and so make the treatment generically applicable. It should be emphasized, however, that PDT remains largely a localized treatment, since the penetration of light in tissues is limited, so that it is not generally useful for systemic diseases.

17.2 Basic Principles and Features of "Standard PDT"

Before discussing the potential novel approaches to PDT, we will summarize the "standard" PDT technique that is characterized by the following:

- High doses of photosensitizer and light, given as a single treatment (even if this is subsequently repeated to achieve complete clinical response).
- Single-photon activation (see Fig. 17.1 and discussion), using continuous-wave (CW) or quasi-CW light sources, and exploiting the Type II photoreaction.
- Photosensitizers that are always activatable.
- Photosensitizers that comprise organic molecules that are directly activated by the light.

Figure 17.1 shows the Jablonski, or energy-level, diagram for PDT. The ground state (S_0) of the photosensitizer absorbs the photon energy and is raised to an electronic excited (S_1) state (lifetime approximately nanosecond). This can de-excite either nonradiatively by fluorescence emission to S_0 or by intersystem crossing to a triplet state (T_1). The $T_1 \rightarrow S_0$ transition is quantum-mechanically forbidden, so that T_1 is long-lived (approximately microsecond) and can exchange energy with ground-state molecular oxygen ($^3O_2$) to generate singlet-state oxygen, $^1O_2$, which is highly reactive and leads to oxidative damage to nearby biomolecules (within 10 s of nanometer). Most often these are cell membrane components because of the photosensitizer microlocalization. Following this photoreaction, S_0 is regenerated, so that the photosensitizer essentially acts as a catalyst. It may eventually be destroyed (photobleached), either directly by the interaction or indirectly through damage due to the $^1O_2$. In addition to the properties common to any therapeutic drug, photosensitizers should have high $^1O_2$ quantum yield, high absorption

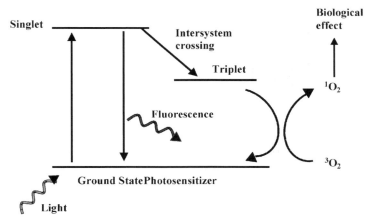

Fig. 17.1. Jablonski diagram for singlet oxygen generation by PDT (Type II photoreaction)

at red/near-infrared wavelength to get good light penetration, high target-to-nontarget uptake ratio in tissues or cells, and low concentration in the skin and eyes to avoid complications from accidental light exposure. Photosensitizers that are not water soluble require formulation into liposomes or lipid emulsions in order to achieve efficient delivery in vivo. The absorption spectra of some PDT photosensitizers are shown in Fig. 17.2, together with a plot showing the general dependence of light penetration in tissues. For example, HPD has its highest absorption in the UVA/blue region and only a very small red absorption (at about 630 nm). This contrasts with most second-generation photosensitizers where the molecule is designed to have the largest absorption peak above 630 nm where the hemoglobin absorption falls rapidly, but below about 800 nm in order to maintain high $^1O_2$ quantum yield and avoid the NIR water absorption bands.

One agent that is used widely is aminolevulinic acid (ALA). While not itself a photosensitizer, administering ALA to cells leads to synthesis of the photosensitizer protoporphyrin IX (PpIX) [8]. The advantage is that there is a degree of intrinsic tissue selectivity that does not just rely on enhanced photosensitizer uptake. ALA-PDT is used in treating tumors, particularly early-stage, and other indications mentioned above. Its fluorescence has also made it useful for early cancer/precancer imaging, either for tumor detection or in guiding tumor surgery, particularly in the brain [9]. As well as selection of good photosensitizer properties (and enough oxygen), successful PDT requires that enough light must be delivered to the target tissue to generate biologically effective levels of $^1O_2$ ($>10^8$ molecules per cell). This can be technologically challenging, since the amount of light required is quite high [10], typically $>100\,\text{J cm}^{-2}$ of red/NIR light incident on the tissue surface. Since this amount of light has to be delivered in at most tens of minutes for clinical practicality

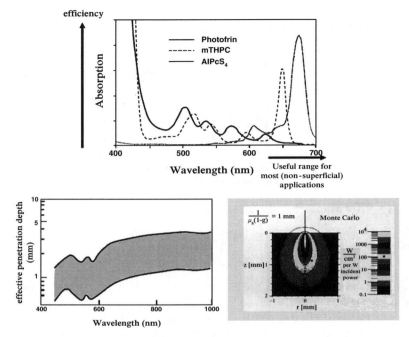

Fig. 17.2. Absorption spectra of some typical photosensitizers (*top*) and the dependence of light penetration in tissues (*bottom*), showing the range of effective penetration ($1/e$) depth spectra in typical tissues (*left*) and a representative Monte Carlo calculation of the distribution of light fluence in tissue of given optical absorption and scattering properties (*right*)

(but see mPDT below), the light sources require typically a few Watts of output power at the right wavelength for the photosensitizer. Suitable light sources are diode lasers, wavelength-filtered lamps, and light emitting diode arrays (LEDs) [3, 11]. Each has its own advantages and limitations. Overall, lasers are preferred when delivery via single optical fibers is required, for example, in endoscopy or when treating a larger tissue volume by multiple interstitial fibers; LED arrays are particularly useful for irradiation of easily accessible tissue surfaces. Lamps have a role when treating very large surface areas and have the advantage of easy (and low cost) change of wavelength to match different photosensitizers.

Different "modes" of target illumination by the light are used: surface, interstitial, or intravascular; external, endoscopic, or intraoperative. In all cases, making the *total system* (light source, power supply, light delivery devices, etc.) ergonomic and economic for specific clinical applications is critical. For the highest efficacy, the spatial distribution of the light in the tissue must be "matched" as far as possible to the volume and shape or the target tissue. This can be very difficult, given (a) the limited penetration of light in tissue (the $1/e$ depth of red/NIR light in tissues is typically only a few millimeter), (b) the high variability optical scattering and absorption properties of tissues, and (c) the limited degree of "tailoring" of the applied light distribution that can be achieved, especially in body locations where access is restricted. A variety of light-delivery devices has been developed for different applications [3, 10]. For example, optical fibers can be modified to produce a cylindrical irradiation volume along the last several centimeter at the tip, and recently it has become possible to make the output distribution deliberately nonuniform to match the target volume profile. Likewise, various balloon devices have been developed to irradiate within body cavities. With multiple fiber delivery, efficient splitting of the output of the light source is needed, and recent developments have included instruments in which nonequal distribution among sources can be achieved, again to match the target geometry in individual patients. These different technologies are still evolving as new clinical applications are investigated. A major challenge in some applications is to make the light source and delivery much cheaper and simpler to use than the established systems where, typically, the source cost is $\sim$\$10 K per Watt and single-use delivery devices are hundreds of dollars.

17.3 Novel PDT Concepts

In this main section we will present several new concepts that are still under development but which, if eventually adopted into clinical practice, would represent "paradigm sifts" in the way that PDT is performed, either in terms of the underlying principles used and/or in the technologies involved.

17.3.1 Two-Photon PDT

The photophysical interaction that has been used in essentially all PDT applications to date is that shown in Fig. 17.1 (although, in some cases, there may be Type 1 contributions to the photobiological effect). Thus, the S_0 state of each molecule is activated by absorbing a single photon. Two-photon (2γ) activation, representing a fundamental alternative to this scheme, is illustrated in Fig. 17.3, in two different implementations both involving the absorption of two photons of light for each activation cycle. Consider *Case A* (referred to here simply as 2γ activation), where a wavelength is used at which the photosensitizer has a nonzero 2γ cross-section, σ. The probability of simultaneous absorption of two photons is then proportional to σI^2, where I is the instantaneous light intensity. Hence, to avoid requiring too much total energy (which could cause thermal damage to the tissue) an ultrashort laser pulse is needed, typically $\sim$100 fs. Since the energy required for the $S_0 \to S_1$ transition is essentially the same as for 1γ activation, the wavelength of the pulsed laser can typically be pushed further into the NIR range, thereby increasing the penetration depth in tissue. This is one of the putative advantages of this approach, although the situation is complex, since it involves a balance between

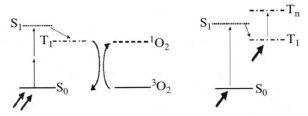

Fig. 17.3. Jablonski diagrams for two different types of 2γ activation involving either (**a**) simultaneous absorption of two photons by S_0 (*left*) or (**b**) sequential absorption of two separate photons by S_0 and then T_1 (*right*)

the reduced tissue attenuation and the quadratic dependence on the activation intensity, which falls off faster with depth than the linearly-dependent 1γ intensity. The second advantage is that, because of the I^2 dependence, focusing the laser beam activates the photosensitizer only within a very small (~femtoliter) tissue volume. This is the same effect as used in 2γ confocal microscopy, where it offers the analogous advantage of reduced photobleaching of the sample above and below the focal plane.

A potential application of this concept is to treat AMD, which is a major application for (1γ) PDT, involving targeting of the abnormal blood vessel growth in the retina. 2γ PDT may be able to achieve vessel closure but minimize any collateral damage caused by activation of photosensitizer outside the blood vessel layer. Recent research indicates that this is possible, in principle [12], but that there are several substantial technical challenges to be overcome, due primarily to the fact that absorption of two photons by S_0 is a very low probability event [13]. As a result, since activation is proportional to σI^2, either σ and/or I must be very high. The former requires the development of new molecules that have σ values several orders of magnitude higher than conventional PDT photosensitizers: in turn, this may also require the use of a "designer" delivery vehicle, since the pharmacological properties may then be difficult to control (e.g., limited water solubility). In principle, the intensity, I, may be arbitrarily high. However, in practice, this is limited by several factors: (a) the need to keep the total energy within sub-thermal range, and thus limiting the pulse energy, which means shorter pulses, (b) the cost of such laser sources, (c) nonlinear processes in the propagation of the laser pulses in the delivery system and the tissue itself, and (d) the onset of photomechanical damage to the tissue even in the absence of photosensitizer, which loses the potential selective advantage of photosensitizer localization.

The practical implementation envisaged for 2γ PDT is to image the region of abnormal blood vessels, or "feeder" vessels, using a confocal scanning laser ophthalmoscope (cslo), as illustrated in Fig. 17.4. This instrument operates on the same principle as confocal microscopy, using a scanning laser beam focused at a specific plane in the object (in this case the neovascular layer).

17 Advanced Photodynamic Therapy 321

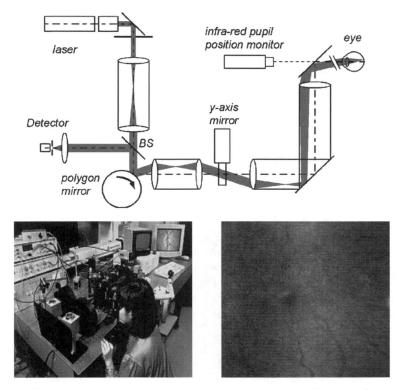

Fig. 17.4. Confocal scanning laser ophthalmoscope under development for image-guided 2γ PDT of age related macular degeneration, showing the optical design (*top*), the system in use, and a typical high-resolution retinal image (*normal*)

The operator would then define the "target" and the laser beam would be switched to "treatment mode," using the high peak-power fs laser. Apart from the fs light source itself, there are several optics challenges in implementing 2γ PDT for AMD, in particular, the following:

- Efficient coupling of the fs laser source into the cslo is complicated by effects such as pulse dispersion that take place as the laser beam propagates through the optical elements
- Achieving diffraction-limited focusing at the back of the eye is restricted by the low numerical aperture of the eye and possibly by scattering (e.g., by cataract) and wavefront distortion (e.g., from astigmatism).

To tackle the second problem, adaptive optics approaches will likely be needed and, in addition, automatic eye tracking will be required to keep the focal spot at the correct location. Both technologies are under development.

Case B, so-called two-photon/two-color activation $(2\gamma/2\lambda)$-PDT, uses two laser pulses of different wavelength and separated in time. The first pulse generates T_1 via the S_1 state. A second pulse, at a wavelength that is strongly absorbed by T_1, then generates higher-order triplet states, T_n. The biological effects are then mediated by the T_n states interacting with nearby biomolecules. The key point is that this process does not require molecular oxygen, and so may avoid the limitations of Type II photosensitizers, which do require adequate free oxygen in the target tissue. Oxygen-independent cell killing by $2\gamma/2\lambda$ PDT has been demonstrated in cells, although using a photosensitizer that is poorly suited to clinical use. However, there are major challenges in moving this concept to the clinic: (a) the photosensitizer must have high absorption in both the S_0 and T_1 states and these must be at useful wavelengths for tissue penetration and (b) a $2\gamma/2\lambda$ laser source is required, preferably with independent control over the intensity and wavelengths of each pulse and pulse–pulse time delay to match the sensitizer photokinetics. Work is in progress to address both challenges. For example, Mir et al. [14] are developing phthalocyanine with specific structural modifications to yield optimum absorption spectra in the ground and excited states, while rapid developments in pulsed laser sources, such as fiber lasers, are likely to yield practical laser technologies in the next few years. (Note that, in principle, this pathway could be activated to operate in CW mode with both wavelengths present simultaneously, although the photoefficiency would be very low.)

17.3.2 Metronomic PDT

At the other end of the light intensity scale is the concept of metronomic photodynamic therapy (mPDT). This is based on the observation that it may be possible to increase the target-specific cytotoxicity by delivering the photosensitizer and the light at very low dose rates over an extended period (hours or days) rather than applying a single, high dose-rate treatment. An example of such specificity is seen in treating brain tumors [15]. In this case, the clinical challenge is that the tumor cells around the edges of a solid tumor infiltrate the normal brain tissue. With standard, high dose-rate treatment, it may not be possible to kill the tumor cells without also destroying normal brain tissue. However, tumor cell-specific programmed cell death (apoptosis) can be achieved by mPDT. This has been demonstrated for ALA. It is not known if the mPDT effect is unique to ALA or applies also to other photosensitizers, or whether the mPDT principle can be applied also to other tumor sites or diseases.

From the biophotonics perspective, a challenge is to develop light sources that could deliver light over a long period and still be practicable, either for clinical use or in preclinical animal models. Figure 17.5 shows an example based on high-efficiency coupling of LEDs into optical fibers, with battery and driver electronics incorporated into a lightweight "backpack." For clinical use, the technology challenges will include (a) how to distribute the light, e.g., in the brain or other site after surgery to remove the bulk tumor tissue, (b) how to integrate the power supply, light source (diode laser or LED), and the

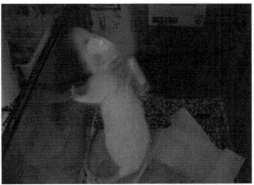

Fig. 17.5. Fiber-coupled LED sources developed for mPDT use in rodent brain, and a source in use to treat brain tumor in a rat model, with the battery and electronics in a wearable backpack

light delivery components, (c) how to maintain sterility and biocompatibility of such an implant and (d) how to provide power, for which one option is to have an external power supply and remote coupling to the source, as is used in some other implantable devices.

The fundamental biological issue with mPDT is whether or not it is possible to kill the target cells faster than their proliferation, while avoiding non-apoptotic killing (to avoid inducing an inflammatory response that could cause secondary damage to normal tissues) and avoiding direct PDT damage to normal tissues.

17.3.3 PDT Molecular Beacons

Molecular beacons were introduced several years ago as a way to enhance the target specificity of molecular imaging agents [16]. In that case the concept was to link a fluorescent reporter molecule (fluorophore) with a molecule

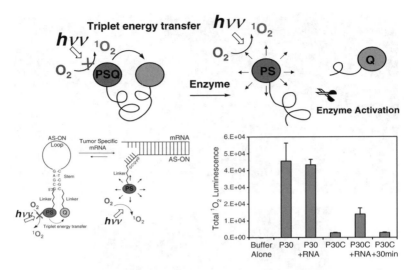

Fig. 17.6. PDT molecular beacons. showing the concept with enzyme cleavage of the linker (*top*), and using an antisense loop (*bottom*), with an example of the low $^1O_2$ luminescence signal (*third bar*) compared to controls (*first* and *second bars*) that increases with unquenching (*fourth bar*) and then decreases with inhibition of the unquenching (*fifth bar*)

(quencher) that suppresses the fluorescence when the two are in close proximity. The quenching efficiency has a very strong ($\sim 1/r^6$) dependence on the quencher-fluorophore distance, r. The linker molecule could then be broken by, for example, interaction with an enzyme that is found in high concentration in the target tissues or cells.

This concept has recently been transferred into PDT using a photosensitizer instead of the fluorophore (Fig. 17.6). Proof-of-principle studies of this concept have been reported [17], showing, for example, that $^1O_2$ can be generated by light exposure at much higher levels after unquenching than before. Selective toxicity has also been shown, both in vitro and in vivo, where only cells/tumors that express the specific enzyme used in the beacon are sensitive to PDT.

There are several possible variations on this idea, in terms of the linker structures. One of the most interesting, but challenging in its synthetic chemistry, is to use a so-called antisense loop, in which the quencher and photosensitizer are held in close proximity when the "hairpin" structure is closed, but then the linker opens up and unquenches the photosensitizer when it hybridizes with a specific nucleotide sequence in the target cells. Antisense beacons have been demonstrated to date in solution. This could have tremendous disease specificity, since it is based on highly defined genetic biomarker characteristics of, for example, tumor cells.

17.3.4 Nanoparticle-Based PDT

There has been great excitement recently in the potential of nanoparticles (NPs) for a wide range of applications, including medical diagnostics and therapeutics [18]. This disruptive technology is based on the special optical, electrical, mechanical, or chemical properties that emerge when materials are formulated at the nanoscale (typically, 1–100 nm) compared to the bulk properties. Optical NPs include the following:

- Metal NPs with light absorption and scattering spectra that are highly dependent on the size, shape, and form (e.g., solids vs. shells, spheres vs. rods)
- Quantum dots (Qdots), semiconductor NPs that are fluorescent in the visible and/or near-infrared, which are very bright, have very low photobleaching, have narrow emission spectra with the peak emission increasing with the Qdot diameter, and have a wide excitation spectrum so that a single light source can excite many different sizes of Qdots
- NPs that are themselves not optically active but that can carry a "payload" of optically active molecules, such as fluorescent dyes or photosensitizers

In addition to these specific novel properties, NPs have the advantage that they can be "decorated" with molecules, such as antibodies or peptide sequences, to target them to specific cells or tissues. The potential of NPs for cancer diagnostics (particularly imaging) and therapeutics has been recently reviewed [19]. An important subset of these techniques is photonics-based, including fluorescence imaging and photothermal and photodynamic treatments.

For PDT there are several potential and distinct ways to utilize NP characteristics, as illustrated in Fig. 17.7, namely (a) as "carriers" of PDT photosensitizers, (b) as photosensitizers themselves, or (c) as "energy transducers."

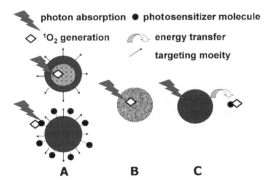

Fig. 17.7. Schematic of three different uses of nanoparticles for PDT: (**a**) as targeted "carriers" of the photosensitizer, (**b**) as direct $^1O_2$ generators, and (**c**) as energy transducers

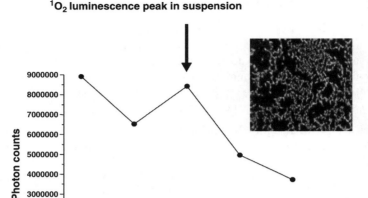

Fig. 17.8. $^1O_2$ signal from porous silicon NPs in water (relative to background): insert shows the ultrastructure of these NPs

For Case A, a variety of NPs can be used that either encapsulate many photosensitizer molecules per NP or have many attached to the surface in order to deliver a high payload of photosensitizer to the target cells/tissues. There are several demonstrations in the literature of this approach. One advantage that PDT has over, say chemotherapeutic drugs, is that generally the toxicity of most photosensitizers is very low in the absence of light activation, so that there is a higher tolerance for nontarget "leakage" in the delivery system.

For Case B, Fig. 17.8 shows an example where singlet oxygen is generated upon light irradiation of porous silicon NPs, without any added molecular photosensitizer. This is believed to be due to direct energy transfer to the molecular oxygen in the liquid that permeates the NPs, which have an extremely large surface-to-mass ratio. It has not yet been shown that this can actually be exploited to produce PDT cell killing: these studies are in progress. Potential barriers are getting the NPs into the cells and having the $^1O_2$ diffuse from the (inner) NP surface to the biological target. (Note that in the case of NPs as photosensitizer delivery vehicles, it may not be necessary for the NPs themselves to penetrate the cell wall if the payload can be released within the target tissue, such as in the interstitial space, from which it would then be taken up by the cells.) In the third scenario, Case C, Qdots are used as the primary light absorber. The energy is then transferred to a photosensitizer molecule that is conjugated to the Qdot, activating it to the S_1 state, as in standard PDT. The generation of $^1O_2$ via Förster Resonant Energy Transfer has been demonstrated in solution and should work in cells/tissues.

Again, the advantage in principle would be the ability to target to Qdots to the desired tissues/cells without compromising the optimum photosensitizer properties. A variation of this approach would be to use it for 2γ PDT, since Qdots are known to have very high σ values ($\sim 10^4$–10^5 GM units compared with ~ 10 GM units for conventional 1γ drugs and $\sim 10^3$ GM units for the best "designer" 2γ molecules to date).

17.4 PDT Dosimetry Using Photonic Techniques

PDT dosimetry, i.e., the measurement of the "dose" that determines the efficacy of the treatment, is complex, due to the following:

(a) The multiple factors involved (photosensitizer, light, oxygen)
(b) The heterogeneity of the local photosensitizer concentration, light fluence rate, and oxygen level in the tissue, which may result in nonuniform deposition of PDT dose throughout the target volume, and
(c) The interdependence and dynamic behavior of the parameters, e.g., $^1O_2$ destroys the photosensitizer molecules (photobleaching), reducing the amount available for activation; the photosensitizer may increase the attenuation of light in the tissue, further limiting the penetration depth; if the photosensitizer concentration and the light intensity (fluence rate) are high enough, this can deplete the ground-state oxygen faster than the blood supply can replenish it.

Three main dosimetry approaches have been used to address this complexity, termed explicit, implicit, and direct [20]. In the first, the attempt is to measure the three basic parameters of light fluence rate, photosensitizer concentration, and molecular oxygen concentration, which are then combined into a predictive estimate of the effective PDT dose. There are many technical details in each of these three measurements [3, 10]. Figure 17.9 illustrates light dosimetry in PDT to destroy prostate cancer by delivering light through multiple diffusing optical fibers to the whole prostate volume.

To optimize the distribution of light throughout the prostate treatment planning is used through a series of steps: 3D MRI is done to determine the volume and shape of the prostate; source fibers are placed at selected positions to cover the prostate volume; the spatial light fluence-rate distribution resulting from delivering selected powers to each of the source fibers is calculated; this is compared with the desired light distribution; and the steps are iterated to optimize the number, length, position, and power from each source. This is analogous to treatment planning in radiation therapy (e.g., brachytherapy using implanted radioactive wires or seeds). For photosensitizer dosimetry, the main technique has been to measure the fluorescent by point spectroscopy or imaging. However, while it is straightforward to make relative measurements, it is more difficult to derive absolute photosensitizer concentration values, since the measured signal depends also on the tissue

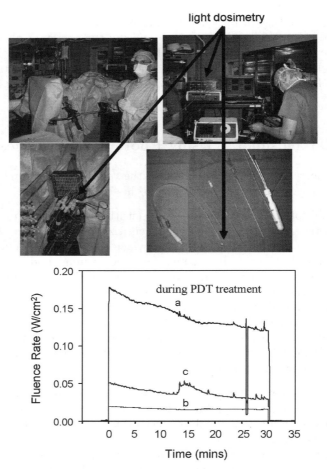

Fig. 17.9. Example of light dosimetry during PDT of prostate cancer, showing the patient set-up, the placement of fibers and detector fibers through a template, the multichannel light dosimeter measurements of the local light fluence rate as a function of time during treatment at three different locations the graph shows: (**a**) in the prostate, (**b**) the urethra and (**c**) the rectum

optical properties. One solution is to measure these properties separately and apply a light propagation model to correct for the attenuation. Alternatively, fiberoptic probes have been developed to reduce this attenuation dependence, for example, by sampling only a very small tissue volume [22].

The measurement of oxygen levels in tissue is well-established, for example, using interstitial microelectrodes or fiberoptic probes. Noninvasive measurements are less well developed and do not directly measure the pO_2, e.g., diffuse reflectance spectroscopy can detect changes in the relative concentration of

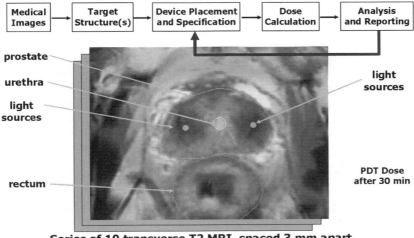

$$\nabla D(\mathbf{x})\nabla\phi(\mathbf{x})-\mu_a(\mathbf{x})\phi(\mathbf{x})=0$$

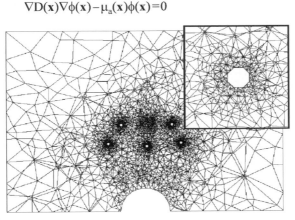

Fig. 17.10. Example of treatment planning for PDT of prostate [21], showing the process for calculating the 3D light distribution in the tissue (*top*, two-source example) and the diffusion-theory equation that is solved on a nonuniform mesh by finite-element analysis (*bottom*, six-source example)

hemoglobin and oxyhemoglobin, from which the oxygen saturation, SO_2, can be derived. (This is also used in studies of brain and muscle physiology [23].)

In implicit dosimetry a surrogate measure of the integrated interplay of light, drug, and oxygen is employed, most commonly the photosensitizer photobleaching. Referring to Fig. 17.1, if the $^1O_2$ can react with and disrupt the photosensitizer molecules such that they become nonphotoactive/nonfluorescent, then this is related to the $^1O_2$ production. This method is technically fairly simple and can be applied at multiple locations in the tissues.

However, interpreting the data rigorously is not simple, since the photobleaching may or may not be $^1O_2$-dependent and there may be non-$^1O_2$ dependent cytotoxic pathways if the tissue pO_2 is low. Nevertheless, under well-controlled conditions, photobleaching can be a very useful PDT dose metric [24].

Lastly, in direct dosimetry, $^1O_2$ itself is quantified by detecting the 1,270 nm luminescence emission from the $^1O_2 \rightarrow ^3O_2$ transition. The signal is very weak, since $^1O_2$ is so reactive in biological media, with only about 1 in 10^8 molecules decaying by luminescence. However, this has been demonstrated recently both in vitro and in vivo and is highly correlated to the PDT response [25–27]. The challenge now is to simplify and reduce the cost and complexity of the technology.

17.5 Biophotonic Techniques for Monitoring Response to PDT

Partly in response to the continuing challenge of PDT dosimetry, but also to gain more direct understanding of the biological effects of PDT treatment, there has been ongoing interest in assessing directly the tissue responses following or even during PDT. The last, in particular, is also motivated by the possibility to use this information for feedback control of the treatment while it is in progress.

A first example that is used for preclinical studies is bioluminescence imaging (BLI). For this the cells of interest are transfected with the luciferase ("firefly") gene, so that, when luciferin is supplied the cells emit visible light through chemiluminescent. Although weak, this can be imaged noninvasively using a high sensitivity photodetector, such as a cooled CCD array. BLI has been used to quantify changes in implanted tumors [28] or the survival of bacteria [5] treated with PDT in animal models. The method can be extended by having the luciferase gene only "turned on" when another gene of interest (e.g., a stress-response gene) is up-regulated by the PDT treatment, so that BLI gives direct in vivo imaging of the response of the target cells at the specific molecular level.

A second example is optical coherence tomography (OCT), the analogue of high-frequency ultrasound imaging, based on interferometry (Fig. 17.11) (explained in detail in the chapter by S. Boppart). Depth scans (A-scans) are produced by detecting the interference signal between the reflected light from a given depth in the tissue and a reference beam: high spatial resolution is obtained by using a light source with a very short coherence length. A 2D image (B-scan) is then generated by scanning the beam across the tissue surface. OCT is approved in ophthalmology and is being widely investigated for a range of other applications, including endoscopic and intravascular imaging using fiberoptic implementation. As with ultrasound, OCT can operate in Doppler mode to image blood flow (by measuring the small shift in phase between successive A-scans caused by movement of the red blood cells or other tissue components).

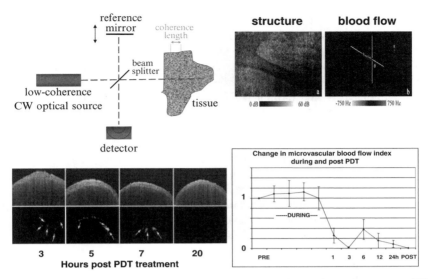

Fig. 17.11. Doppler OCT for monitoring vascular response of tumor tissue to PDT treatment. (*Top*) Principle of OCT and example of cross-sectional imaging in vivo. (*Bottom*) Structural and DOCT images at different time points following PDT treatment showing blood vessel shut down (*left*) and plot of volume-averaged measure of blood flow in the tissue during and following Photofrin-PDT treatment of melanoma

For monitoring the changes to tissue from PDT, one can look at either the structural or the Doppler images. The latter have proved to be more useful to date, since the changes in tissue structure are rather subtle. Figure 17.11 also shows examples of the changes in blood flow in individual microvessels due to PDT. Such changes can be seen in some cases even during the light treatment [29], so that there is the possibility to use the DOCT information to modify the treatment "on line." A limitation with OCT is that the depth of imaging is small ($\sim$1–2 mm). Recently, we reported an OCT probe within a small-diameter needle that can be placed interstitially [30], e.g., at the tumor base to ensure that adequate PDT dose is delivered.

17.6 Biophotonic Challenges and Opportunities in Clinical PDT

We end with a brief overview of some of the major outstanding limitations of PDT and how these might be overcome, at least in part by applying biophotonic science and technologies.

Certainly PDT can be both safe and effective. However, there can be significant variations in the response. One of the obvious problems is that there is little or no "tailoring" of the treatment parameters used to the individual patient or lesion. The various dosimetry methods outlined above could improve this situation, but they are rarely used in routine clinical practice at this time, in part because they are still rather complicated to use and interpret. A major technical challenge is to devise the next generation of dosimetry devices that will be simple and fast to use, minimally invasive, inexpensive, and reliable. Until this happens, PDT will continue to evolve largely as an empirical procedure with optimization based on dose ranging in clinical trials.

A second issue is the continuing high cost of light sources. Using existing technologies it is not clear how this can be significantly reduced. Certainly, part of the cost comes from dealing with medical devices where significant investments are required to ensure safety and reliability. It is becoming clear that, for some potentially important applications, it is the cost of the light source that is limiting, not the cost of the photosensitizer. Hence, there are significant opportunities for optical engineers and physicists to devise new devices.

A further major limitation is that the specificity of photosensitizer targeting of disease has not been high enough to use unlimited light safely. Attempts have been made to improve this, for example, by linking the photosensitizer to antibodies to target tumors cells, but these have not been particularly successful to date. PDT beacons could change this significantly. However, it will be challenging to get these into clinical trials and then through regulatory approvals, so that the preclinical results in vivo will have to be outstanding to make the effort worthwhile.

With respect to tissue response monitoring, there are many novel biophotonic imaging and spectroscopic techniques being developed for other applications that could also be valuable in PDT: fluorescence imaging is a prime example, while others include Raman spectroscopy (to report biochemical changes) and intravital second harmonic generation imaging (changes in tissue architecture). PDT monitoring will not drive the development of these techniques but will benefit from them.

Three other major aspects of PDT are worth mentioning:

– The parallel development of "photodynamic diagnostics," particularly for tumor detection and localization [23]
– The combination of PDT with other therapeutic modalities, e.g., (fluorescence guided) tumor resection followed by PDT to eliminate residual minimal disease
– The extension of PDT into new applications, particularly to modify cells or tissue function rather than destroying them: recent examples are for modifying bone growth and for targeting epilepsy.

In addition to these clinical applications, the principle of light-activated drugs may also be useful for basic life sciences research. One example is the use of 2γ PDT to uncage growth factors in order to direct the growth of neurons in tissue engineering [31].

17.7 Conclusions

Although PDT is over a century old, it is far from a mature science. Addressing some of the challenges in PDT, particularly in the optical technologies and dosimetry, has been a major driving force in biophotonics as a whole, with substantial spin-off into other applications. For example, much of the early work in tissue optics was motivated by the need to understand and control the distribution of light in PDT of solid tumors. In turn, this laid the foundation for diffuse optical tomography and other therapeutic and diagnostic techniques. In the same way, studies of photosensitizer fluorescence led directly to imaging based on tissue autofluorescence: subsequently, attempts to improve the image contrast stimulated work on targeted fluorophores, including most recently quantum dots.

It is reasonable to expect that PDT will continue, both academically and commercially, especially with the extension into a diverse range of new applications and using some of the new concepts outlined here. There is still certainly much to be done that will require the efforts of many different disciplines.

Acknowledgments

The author thanks the following agencies for support of work illustrated here: the National Cancer Institute of Canada ($^1O_2$ dosimetry and DOCT monitoring), the National Institutes of Heath, US (prostate PDT and mPDT [(CA-43892)], beacons), the Canadian Institute for Photonic Innovations (2-photon PDT), and Photonics Research Ontario (PDT technology development). Also, thanks to many colleagues and students in the PDT program at the Ontario Cancer Institute and to our clinical and industry collaborators.

References

1. T. Patrice (ed.), *Photodynamic Therapy*, (Royal Society of Chemistry, UK, 2003)
2. R. Ackroyd et al., Photochem. Photobiol. **74**, 656 (2001)
3. B.C. Wilson, S.G. Bown, in *Handbook of Lasers and Applications*, ed. by C. Webb, J. Jones (IOP Publ, UK, 2004), p. 2019
4. T.J. Dougherty, J. Clin. Laser Med. Surg. **20**, 3 (2002)
5. T.N. Demidova, M.R. Hamblin, Int. J. Immunopathol. Pharmacol. **17**, 245 (2004)
6. B.W. Henderson, T.J. Dougherty, Photochem. Photobiol. **55**, 145 (1992)
7. N.L. Oleinick, H.H. Evans, Radiat. Res. **150**, S146 (1998)
8. Q. Peng et al., Photochem. Photobiol. **65**, 235 (1997)
9. W. Stummer et al., Acta Neurochir. Suppl. **88**, 9 (2003)
10. M.S. Patterson, B.C. Wilson, in *Modern Technology of Radiation Oncology*, ed. by J. Van Dyke (Medical Physics Publishing, Madison, WI, USA, 1999), p. 941

11. L. Brancaleon, H. Moseley, Lasers Med. Sci. **17**, 173 (2002)
12. K.S. Samkoe, D.T. Cramb, J. Biomed. Opt. **8**, 410 (2003)
13. A. Karotki et al., Photochem. Photobiol. **82**, 443 (2006)
14. Y. Mir et al., Photochem. Photobiol. Sci. **5**, 1024 (2006)
15. S.K. Bisland et al., Photochem. Photobiol. **80**, 2 (2004)
16. U. Mahmood, R. Weissleder, Mol. Cancer Ther. **2**, 489 (2003)
17. J. Chen et al., J. Am. Chem. Soc. **126**, 11450 (2004)
18. M. Ferrari, Nat. Rev. Cancer **5**, 161 (2005)
19. B.C. Wilson, in *Photon-Based NanoScience and Technology*, ed. by J. Dubowski, S. Tanev (Springer, Netherlands, 2006), p. 121
20. B.C. Wilson et al., Lasers Med. Sci. **12**, 182 (1997)
21. R.A. Weersink et al., J. Photochem. Photobiol. **79**, 211 (2005)
22. B.W. Pogue, G. Burke, Appl. Opt. **37**, 7429 (1998)
23. B.C. Wilson, in *Handbook of Lasers and Applications*, ed. by C. Webb, J. Jones (IOP, UK, 2004), p. 2087
24. J.S. Dysart, M.S. Patterson, Photochem. Photobiol. Sci. **5**, 73 (2006)
25. M. Niedre et al., Photochem. Photobiol. **75**, 382 (2002); **81**, 941 (2005)
26. M. Niedre et al., Cancer Res. **63**, 7986 (2003)
27. M. Niedre et al., Br. J. Cancer **92**, 298 (2005)
28. E. Moriyama et al., Photochem. Photobiol. **80**, 242 (2004)
29. H. Li et al., Lasers Surg. Med. **38**, 754 (2006)
30. V.X. Yang et al., Opt. Lett. **30**, 1791 (2005)
31. Y. Luo, M.S. Stoichet, Nat. Mater. **3**, 249 (2004)

Index

1D photonic crystal, 103
2D photonic crystal, 103

Active pixel sensor, 239
Avidin, 200

Biosensor, 107, 199
BSA, 120

Carotenoids, 3, 31, 42
CARS:Coherent anti-Stokes Raman, 48
CCD, 239
Chlorophylls, 3, 29
Chloroplasts, 2

DHM, 164
DHM Life Cell Imaging, 171
DHM numerical focus, 170
DHM resolution, 170
Digital Holographic Microscopy, 164
Digital holographic reconstruction, 166
Distal endoscopic ESPI, 158
DNA, 114
Double exposure subtraction ESPI, 152

Effective dielectric constant, 110
Electronic Speckle Pattern Interferometry, 152
ESPI, 152

Fluorescence, 49
Fluorescent dye molecule, 87
Functionalization, 111

Genetic engineering, 202

Genome, 177
Genomics, 219
Gram bacteria, 114
Greenhouse effect, 15

Hydrogenases, 19

IgG, 117

Keratinocytes, 301

Label-free, 88
Label-free optical biosensing, 109
Laser tissue welding, 275
Light cone, 104
Light harvesting complexes, 6

Mach–Zehnder interferometer, 128
Michelson-type interferometer, 128
Microcavity, 94, 105, 110, 111, 121
Microscopic Speckle Interferometry, 161
Multi-photon microscopy, 47
Mutant Type protein, 195

NDRM, 166
Non diffractive reconstruction, 166

Ophthalmology, 127
Optical tweezers, 249
Optrodes, 204, 207

Phosphorescennce, 49
Photonic bandgap, 105
Photonic crystals, 101
Photosynthesis, 1, 17
Plasmid, 194

Porous silicon, 107
Proteomeics, 219
Proteomics, 221
Proximal endoscopic ESPI, 156

Quantum dots, 178

Second harmonic generation, 48
Sensitivity vector, 153
Setups for DHM, 165
Silanization, 114
Single photon avalanche diode, SPAD, 239

Singlet oxygen, 30
Spatial phase shifting, 153
SPS, 153
Stable transfection, 194
Streptavidin, 115

Thylakoids, 2, 6
Transfection, 194
Transient transfection, 194
Two-photon cross-section, 58

Vector, 194